Photonics
Modelling
and Design

OPTICAL SCIENCES AND APPLICATIONS OF LIGHT
Series Editor
James C. Wyant

Charged Particle Optics Theory: An Introduction, *Timothy R. Groves*

Photonics Modelling and Design, *Slawomir Sujecki*

Nonlinear Optics: Principles and Applications, *Karsten Rottwitt and Peter Tidemand-Lichtenberg*

Numerical Methods in Photonics, *Andrei V. Lavrinenko, Jesper Lægsgaard, Niels Gregersen, Frank Schmidt, and Thomas Søndergaard*

*Please visit our website **www.crcpress.com** for a full list of titles*

Photonics Modelling and Design

Sławomir Sujecki
The University of Nottingham, UK

CRC Press
Taylor & Francis Group
Boca Raton London New York

CRC Press is an imprint of the
Taylor & Francis Group, an **informa** business

CRC Press
Taylor & Francis Group
6000 Broken Sound Parkway NW, Suite 300
Boca Raton, FL 33487-2742

First issued in paperback 2017

© 2015 by Taylor & Francis Group, LLC
CRC Press is an imprint of Taylor & Francis Group, an Informa business

No claim to original U.S. Government works

ISBN-13: 978-1-4665-6126-7 (hbk)
ISBN-13: 978-1-138-80938-3 (pbk)

Dedication

To my family and friends

Contents

Preface

The intention of this book is to introduce an engineer or applied physicist to the modelling and design of photonic devices. This book was written on the basis of both teaching and research carried out by the author. It is therefore written in such a way that it contains material suitable for both undergraduate and master's students but is also of benefit to PhD students and researchers interested in the modelling and design of photonic devices.

The author has tried to avoid using specialised mathematical and quantum mechanical language to make this book approachable to a wider audience. Further, the book contains a strong "hands on" element that is backed up by several, relatively simple illustrative examples of software developed within the MATLAB® environment.

The contents of this book are organised as follows; first, Chapters 2–4 present an analysis of light propagation in dielectric media. Then the topics of heat diffusion and carrier transport are discussed in Chapters 5 and 6, respectively. In Chapters 7 and 8, the theory presented in Chapters 2–6 is used to develop fibre and semiconductor laser models. Finally, in Chapter 9 the propagation of short optical pulses in optical fibres is discussed.

To a large extent, this book is self-contained. However, the reader is expected to have a fair understanding of photonics in general and a good understanding of the operating principles of fibre and semiconductor lasers. If there is a need for familiarisation with the basic theory prior to the discussion of a particular chapter, suitable literature is suggested in the introductory sections of the relevant chapter.

The completion of this book would not have been possible without scientific collaboration that occurred within the EU projects: Ultrabright, Bright.EU, Brighter.EU, FastAccess, Copernicus and MINERVA. A significant contribution was also provided by the discussion platform that was created by the NUSOD and ICTON conferences. I am therefore grateful to Dr J. Piprek (NUSOD Institute) and Prof. M. Marciniak (National Institute of Telecommunications, Warsaw, Poland) who have initiated and run the NUSOD and ICTON conferences, respectively. I am also grateful to the researchers that I collaborated closely with on the EU projects: Prof. I Esquivias and Dr L. Borruel from Universidad Politécnica de Madrid, Spain; Dr B. Sumpf, Dr H. Wenzel and Dr G. Erbert from Ferdinand Braun Institut für Höchstfrequenztechnik, Germany; Dr N. Michel and Dr M. Krakowski from Thales Group, III-V Lab, France; Prof. A. Larsson from Chalmers University of Technology, Sweden; Dr P. Uusimaa from Modulight Inc., Finland; Dr. G. Lucas-Leclin, Dr N. Dubreuil and Dr G. Pauliat from Laboratoire Charles Fabry de L'Institut d'Optique, France; Dr M. Dumitrescu and Prof. M. Pessa from the Tampere University of Technology, Finland; Dr Birgitte Thestrup, Prof. Paul Michael Petersen and Prof. O. Bang from TU Denmark; and Dr S. Lamrini and Dr P. Fürberg from LISA laser products OHG, Germany. I also owe thanks to my colleagues: Prof. A.B. Seddon, Prof. E.C. Larkins, Prof. P. Sewell, Dr A. Vukovic, Dr D. Furniss, Dr Z. Tang, Dr R. MacKenzie and especially Prof. T.M. Benson for the careful reading of the manuscript. Finally, I thank my PhD student Mr Ł. Sójka for his contributions to Chapter 7.

The MathWorks, Inc.
3 Apple Hill Drive
Natick, MA 01760-2098 USA
Tel: 508-647-7000
Fax: 508-647-7001
Email: info@mathworks.com
Web: www.mathworks.com

Biography

Slawomir Sujecki graduated from the Faculty of Electronic and Information Technology of the Warsaw University of Technology, Poland, in 1993. He obtained his PhD (1997) and DSc (2010) degrees, specialising in modelling and design of photonic devices, also from the Warsaw University of Technology. In 1998, he was appointed as a lecturer at the Kielce University of Technology, Poland, and in 1999 he became a researcher at the National Institute of Telecommunications, Warsaw, Poland. In 2000, Dr Sujecki joined the University of Nottingham, UK, as a research assistant and was appointed a lecturer in 2002, and then promoted to associate professor in 2012. His main research area is laser design, especially laser diodes and fibre lasers. Dr Sujecki has participated in several research projects funded by the European Community: Ultrabright, Bright.EU, Brighter.EU, FastAccess, Copernicus and MINERVA, where his main area of activity was the modelling and design of photonic devices.

Dr Sujecki was awarded fellowships by Deutscher Akademischer Austauschdienst, British Council, Royal Society and Wolfson Foundation. He is a senior member of IEEE, a life member of OSA, and a member of the Program Committee of the NUSOD Conference (Numerical Simulation of Optoelectronic Devices).

1 Introduction

Modelling and design play a very important role in photonics and, with progress in computer science and technology, gradually more and more sophisticated tools have become available for the development of novel photonic devices. To focus the discussion on a particular representative example, let us consider a semiconductor laser diode (LD) using a C-mount packaging technology that can be readily purchased from a number of vendors (Figure 1.1). During its normal operation, the LD is electrically pumped from a current source and emits a coherent light beam (Figure 1.2). Therefore, the cathode and anode of the LD have to be connected to electrical wires, typically using wire bonding technology, so that the LD can be conveniently driven by a current source. The light can be collected from the front facet of the LD, which is often coated with an antireflection material. The processes of the current conduction and light emission in LD are accompanied by heat generation and the creation of stress and strain fields. Therefore, the LD chip is soldered onto a submount, which in turn is mounted onto a heat sink. The heat sink facilitates the heat dissipation whereas the stress and strain are reduced by the submount, with an appropriate soldering technique. This simple introductory discussion reveals that a comprehensive model of a LD should contain an optical model that will accurately describe the light propagation, a model of the active medium that will describe the process of the carrier transport within the device and the conversion of the pumping current into light, and a thermal model that can predict the device's operating temperature and facilitate the optimisation of the heat extraction (Figure 1.3). Additionally, a stress and strain model could be included to calculate the distribution of stress and strain tensor fields. More generally, the description presented of an LD package example obviates the following statement: The modelling and design of photonic devices is, in principle, a multidisciplinary field and requires familiarisation with four areas of physics, namely the theory of light propagation, solid state physics (quantum mechanics), heat diffusion theory, and elasticity theory.

OPTICAL MODEL

The optical models of modern photonic design tools are based on Maxwell's electromagnetic field theory. The electromagnetic field theory predicts that light can be approximated by an electromagnetic wave. In fact, the concept of light in the wave form was initially introduced a century earlier by Thomas Young. Interestingly, the first theory of light developed in the modern era, namely the geometric optics, predicted that light is not a wave but a stream of corpuscles. This postulate was justified by several experiments that were carried out in the late sixteenth and seventeenth centuries and, even though it was strongly contested by Christiaan Huygens, geometric optics established itself as the theory of light back in the eighteenth

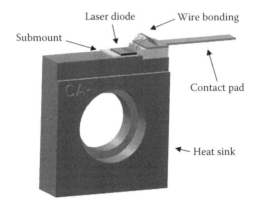

FIGURE 1.1   The schematic of a laser diode on a C-mount.

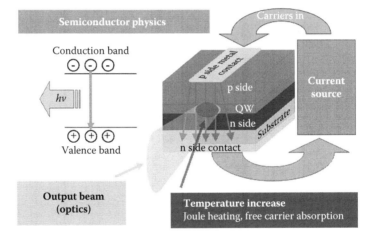

FIGURE 1.2   Physical processes accompanying the light emission process in a laser diode.

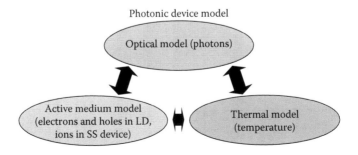

FIGURE 1.3   A schematic diagram of a photonic device model.

century. The geometrical optics allowed the phenomena of light reflection and refraction to be explained and was successfully applied to perfect the design of telescopes through the seventeenth century. In the late eighteenth and early nineteenth centuries, a growing number of experiments demonstrated the wave nature of light. This led to the development of wave optics. Namely, a theory that postulates that light is a wave that can be described by a scalar function that is a solution of the wave equation:

$$\Delta u - \frac{1}{c^2} \frac{\partial^2 u}{\partial t^2} = 0 \qquad (1.1)$$

In Equation 1.1, u is a real function of position and time, Δ is the Laplace operator, which in the Cartesian coordinate system x, y, z can be expressed as:

$$\Delta u = \frac{\partial^2 u}{\partial^2 x} + \frac{\partial^2 u}{\partial^2 y} + \frac{\partial^2 u}{\partial^2 z}$$

The constant c denotes the speed of light in the medium.

Wave optics allowed the phenomena of light diffraction and interference to be explained. It failed, however, to provide an explanation for the light birefringence phenomenon—a physical phenomenon that was observed first as early as the seventeenth century. This subsequently led to the postulate that light is an electromagnetic wave. Namely, a wave that can be fully described by four vector fields, namely the electric field, magnetic field, electric flux field, and magnetic induction field. These four fields can be obtained by solving Maxwell's equations that, for an isotropic medium, have the form:

$$\nabla \times H = \frac{\partial D}{\partial t}$$
$$\nabla \times E = -\frac{\partial B}{\partial t} \qquad (1.2)$$

In the absence of free electric charges Equation 1.2 is complemented by the following conditions that are used to eliminate spurious solutions:

$$\nabla \cdot D = 0$$
$$\nabla \cdot H = 0$$

The number of the unknown fields in Equation 1.2 can be reduced by adding the material constitutive equations:

$$D = \varepsilon E$$
$$B = \mu_0 H \qquad (1.3)$$

where ε is the dielectric constant and μ_0 represents the magnetic permeability of the free space. Even though it is known that the electromagnetic wave theory fails to explain the light quantisation phenomenon, it is, nowadays, the most established theory of light that is used for the development of photonic design and modelling tools. The more sophisticated theory based on quantum electrodynamics has been advanced considerably over the last decades; however, it is still too complicated for a direct application in photonic design tools. Hence, an approach based on a phenomenological incorporation of the main light characteristics that come from quantum electromagnetics into the classical theory is preferred. We gradually develop the theory of light propagation in Chapters 2 through 4 and apply it to the modelling of active devices in Chapters 7 and 8, and to the modelling of light propagation in the nonlinear media in Chapter 9.

ACTIVE MEDIUM MODEL

The active medium models are based in principle on quantum mechanics. In the case of semiconductor devices two main phenomena need to be taken into account namely, the current conduction and the interaction of the photons with the current-carrying electrons. Semiconductor crystals are made up of fairly large atoms. The atom of gallium has for instance 31 electrons. However, according to semiconductor physics, only the electrons of the outer electron shell contribute to the current flow and effectively exchange energy with the optical field. In a semiconductor crystal, the outer electron shell electrons are not bound to a specific atom. Further, their energies are closely spaced and can be grouped into two bands: the conduction band and the valence band. The energy band gap between these two top energy bands plays a crucial role in determining the optical properties of a particular semiconductor. This is because one of the main modes of interaction between electrons and photons consists of an electron falling from the conduction band to the valence band (an interband transition), while releasing the energy in a form of a photon. To preserve energy the photon frequency multiplied by Planck's constant has to be equal to the band gap energy. Because the band gap energy is an intrinsic property of a particular semiconductor, there is a correlation between a particular semiconductor material used and the operating wavelength of the device. For instance, GaN LDs emit light in the blue wavelength range whereas the GaAlAs based LDs are used for designing light sources that operate near 1 μm wavelength.

As mentioned earlier, only the conduction and valence band electrons participate in the current conduction in a semiconductor. In the case of the valence band, it is more convenient to model the electrical transport of fictitious positive charges (holes) that correspond to a missing electron rather than all the valence electrons. Therefore, the simplest comprehensive model for studying the current flow in a semiconductor includes an equation for the electrons in the conduction band and a separate equation for holes in the valence band. Because the spatial variations in electron and hole concentrations within a semiconductor material may source a local electric field, to obtain a self-consistent model it is necessary to include a third equation that will allow the spatial distribution of the locally induced electric field to be calculated. The set of three equations that describe the electron and hole transport in a semiconductor material, and known in the literature as the drift-diffusion equations, are:

$$\nabla \cdot \mathbf{J_n} = qR$$

$$\nabla \cdot \mathbf{J_p} = -qR \tag{1.4a}$$

$$\nabla \cdot (\varepsilon_w \nabla \phi) + q(p - n + N_D^+ - N_A^-) = 0 \tag{1.4b}$$

Vectors $\mathbf{J_n}$ and $\mathbf{J_p}$ in Equation 1.4a stand for the electron and hole current density respectively. R denotes the net recombination rate, whereas q is the electron charge. In Equation 1.4b, ϕ denotes the electric potential distribution, p and n are the distributions of the electron and hole density, respectively. N_D^+ and N_A^- stand for the concentration of the donors and acceptors whereas ε_w is the static electric permittivity. Because the solution of Equation 1.4 is not trivial, when modelling the LDs, Equation 1.4 sometimes approximated by a unipolar diffusion equation:

$$D_n \nabla^2 N - R + \frac{J}{qd} = 0 \tag{1.5}$$

In Equation 1.5, D_n denotes a diffusion constant, N is the carrier concentration, J stands for the injection current density, and d is the width of the quantum well. The current conduction in semiconductors will be discussed in Chapter 6 whereas the interaction of electrons and photons in LDs will be studied in Chapter 8.

Several important photonic devices rely on the interaction of photons with the electronic energy levels of lanthanide atoms. Studying the properties of the lanthanide atoms falls mostly into the domain of atomic spectroscopy. The currently established model of infrared and visible light interaction with the lanthanide atoms has been formulated by B. R. Judd and G. S. Ofelt in the second half of the twentieth century [1]. Their theory postulates that the photons interact with the electrons in the partially filled 4f shell of a lanthanide atom. The interaction of photons and lanthanide atoms is discussed in Chapter 7.

HEAT DIFFUSION AND STRESS–STRAIN MODELS

The heat diffusion in solid state photonic devices is studied to optimise the thermal management within the device. The most established theory at the moment assumes that the heat diffuses within the device via the heat conduction process. The heat flux measured in Watts of thermal power over unit surface is assumed to be proportional to the temperature gradient. The balance of the net heat conducted into a unit volume, the heat generated, and the thermal energy stored within the unit volume yields the heat conduction equation:

$$\nabla \cdot \kappa \, \nabla T + H = c \frac{\partial T}{\partial t} \tag{1.6}$$

In Equation 1.6, T stands for the temperature distribution, H is the heat generation rate, c is the volume heat capacity, and κ is the heat conductivity. The accurate prediction of the temperature distribution within a photonic device can only be performed if a fairly good estimate of all the constants and the heat generation rate is available. It must be stressed that the accurate evaluation of the heat generation rate within a semiconductor device is not trivial and needs to be carried out carefully. Otherwise, the developed device model might not fulfil the energy conservation principle. Heat diffusion modelling will be discussed in Chapter 5.

Finally, we note that the stress and strain models typically do not form a part of the device model. It is generally recognised that the stress and strain fields are usually not directly coupled with the process of light generation in the photonic device. This does not mean, however, that the stress and strain distributions do not affect the operation of a photonic device. For instance, strained quantum wells are used in LDs [2] whereas strain-induced optical waveguides are of importance for integrated optics [3]. When including stress and strain in photonic device modelling, it is usual practice to calculate the distribution of stress and strain tensor fields separately using standard professional tools, for example, ANSYS™. Alternatively, if available, approximate analytical expressions can be used [4]. The methods for the calculation of stress and strain tensor field distributions are not discussed in further detail in this book because there is abundant literature on this topic [5].

PHOTONIC DEVICE MODELS

The discussion presented in first three sections of this chapter shows that design and modelling in photonics requires finding solutions to fairly complex sets of partial differential equations. In a general case, such solutions can only be obtained using numerical methods and sufficiently fast computers. During the last two decades, there has been significant progress in computer technology. Nowadays, commercially available standard personal computers have several giga bytes of memory and multicore processors operating at clock speeds of over 1 GHz. This gives sufficient computational power for tackling fairly complex problems involving partial differential equations on a simple PC. So unlike the last decade of the twentieth century, quite a significant proportion of the design work in photonics can now be accomplished without using expensive supercomputers. As a result, the customer basis for the software companies that provide professional design tools for photonic devices significantly widened improving their profitability and ability to survive in the market. There are currently several companies that provide software for the design of semiconductor lasers, planar optical waveguides, optical fibres, and bulk optics elements, for example, Optiwave, Photon Design, RSoft, Crosslight, or TimberCad. The scientific community that researches into novel modelling and design techniques has also grown over the last two decades. Special terminology was developed that facilities specifying various aspects of the numerical models. Because this terminology will be used throughout the book, we give a short account of it as follows.

One of the primary characteristics of a photonic device model is its dimensionality. A three-dimensional model (a 3D model) for instance, is based on the numerical solution of the partial differential equations that considers all three spatial dimensions. In principle, a 3D model provides a rigorous solution of the partial differential equations. The larger the number of dimensions that are included directly in the computations the better are the accuracy of the results. This, however, is usually achieved at the expense of a larger computational overhead (Figure 1.4). The primary reason for using models with a dimensionality of less than 3 is their superior computational efficiency. Models with a large computational efficiency are typically used to search thoroughly the design parameter space. The full 3D model is applied once the design parameter space is sufficiently narrowed for the slow but accurate design tools to be used for fine-tuning the design parameters and more in-depth understanding of the photonic device operating principles. This combined approach allows achieving a shorter design time, which means also a faster introduction of a new photonic device into the market and a lower design cost (less computer resources and design engineer time). Thus typically in the design process several design tools with various dimensionalities are used concurrently to reduce the overall design effort.

Two other important concepts that are used in photonic engineering are the first principles model and the phenomenological model. A first principles model is based on the solution of the fundamental equations that govern the particular physical theory that forms the basis for the model. When studying the interactions between light and matter the most fundamental theory that is currently established in physics is quantum electrodynamics [6,7]. However, this theory is too complicated for a direct application in the photonic design and modelling. In fact, even the direct solution of the Schrödinger equation for the (t electron) helium atom is challenging, whereas many of the atoms or ions that find application in photonics have more than 30 electrons, for example, gallium (31 electrons) or a trivalent erbium ion (65 electrons). This fact makes the development of strictly first principles models in photonics virtually impossible. Nonetheless, the concept of the first principles model is used with a reference to some existing models and it usually means a model that uses the most fundamental equations and only a small number of arbitrary adjustable fitting parameters. This definition is too some extent inherently ambiguous and hence claiming a first principles model should be always followed by an explanation of what exactly is meant. A phenomenological model, on the other hand, is a model that is derived from experimental observation using the intuition of the person who

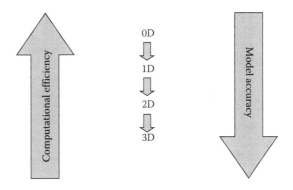

FIGURE 1.4   Trade-offs between the 0D, 1D, 2D, and 3D models.

developed the model. A phenomenological model relies on a large number of arbitrary adjustable parameters. Developing a phenomenological model that can turn into a useful design tool is not a simple task and requires both a very good understanding of photonics in general and of the particular design problem.

Another important concept that is used in the photonic modelling and design is one of a predictive model. A predictive model can be used to calculate the change in the device characteristics subject to a change of either a single parameter or a set of parameters. It is therefore good practice when discussing predictive models to specify for which parameters and device characteristics the particular model is predictive. Further, a predictive model can be predictive only within a particular parameter range. For instance, a high power LD model can calculate correctly the output power only for moderate values of the bias current but fails to predict correctly the maximum output power (i.e., to predict the light–current characteristic within the entire range of the bias current values). Predictive models are of paramount importance in the design of photonic devices. Accurate, (predictive) design tools can significantly shorten product development time and cost. This, however, does not mean that the modelling tools that are not predictive are not useful for the development of novel photonic devices. In fact, in many cases, very useful information can be gained from models that can only reproduce the experimentally observed results. Such models typically play a crucial role at the early stage of novel photonic device development, when the behaviour of the device is not yet fully understood. A model that successfully reproduces the experimental results can identify the origin of the observed device behaviour and be the first step on the way to the development of a predictive design tool.

In computational electromagnetics, the models fall into one of two categories, namely time domain models and frequency domain models. Time domain models are based on the solution of the equations in the time domain whereas the frequency domain models use the Fourier transform and provide the solution in the frequency domain. This classification has been also used in the analysis of light propagation in optical waveguides. However, when modelling more complicated structures, for example, LDs, a different classification is used. Namely, the models used for LD design are divided into the following three categories: time domain models, continuous wave (CW) models, and multiwavelength (spectral) models. A CW model allows device characteristics to be studied in the steady state only. The time domain models on the other hand allow for modelling of transients and also for performing steady state analysis. CW spectral models cannot be used for simulating transients; however, they can predict (or reproduce) the laser spectrum. In fact, a spectral model is a frequency domain model that performs the analysis within a wide range of frequencies. The difference between these three types of models will become more obvious in Chapter 8.

For the sake of completeness, it should be also mentioned that the models can be classed as either analytical models or numerical models. An analytical model is based on the analytical solution of the equations that describe the device operation whereas numerical models use numerical methods to obtain the solution. Analytical models, although important, have a very limited application in photonic design. Most of the problems faced by a photonic design engineer require an application of numerical modelling techniques, which are discussed in this book.

The progress in computer technology has been accompanied by the development of the computer programming environment. Initially, most of the software in photonics was developed using standard programming languages, for example, FORTRAN or C++. More recently increased use is made of specialised engineering software environments, like MATLAB®, which has been used throughout this book for writing illustrative examples, or Python.

REFERENCES

1. Judd, B.R., *Operator Techniques in Atomic Spectroscopy*. 1963, New York: McGraw-Hill.
2. Erbert, G., et al., High-power tensile-strained GaAsP-AlGaAs quantum-well lasers emitting between 715 and 790 nm. *IEEE Journal of Selected Topics in Quantum Electronics*, 1999. 5(3): p. 780–784.
3. Saitoh, K., M. Koshiba, and Y. Tsuji, Stress analysis method for elastically anisotropic material based optical waveguides and its application to strain-induced optical waveguides. *Journal of Lightwave Technology*, 1999. 17(2): p. 255–259.
4. Chow, W.W. and S. Koch, *Semiconductor-Laser Fundamentals: Physics of the Gain Materials*. 1999, Berlin: Springer.
5. Sadd, M.H., *Elasticity, Theory, Applications and Numerics*. 2009, Oxford: Academic Press.
6. Sargent, M., M.O. Scully, and W.E. Lamb, *Laser Physics*. 1974, London: Addison-Wesley.
7. Marcuse, D., *Principles of Quantum Electronics*. 1980, London: Academic Press.

2 Light Propagation in Homogenous Media

In this chapter, we discuss the numerical analysis of light propagation in homogenous media. It is assumed that the reader is familiar with wave optics, electromagnetic optics, and Fourier optics (see [1,2]). In particular, we discuss the application of the Fourier method. In the first section, we discuss the application of the Fourier method to study optical beam propagation in homogenous media. In the second section, we extend the Fourier method and study optical beam reflection and refraction. In the third section, we discuss the application of paraxial and wide angle approximations. Finally, in the last section, we discuss the modelling of optical systems that include thin bulk optical elements. In this chapter, we also introduce several basic concepts that are further explored in the remaining chapters of this book. A description of the application of ray optics methods can be found in Poon and Kim [2].

FOURIER METHOD

The Fourier method provides a simple and an efficient way of calculating the evolution of an optical beam envelope function when propagating in a homogenous medium. To explain how the Fourier method works, we introduce the Cartesian coordinate system (x,y,z) and consider the simplest case, namely an optical wave that propagates with the propagating vector in the plane (x,z). For such a case, there is no dependence of the wave function u on the spatial variable y and hence the wave equation reduces to

$$\frac{\partial^2 u}{\partial x^2} + \frac{\partial^2 u}{\partial z^2} - \frac{1}{c^2}\frac{\partial^2 u}{\partial t^2} = 0 \tag{2.1}$$

To find a solution for Equation 2.1 when the time dependence of $u(x,z,t)$ is harmonic, we follow the standard steps, that is, we introduce the complex wave amplitude $\tilde{U}(x,z)$ that is defined through the equation:

$$u(x,z,t) = u(x,z)\cos(\omega t + \varphi) = \frac{1}{2}\left[\tilde{U}(x,z)e^{j\omega t} + \tilde{U}^*(x,z)e^{-j\omega t}\right] \tag{2.2}$$

whereby the phase φ was incorporated into the complex wave amplitude $\tilde{U}(x,z) = u(x,z)e^{j\varphi}$. Substituting 2.2 into 2.1, multiplying Equation 2.1 by $\exp(-j\omega t)$ and integrating over the period $T = 2\pi/\omega$ yields the following equation for the complex wave amplitude $\tilde{U}(x,z)$:

$$\frac{\partial^2 \tilde{U}}{\partial x^2} + \frac{\partial^2 \tilde{U}}{\partial z^2} + k^2 \tilde{U} = 0 \qquad (2.3)$$

whereby k is equal to $2\pi/\lambda_{med}$, and λ_{med} gives the wavelength measured in the medium. Alternatively, k can be defined as $2\pi n/\lambda$ where λ is the wavelength measured in the free space and n is the refractive index of the medium.

In a homogenous medium, that is when $k(x,z) = $ constant, a solution of the partial differential equation 2.3 can be found easily using the method of variable separation. This yields $\tilde{U}(x,z)$ in the following form:

$$\tilde{U}(x,z) = A\,e^{-jk_x x}e^{-jk_z z} \qquad (2.4)$$

The separation constants k_x and k_z can assume any value between $+\infty$ and $-\infty$ and are interdependent, whereas A is an arbitrary constant. The relation between k_x and k_z can be obtained by substituting $\tilde{U}(x,0)$ back into Equation 2.3. This yields the following equation:

$$k_x^2 + k_z^2 = k^2 \qquad (2.5)$$

Therefore, each solution of Equation 2.3 in the form 2.4 can be identified by specifying either k_x or k_z. The other constant needed in 2.4 can be obtained from 2.5. The pair of constants: k_x and k_z can also be understood as two components of a vector (wave vector) along the x and z directions of the Cartesian coordinate system, respectively (Figure 2.1). From 2.5 it follows that the absolute value of the wave vector $\vec{k}$ is equal to k. Finally, once the wave vector is specified the spatial and time dependence of a particular solution of the original Equation 2.1 can be obtained from 2.2.

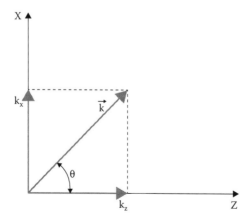

FIGURE 2.1 A graphical representation of the propagation vector in the rectangular coordinate system.

The character of the solution of Equation 2.1 depends critically on the value of the constants k_x and k_z. When k_x is equal to zero then at any time instant t the dependence of $u(x,z,t)$ on the x variable is given by a constant function, whereas the dependence on the z variable is given by a cosine function. This case clearly corresponds to a wave that propagates along the z-axis. When k_x is not equal to zero then at any time instant t, the dependence of $u(x,z,t)$ on the x variable is given by a cosine function, whereas the dependence on the z variable depends on the value of k_x. If k_x is less than k, then the dependence on the z variable is given by a cosine function. This case corresponds to a wave propagating at an angle to the z-axis. If $k_x = k$, then the dependence on the z variable is given by the constant function. Whereas, for $k_x > k$, the z dependence is given by an exponential function. The former case corresponds to a wave propagating along the x-axis and the latter one to an evanescent wave that decays along the z-axis. We illustrate these different cases in Figure 2.2.

After this introductory discussion we can pose the problem namely, how to calculate the distribution of $\tilde{U}$ at the position $z = L$ if the distribution of $\tilde{U}$ at the position $z = 0$, $\tilde{U}(x,0)$ is known (Figure 2.3).

First we observe that if $\tilde{U}$ is given by 2.4 then the calculation of $\tilde{U}(x,L)$ subject to known $\tilde{U}(x,0)$ is straightforward:

$$\tilde{U}(x,L) = \tilde{U}(x,0)e^{-jk_z L} \tag{2.6}$$

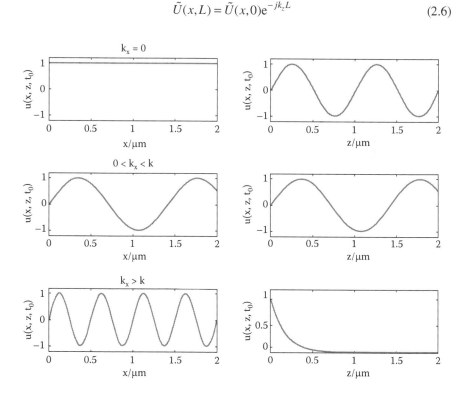

FIGURE 2.2 Dependence of $u(x,z,t)$ at a time instant t_0 on x and z for three selected values of k_x.

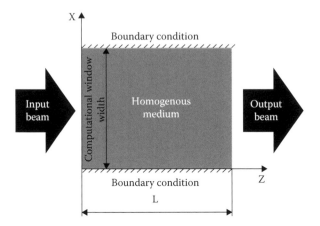

FIGURE 2.3 Schematic diagram illustrating the problem of the optical beam propagation through a section of a homogenous medium.

For an arbitrary $\tilde{U}(x,0)$ this observation leads to the following conclusion: if we could represent $\tilde{U}(x,0)$ as a sum of $e^{-jk_x x}$ functions with some weighting coefficients (a linear combination) then we could easily calculate $\tilde{U}(x,L)$ for each function and then sum them up to obtain the final result. The representation of an arbitrary wave function $\tilde{U}(x,0)$ by a sum of $e^{-jk_x x}$ functions is in fact nothing else but expressing $\tilde{U}(x,0)$ using the Fourier series. In the limit, when k_x varies continuously, the representation of $\tilde{U}(x,0)$ in terms of $e^{-jk_x x}$ functions can be obtained by calculating the Fourier transform of $\tilde{U}(x,0)$ with respect to x:

$$\tilde{U}(k_x,0) = \int_{-\infty}^{\infty} \tilde{U}(x,0)e^{-j2\pi f_x x}dx \tag{2.7a}$$

where $2\pi f_x = k_x$. The distribution of the wave function in the Fourier domain at $z = L$ is obtained from:

$$\tilde{U}(k_x,L) = \tilde{U}(k_x,0)e^{-jk_z L} \tag{2.7b}$$

whereas the result of the final summation can be obtained by performing an integration by calculating the inverse Fourier transform of 2.7b:

$$\tilde{U}(x,L) = \int_{-\infty}^{\infty} \tilde{U}(k_x,L)e^{j2\pi f_x x}df_x \tag{2.7c}$$

The technique given by Equation 2.7 can be used to calculate an optical beam evolution analytically. However, the use of such technique is limited to the optical beam shapes for which the relevant integrals can be analytically evaluated (see [1,3–5]). For an arbitrary optical beam shape, the application of numerical methods is more convenient. An algorithm that could be easily implemented on a personal computer can be obtained using a discrete Fourier transform (DFT) instead of the Fourier transform in 2.7. For this purpose $\tilde{U}(x,0)$ has to be sampled at a set of equidistantly spaced points x_l within a window of width $M\Delta x$ where Δx is the sampling interval and M is the total number of samples. The set of samples $\tilde{U}(x,0)$ can be used to calculate the DFT values at a set of spatial frequencies $k_{xm} = l/M\Delta x$. Each discrete harmonic is then propagated using 2.6, and the final summation is performed using the inverse DFT. Both the DFT and the inverse DFT can be easily and efficiently implemented on a computer using fast Fourier transform (FFT) algorithms that are readily available from most numerical modelling libraries and from standard reference books on numerical modelling [6]. The proposed method can be therefore summarised by replacing Equation 2.7 with the following set of equations:

$$\tilde{U}(k_{xm},0) = \Delta x \sum_{l=0}^{M-1} \tilde{U}(x_l,0)e^{-j2\pi lm/M} \qquad (2.8a)$$

$$\tilde{U}(k_{xm},L) = \tilde{U}(k_{xm},0)e^{-jk_{zn}L} \qquad (2.8b)$$

$$\tilde{U}(x_l,0) = \frac{1}{M} \sum_{m=0}^{M-1} \tilde{U}(k_{xm},0)e^{j2\pi im/M} \qquad (2.8c)$$

The preceding discussion also allows us to introduce the concepts of the basis function and the basis function set. A basis function set is introduced to facilitate the solution of the problem using numerical methods and allow convenient implementation of the resulting algorithm using a standard programming language. In the example considered, $e^{-jk_x x}$ is a basis function and all functions $e^{-jk_x x}$ for k_x varying from $-\infty$ to $+\infty$ form the basis function set. The basis set is introduced because it is not straightforward to calculate numerically $\tilde{U}(x,L)$ for an arbitrary $\tilde{U}(x,0)$. However, such calculations become straightforward if $\tilde{U}(x,0)$ is either a $e^{-jk_x x}$ function or a sum of such functions. Thus using $e^{-jk_x x}$ basis functions allows for simplifying the numerical calculations. A problem that is inherently linked with a selection of a basis function set is that of determining what class of functions can be represented by the selected basis function set (or in which function set the basis function set forms a dense subset as the problem is formulated in the functional analysis [7]). A basis set cannot usually represent an arbitrary function. The set of functions that can be represented by a linear combination of the basis functions in the example considered is known from the Fourier series theory and corresponds to the class of all L^2 integrable functions [7]. This class of functions contains practically any physically relevant

---

ALGORITHM 2.1   FOURIER METHOD FOR OPTICAL BEAM PROPAGATION IN A HOMOGENOUS MEDIUM

1. Start
2. Set the free space wavelength λ
3. Select N for $2N$ equidistant sampling points for FFT
4. Initialise: at $z = 0$, $\widetilde{U}(x,0)$
5. Calculate FFT of $\widetilde{U}(x,0)$: $\widetilde{V}(k_x,0) = \Im\left(\widetilde{U}(x,0)\right)$
6. Calculate $\widetilde{V}(k_x,L) = \widetilde{V}(k_x,0)\,e^{jk_zL}$
7. Obtain $\widetilde{U}(x,L)$ by calculating inverse FFT of $\widetilde{V}(k_x,L)$
8. Stop

---

optical beam shape, which makes the proposed Fourier method a very useful tool for studying optical beam propagation in a homogenous medium. Finally, it is noted that it is not necessary for the reader to be familiar with functional analysis to understand the presented material; however, further information on this topic can be found in a number of textbooks [7,8].

After the preceding discussion, we finally propose Algorithm 2.1 as an implementation of the Fourier Method discussed.

Before Algorithm 2.1 can be implemented using a programming language, the values of k_z at each sampling point for $\widetilde{V}(k_x,0)$ have to be calculated. The values of k_z can be obtained from 2.5 as:

$$k_z = \pm\sqrt{k^2 - k_x^2} \tag{2.9}$$

The dependence of k_z on k_x is shown in Figure 2.4. For $k_x > k$, k_z is purely imaginary and may lead to either an exponential decay (Figure 2.2) or growth. The selection of the appropriate sign in 2.9 is therefore crucial for preserving the stability of Algorithm 2.1. The sign should be selected so that the evanescent waves are attenuated. The opposite value of the sign will yield spurious amplification of the evanescent waves and render Algorithm 2.1 unstable. After this discussion it is fairly straightforward to implement Algorithm 2.1 in MATLAB script:

```
% Fourier Method propagation for a Gaussian shape optical beam
clear % clears variables
clear global % clears global variables
format long e
% initial constants
I = sqrt(-1);% remember not to overwrite pi or I !
pi = 3.141592653589793e+000;

lam = 1.06;% wavelength [mi]
n = 1.0;% refractive index of the medium
k = 2*pi*n/lam;
```

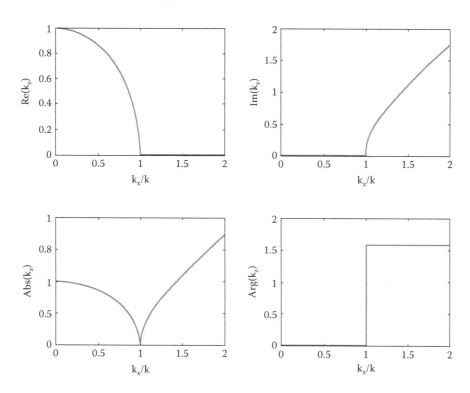

FIGURE 2.4 Dependence of k_z on k_x for formula (2.9).

```
wind_width = 600;% numerical window width[mi]
NoSamp = 2^12;% number of samples
dx = wind_width/NoSamp;
x_init = dx*(0:NoSamp-1);% x values
x = x_init-wind_width/2;

% calculating the initial beam distribution
beam_width = 30;% beam half width[mi]
teta = pi/24;%beam tilt angle
E = exp(-x.*x/(beam_width*beam_width)).*exp(-I*k*x*sin(teta));
Z(1,:) = abs(E).*abs(E);
X = x;
Y(1) = 0;

dist_step = 60.;%propagation distance [mi]
for m = 1:50
kx = 2*pi*(-NoSamp/2:NoSamp/2-1)/(NoSamp*dx);
% calculating the values of kx
%calculating kz and changing the sign of the imaginary part to
%keep stable
kz = sqrt(k^2-kx.^2);
mask1 = imag(kz)>0;
```

```
mask2 = imag(kz)< = 0;
kz = -mask1.*kz+mask2.*kz;
%calculating the field at the next z position
fE = (fft(E));
fE = fftshift(fE);
fE1 = fE.*exp(-I*dist_step*kz);
fE1 = ifftshift(fE1);
E1 = ifft(fE1);% field at position z = dist
E = E1;
Z(m+1,:) = abs(E).*abs(E);
Y(m+1) = m*dist_step;
end

contourf(X,Y,Z,'EdgeColor','none')
xlabel('x/micrometers','Fontsize',20)
ylabel('z/micrometers','Fontsize',20)
colormap cool
```

Note that `fftshift` command was used to shift the zero frequency component to the centre of the spectrum. This corresponds to calculating DFT within the interval stretching from $-M/2$ to $M/2$ rather than from 0 to M, which is the way that the MATLAB FFT algorithm calculates the DFT. Introducing this shift simplifies the calculation of the k_x values that correspond to the values of the DFT calculated by the FFT algorithm. Once the shift is performed, k_x values that grow linearly from $-\pi/M$ dx to π/M dx can be used. Furthermore, we inverted the sign of the imaginary part of k_z to keep the evanescent modes attenuated. In the presented script, we split the distance of 3 mm into 50 steps. Although such splitting of the distance is not necessary and results in a longer calculation time, it allows for analysing the beam evolution using the 3D plotting functions of MATLAB.

The last points left to discuss are related to the selection of the numerical window size and the sampling rate. However, because these are the standard problems related to an application of FFT we refer the reader to standard textbooks [6]. Here, we only consider several simple illustrating examples that show the main difficulties in applying the Algorithm 2.1 and discuss them in the context of the numerical analysis of optical beam propagation.

As an illustrating example, we consider a plane wave with a Gaussian envelope that propagates at an angle θ with respect to the z-axis:

$$\tilde{U}(x,L) = e^{\left(\frac{x}{w}\right)^2} e^{-jkx\sin(\theta)} \tag{2.10}$$

The Gaussian envelope beam 2.10 is propagated at an angle α over a distance L in a homogenous medium (Figure 2.5) [9]. The initial half-width of the Gaussian envelope w is measured at the $1/e$ point, at $z = 0$ μm, along the x-axis. The wavelength used in the simulations is 1.06 μm and corresponds to a typical operating wavelength of an Nd:YAG laser, whereas the reference refractive index was assumed equal to unity.

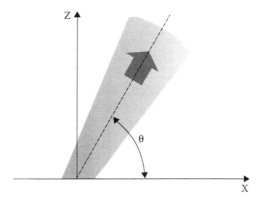

FIGURE 2.5   Schematic diagram of an off-axis propagating optical beam.

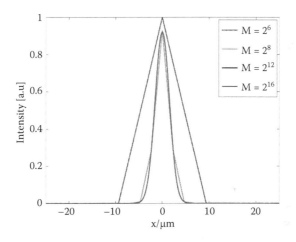

FIGURE 2.6   Light intensity distribution at $z = 10$ µm calculated by Algorithm 2.1. The initial beam width equals 2.828 µm while the tilt angle $\theta = 0°$. The calculation window is 600 µm.

Figure 2.6 shows the light intensity distribution at $z = 10$ µm calculated by Algorithm 2.1 using a variable number of sampling points. The initial beam half-width equals 2.828 µm while the tilt angle $\theta = 0°$. The computational window width is 600 µm (Figure 2.3). With a relatively small number of sampling points, a crude approximation of the beam shape is obtained. However, as the number of sampling points increases, a more accurate solution is obtained. Figure 2.6 shows that no further improvement of the accuracy can be observed when increasing the number of sampling points beyond 2^{12}.

A more complicated behaviour of the calculated solution on the number of sampling points is observed when the tilt angle is different than zero. Figure 2.7 shows the light intensity distribution at $z = 10$ µm calculated by Algorithm 2.1 using a variable number of sampling points. The computational window width and the initial beam half-width

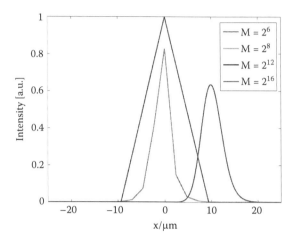

FIGURE 2.7 Light intensity distribution at $z = 10$ μm calculated by Algorithm 2.1. The initial beam width equals 2.828 μm while the tilt angle θ = 45°. The calculation window is 600 μm.

w are kept the same as in the previous example. However, the tilt angle θ = 45°. In this case, the solution obtained with a low sampling rate predicts that, despite the fast phase variations introduced by the large tilt angle, the beam propagates approximately along the z-axis. This is a result that is qualitatively wrong. Only with the sufficiently large sampling rate, the correct direction of the optical beam propagation is obtained. To explain the origin of this phenomenon we plot the light intensity distribution of the initial field at $z = 0$, its DFT and the field distribution at $z = L$ for 2^{12} and 2^8 sampling points, respectively in Figures 2.8 and 2.9. The vertical lines mark the boundary between the values of k_x that correspond to the propagating waves and the values of k_x that correspond to the evanescent waves. The vertical lines allow for an easier location of the calculated spectrum in the Fourier domain. Figure 2.7 shows that if the number of sampling points is increased beyond 2^{12} no further improvement of the accuracy can be observed. For 2^{12} sampling points, Figure 2.8 shows that the DFT of the initial field has nonzero values for k_x between 3/μm and 5/μm. This agrees with expectations, since the k_x value of a plane wave that propagates at 45° is equal to 4.2/μm. However, the DFT calculated with 2^8 sampling points has no components beyond $k_x = 2$/μm. This results from the fact that the width of the computational window for DFT is linked with the sampling rate. A low sampling rate intrinsically imposes a narrow computational window in the Fourier domain. In the example considered, selecting 2^8 sampling points results in a computational window in the Fourier space that is limited to 2/μm, which is insufficient for analysing beam propagation at an angle of 45° since the k_x value of a plane wave that propagates at 45° is equal to 4.2/μm. We note that alternatively one could arrive at the same conclusions by applying Nyquist theorem. In the example considered, it was fairly easy to notice that the result calculated using 2^8 samples was incorrect. However, Algorithm 2.1 is known to produce results that seem correct but which prove wrong under a more detailed investigation. One should therefore use this

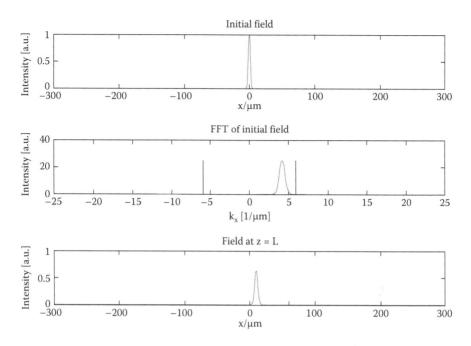

FIGURE 2.8 Light intensity distribution calculated by Algorithm 2.1. The initial beam width equals 2.828 µm while the tilt angle $\theta = 45°$. The calculation window is 600 µm while the number of samples is 2^{12}.

algorithm with caution and always check if the number of sampling points is sufficient for the problem considered.

In the next example, we study another phenomenon, namely the prediction by Algorithm 2.1 of a numerical artefact. Figure 2.10 shows the light intensity distribution calculated by Algorithm 2.1. The initial beam half-width equals 30 µm, whereas the tilt angle $\theta = 7.5°$. The calculation window is 600 µm. The number of samples is 2^8 and the beam is propagated for 3000 µm. To allow for a simpler understanding of the observed results, instead of calculating the field directly at $z = L$, we applied the algorithm iteratively using 500 steps along the z-axis. By collating together the results, we obtained a two-dimensional (2D) distribution of the light intensity. These results show that once the beam reaches the boundary of the computational window it spuriously re-enters the computational window at the other side. Such a phenomenon is a pure numerical artefact and can be avoided by either using a sufficiently large window or using special transparent boundary conditions, which will be discussed in further detail in subsection "Boundary Condition" in the context of the beam propagation algorithms.

The last example that we study in this section is designed to explore the numerical stability and reciprocity of Algorithm 2.1. For Algorithm 2.1 to be a useful tool for studying the optical beam propagation it has to be numerically stable, that is, not to produce any uncontrolled growth of numbers representing the solution that can result in a numerical underflow or overflow. Furthermore, it should be reciprocal, that is, if

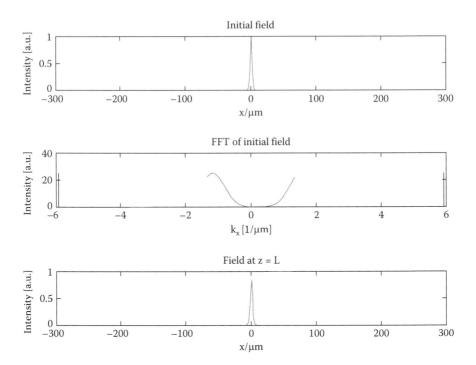

FIGURE 2.9 Light intensity distribution calculated by Algorithm 2.1. The initial beam width equals 2.828 μm while the tilt angle θ = 45°. The calculation window is 600 μm while the number of samples is 2^8.

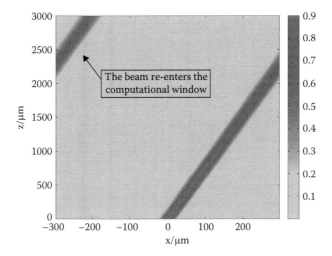

FIGURE 2.10 Light intensity distribution calculated by Algorithm 2.1. The initial beam width equals 30 μm while the tilt angle θ = 7.5°. The calculation window is 600 μm while the number of samples is 2^8 while the beam is propagated for 3000 μm.

the algorithm, for a given initial field distribution at $z = 0$ calculates a particular field distribution at $z = L$, then reversing steps and using the field at $z = L$ as an initial field should yield back the field at $z = 0$. Figure 2.11 shows the light intensity distribution calculated by Algorithm 2.1. The initial beam width equals 30 µm, whereas the tilt angle $\theta = 7.5°$. The calculation window is 600 µm, whereas the number of samples equals 2^{12} and the beam is propagated for 1.8 µm. The arrow in Figure 2.11 indicates the direction of the optical beam propagation. Again, as in the previous example, we obtain the optical beam shape at $z = 1.8$ µm by calculating iteratively the final beam distribution in 225 steps. We first calculate the optical beam intensity distribution at $z = 1.8$ µm and then use this distribution as an initial beam distribution and try to calculate the beam distribution at $z = 0$ µm using the negative value of the propagating step. If the algorithm was reciprocal, the initial field at $z = 0$ µm should be recovered. However, as can be observed in Figure 2.11 this is not the case. The algorithm not only fails to give the initial field distribution at $z = 0$ µm but also becomes unstable. This behaviour follows from the fact that Algorithm 2.1 used with a negative step, according to 2.8b, provides exponential amplification to all frequency components present in the spectrum that correspond to the evanescent waves. When carrying the calculations analytically using 2.7 this phenomenon does not exist since all operations are performed exactly. However, when using a numerical algorithm, round off errors produce spurious spectral components at the frequencies corresponding to the evanescent modes, which get amplified exponentially according to 2.8b. On the other

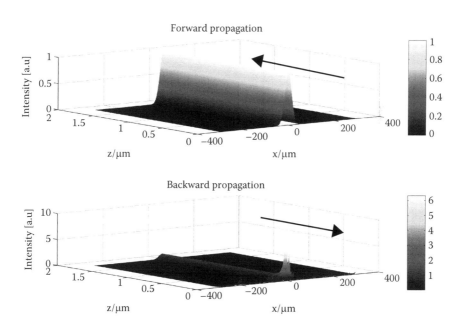

FIGURE 2.11 Light intensity distribution calculated by Algorithm 2.1. The initial beam width equals 30 µm while the tilt angle $\theta = 7.5°$. The calculation window is 600 µm while the number of samples is 2^{12} while the beam is propagated for 1.8 µm. The arrow indicates the direction of the optical beam propagation.

hand, if the evanescent modes are not included in the computation, for example, when using a coarser sampling, Algorithm 2.1 is fully reciprocal (Fig. 2.12).

Another problem that is important to consider when analysing the accuracy of the results obtained by Algorithm 2.1 is that of the power conservation. Any algorithm that is used for solving a physical problem should, in principle, give results that do not violate the basic laws of physics. One of the fundamental principles in physics is that of energy conservation. From this principle it follows that if in the considered region (Figure 2.2) there are no sources of optical waves, then the total amount of power that enters the region at the point $z = 0$ should be equal to the total power leaving the region at $z = L$. In wave optics, it is postulated that the light intensity is proportional to $\left|\tilde{U}(x,z)\right|^2$ [1]. The light intensity has got units of W/m² hence the integral of the light intensity calculated over the entire transverse plane gives, in principle, the total amount of energy that crosses a particular plane within a unit time. It would therefore be expected that the integral of $\left|\tilde{U}(x,0)\right|^2$ along the x-axis should be equal to that calculated using $\left|\tilde{U}(x,L)\right|^2$. Strictly, such a calculation gives us the conservation of power density since we only consider a two-dimensional problem. In theory, numerical algorithms that preserve a particular physically meaningful quantity are referred to as symplectic algorithms [10,11]. Therefore, we would like to establish if Algorithm 2.1 is symplectic with respect to the energy conservation. First of all, we would like to make an observation that proving the energy conservation of the problem 2.7 does not in a general case necessarily imply that 2.8 also preserves energy. We will therefore consider 2.7 first. From Parseval's theorem it follows that

$$\int\limits_{-\infty}^{\infty} \left|\tilde{U}(x,0)\right|^2 dx = \int\limits_{-\infty}^{\infty} \left|\tilde{U}(f_x,0)\right|^2 df_x$$

and

$$\int\limits_{-\infty}^{\infty} \left|\tilde{U}(x,L)\right|^2 dx = \int\limits_{-\infty}^{\infty} \left|\tilde{U}(f_x,L)\right|^2 df_x$$

However, we can also observe that

$$\left|\tilde{U}(f_x,L)\right|^2 = \tilde{U}(f_x,L) * \tilde{U}(f_x,L)^* = \tilde{U}(f_x,0)\, e^{-j\sqrt{k^2 - 2\pi f_x^2}} * \tilde{U}(f_x,0)^* e^{-\left(j\sqrt{k^2 - 2\pi f_x^2}\right)^*}$$

If f_x corresponds to a propagating wave, then $j\sqrt{k^2 - 2\pi f_x^2}$ is purely imaginary and hence $\left|\tilde{U}(f_x,L)\right|^2 = \left|\tilde{U}(f_x,0)\right|^2$. However, for the evanescent waves, $j\sqrt{k^2 - 2\pi f_x^2}$ is purely real and hence $\left|\tilde{U}(f_x,L)\right|^2 \neq \left|\tilde{U}(f_x,0)\right|^2$. Consequently, if the spectrum of an optical beam contains only propagating waves, the power is conserved but in a general case the power is not conserved. Similarly, for 2.8 Parseval's theorem implies that:

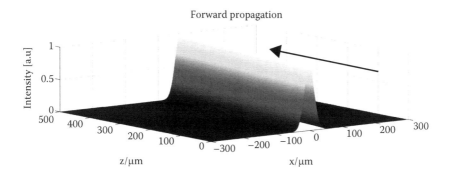

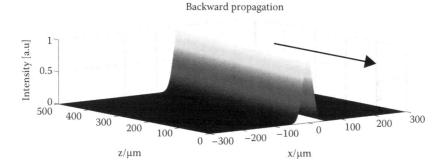

FIGURE 2.12 Light intensity distribution calculated by Algorithm 2.1. The initial beam width equals 30 μm while the tilt angle θ = 7.5°. The calculation window is 600 μm while the number of samples is 2^8 while the beam is propagated for 500 μm. The arrow indicates the direction of the optical beam propagation.

$$\sum_{l=0}^{M-1}\left|\tilde{U}(k_{xm},0)\right|^2 = \frac{1}{M}\sum_{l=0}^{M-1}\left|\tilde{U}(x_l,0)\right|^2$$

and

$$\sum_{l=0}^{M-1}\left|\tilde{U}(k_{xm},L)\right|^2 = \frac{1}{M}\sum_{l=0}^{M-1}\left|\tilde{U}(x_l,L)\right|^2$$

Similarly, as in the previous case, it can be shown that $\left|\tilde{U}(f_x,L)\right|^2 = \left|\tilde{U}(f_x,0)\right|^2$ for propagating waves and $\left|\tilde{U}(f_x,L)\right|^2 \neq \left|\tilde{U}(f_x,0)\right|^2$ for the evanescent waves. Therefore, as in the previous case, if the optical beam spectrum contains propagating waves only the power is conserved, that is:

$$\sum_{l=0}^{M-1}\left|\tilde{U}(k_{xm},0)\right|^2 = \sum_{l=0}^{M-1}\left|\tilde{U}(k_{xm},L)\right|^2$$

The power conservation for an algorithm is closely linked with the reciprocity. In fact, it can be observed that the presence of the evanescent modes is linked with both the lack of reciprocity and the lack of power conservation of Algorithm 2.1. This might seem a fairly disappointing conclusion. However, we will discuss the nature and origin of evanescent modes in the next section, which will provide a better understanding of this particular problem.

The final issue that is yet to be considered is that of the theoretical rigour of the approach. Algorithm 2.1 allows numerical calculation of the solution of Equation 2.1. However, Equation 2.1 itself is already an approximation of the rigorous theory, which in this case is represented by the Maxwell's equations. It is therefore necessary to assess how accurate the predictions of an algorithm based on Equation 2.7 are. For this purpose, in the following we consistently derive the solution of the Maxwell's equations for an in-plane propagation in a homogenous medium. We will consider an isotropic case only.

From Maxwell's equations (1.2) and material equations (1.3) we obtain:

$$\begin{cases} \left(\dfrac{\partial H_z}{\partial y} - \dfrac{\partial H_y}{\partial z}\right)\vec{i}_x + \left(\dfrac{\partial H_x}{\partial z} - \dfrac{\partial H_z}{\partial x}\right)\vec{i}_y + \left(\dfrac{\partial H_y}{\partial x} - \dfrac{\partial H_x}{\partial y}\right)\vec{i}_z = j\omega\varepsilon\left(\vec{i}_x E_x + \vec{i}_y E_y + \vec{i}_z E_z\right) \\[2mm] \left(\dfrac{\partial E_z}{\partial y} - \dfrac{\partial E_y}{\partial z}\right)\vec{i}_x + \left(\dfrac{\partial E_x}{\partial z} - \dfrac{\partial E_z}{\partial x}\right)\vec{i}_y + \left(\dfrac{\partial E_y}{\partial x} - \dfrac{\partial E_x}{\partial y}\right)\vec{i}_z = -j\omega\mu_0\left(\vec{i}_x H_x + \vec{i}_y H_y + \vec{i}_z H_z\right) \end{cases}$$

$$(2.10)$$

where $\vec{i}_x, \vec{i}_y, \vec{i}_z$ are the unit vectors along the x, y, and z axes. The vector fields on both sides of Equation 2.10 are equal if each of the components of these vector fields is equal. This splits Equation 2.10 into a set of six coupled equations for each of the components of the vector fields standing on both sides of 2.10. For in-plane propagation, we also have that $\partial/\partial y = 0$ for all field components. This observation leads to a split of the six coupled equations into two sets of three coupled partial differential equations:

$$\frac{\partial E_y}{\partial z} = j\omega\mu_0 H_x$$

$$\frac{\partial E_y}{\partial x} = -j\omega\mu_0 H_z$$

$$\frac{\partial H_x}{\partial z} - \frac{\partial H_z}{\partial x} = j\omega\varepsilon E_y \qquad (2.11a)$$

$$-\frac{\partial H_y}{\partial z} = j\omega\varepsilon E_x$$

$$\frac{\partial H_y}{\partial x} = j\omega\varepsilon E_z$$

$$\frac{\partial E_x}{\partial z} - \frac{\partial E_z}{\partial x} = -j\omega\mu_0 H_y \qquad (2.11b)$$

The set of three equations 2.11a links together the E_y, H_x, and H_z components of the electromagnetic fields, whereas that of 2.11b links together the E_x, E_z, and H_y components. It is clear that the set of equations 2.11a corresponds to a y polarised wave, whereas the 2.12b to an x-polarised wave. A wave with an arbitrary polarisation can be obtained by a superposition of these two waves. By eliminating H_x and H_z from 2.11a one obtains:

$$\frac{\partial^2 E_y}{\partial x^2} + \frac{\partial^2 E_y}{\partial z^2} + k^2 E_y = 0 \qquad (2.12a)$$

Similarly, eliminating H_y and E_z from 2.11b yields:

$$\frac{\partial^2 E_x}{\partial x^2} + \frac{\partial^2 E_x}{\partial z^2} + k^2 E_x = 0 \qquad (2.12b)$$

Equations 2.12a and 2.12b have the same form as Equation 2.3 and hence can be solved using Algorithm 2.1 without any modifications. This leads to the conclusion that Algorithm 2.1 provides a rigorous method of solving Maxwell's equations. Equation 2.12 shows that the function $\tilde{U}(x,z)$ can be interpreted as either E_x or E_y depending on the assumed initial polarisation of the wave. Once $\tilde{U}(x,L)$ is calculated using Algorithm 2.1, the other components of the electromagnetic field can be obtained from either 2.11a or 2.11b for the assumed polarisation of the wave. The distinction between two polarisations becomes distinct when an optical wave impinges upon an interface between two homogenous media. This case we will consider in detail in the next section [12].

OPTICAL BEAM REFLECTION AND REFRACTION

The differentiation between the wave optics and electromagnetic optics becomes more distinct when an optical beam is incident at a plane boundary between two homogenous media. The Fourier method can be fairly easily extended to analyse the optical beam propagation in a piecewise homogenous medium consisting of two homogenous regions as defined by Figure 2.13. This is the case because the analytical expressions for the reflection and transmission coefficients can be easily derived when the incident wave is a plane wave. It is well known that a plane wave incident at the boundary between the regions 1 and 2 undergoes partial reflection and refraction (Figure 2.14), for example, see [1,12]. When the incident wave is given by 2.4 it is fairly easy to calculate the reflection and transmission coefficients, which are defined as:

$$R = \frac{A_r}{A_i} \qquad (2.13a)$$

$$T = \frac{A_t}{A_i} \qquad (2.13b)$$

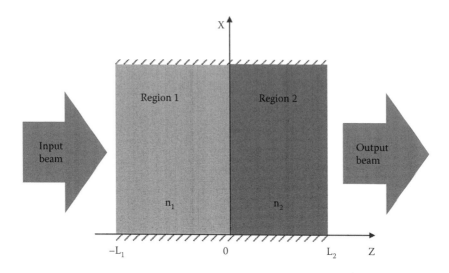

FIGURE 2.13 Schematic diagram illustrating the problem of the optical beam propagation through a medium consisting of two homogenous regions.

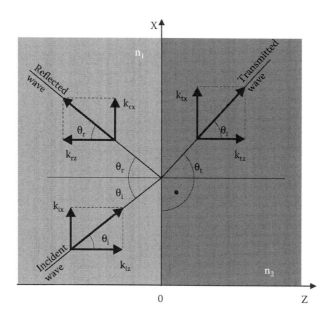

FIGURE 2.14 Reflection and refraction at the boundary between two homogenous regions.

where A_i, A_r, and A_t are the amplitudes of the incident, reflected, and transmitted waves, respectively. To derive the coefficients R and T, let us first write explicitly the expressions for the incident, reflected, and transmitted waves:

$$\tilde{U}_i(x,z) = A_i e^{-jk_{ix}x} e^{-jk_{iz}z}$$

$$\tilde{U}_r(x,z) = A_r e^{-jk_{rx}x} e^{jk_{rz}z}$$

$$\tilde{U}_t(x,z) = A_t e^{-jk_{tx}x} e^{-jk_{tz}z} \qquad (2.14)$$

Because the problem of the wave reflection at the boundary between two dielectric media has been studied in detail in many books [1,12], we will only briefly retrace the derivation of the reflection and transmission coefficients. With respect to the plane boundary between the regions 1 and 2, two distinct orientations of the wave are possible. Either the electric or the magnetic field vector can be parallel to the boundary. The former case is referred to in the literature as the perpendicular [12] or transverse electric (TE) [1] polarisation, whereas the latter one is typically named as the parallel or transverse magnetic (TM) case. Clearly the TE case is described by Equation 2.11a, whereas the latter one by Equation 2.11b. We start the derivations by considering the TE case first. In this case, it is most convenient to derive the reflection and transmission coefficients in terms of the electric field because the electric field has only an E_y component and is therefore invariant under any rotation in the x–z plane. For the TE polarisation the tangential component of electric field is continuous at the boundary between two media, therefore E_y is continuous. Furthermore, the H_x component of the magnetic field is continuous. Therefore from 2.11a, it follows that $\partial E_y / \partial z$ is continuous. Thus, imposing the continuity of $\tilde{U}(x,z)$ and $\partial \tilde{U}(x,z)/\partial z$ for 2.14 at $z = 0$ yields two equations:

$$\begin{cases} A_i A + A_r = A_t \\ -k_{iz}A_i + k_{iz}A_r = -k_{tz}A_t \end{cases}$$

From which upon substituting 2.13 and noticing that with the definitions 2.14 $k_r z = k_i z$, one obtains the reflection and transmission coefficients expressed in terms of the z-component of the wave vector:

$$R = \frac{k_{iz} - k_{tz}}{k_{iz} + k_{tz}} \qquad (2.15a)$$

and

$$T = \frac{2k_{iz}}{k_{iz} + k_{tz}} \qquad (2.15b)$$

Commonly, the transmission and reflection coefficients and 2.14 are expressed in terms of the incidence angle θ_i and the refractive indices n_1 and n_2 in both media. To recast 2.14 and 2.15 in such a form, one needs to observe that the continuity of the tangential electric field component implies that $k_i x = k_r x$ (which is equivalent to stating that the angle of reflection is equal to the angle of incidence and consequently, $k_i z = -k_r z$, since $k_{ix}^2 + k_{iz}^2 = k_{rx}^2 + k_{rz}^2 = k_1^2$). The continuity of the tangential electric field component implies also that $k_i x = k_t x$ and hence $n_1 \sin\theta_i = n_2 \sin\theta_t$, which is the Snell's law that allows for expressing θ_t in terms of n_1, n_2, and θ_i. Finally, from Figure 2.15 we can extract the following formulae: $k_i z = k_1 \cos\theta_i$, $k_i x = k_1 \sin\theta_i$, $k_t z = k_2 \cos\theta_t$, and $k_t x = k_2 \sin\theta_t$. Therefore, 2.14 and 2.15 can be expressed in the following way:

$$\tilde{U}_i(x,z) = A_i e^{-jkn_1 \sin\theta_i x} e^{-jkn_1 \cos\theta_i z}$$

$$\tilde{U}_r(x,z) = A_r e^{-jkn_1 \sin\theta_i x} e^{jkn_1 \cos\theta_i z}$$

$$\tilde{U}_t(x,z) = A_t e^{-jkn_2 \sin\theta_2 x} e^{-jkn_2 \cos\theta_2 z} \qquad (2.16)$$

and

$$R = \frac{n_1 \cos\theta_1 - n_2 \cos\theta_2}{n_1 \cos\theta_1 + n_2 \cos\theta_2} \qquad (2.17a)$$

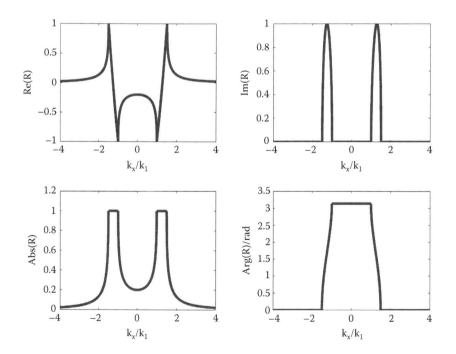

FIGURE 2.15 Dependence of the reflection coefficient R on k_x for TE polarisation when $n_1 = 1.0$ and $n_2 = 1.5$.

$$T = \frac{2n_1 \cos\theta_1}{n_1 \cos\theta_1 + n_2 \cos\theta_2} = 1 + R \qquad (2.17b)$$

Whereas the term dependent on θ_2 can be expressed in terms of the incidence angle θ_1 and the refractive indices n_1 and n_2 in both media using the Snell's law: $n_2 \cos\theta_2 = \sqrt{n_2^2 - n_1^2 \sin^2\theta_1}$. However, for use with a Fourier transform based algorithm, it is more convenient to express R and T in terms of k_x, k_1, and k_2. This can be easily done by observing that $k_{iz} = \sqrt{k_1^2 - k_x^2}$, $k_{tz} = \sqrt{k_2^2 - k_x^2}$ while due to the continuity of the tangential electric field component $k_x = k_{xi} = k_{xr} = k_{xt}$. Hence, we obtain R and T in the following form:

$$R(k_x) = \frac{\sqrt{k_1^2 - k_x^2} - \sqrt{k_2^2 - k_x^2}}{\sqrt{k_1^2 - k_x^2} + \sqrt{k_2^2 - k_x^2}} \qquad (2.18a)$$

and

$$T(k_x) = \frac{2\sqrt{k_1^2 - k_x^2}}{\sqrt{k_1^2 - k_x^2} + \sqrt{k_2^2 - k_x^2}} \qquad (2.18b)$$

This derivation can be repeated for the TM polarisation with the only modification coming from the fact that the z partial derivative of H_y component is equal to $-j\omega\varepsilon E_x$ (2.11b). Thus in this case, the continuity conditions at $z = 0$ are to be imposed for $\tilde{U}(x,z)$ and $1/\varepsilon\ \partial\tilde{U}(x,z)/\partial z$. Following the same steps as in the TE case we obtain two equations:

$$\begin{cases} A_i + A_r = A_t \\ -\dfrac{k_{iz}}{n_1^2} A_i + \dfrac{k_{iz}}{n_1^2} A_r = -\dfrac{k_{tz}}{n_2^2} A_t \end{cases}$$

which yield:

$$R = \frac{k_{iz}/n_1^2 - k_{tz}/n_2^2}{k_{iz}/n_1^2 + k_{tz}/n_2^2} \quad \text{and} \quad T = \frac{2k_{iz}/n_1^2}{k_{iz}/n_1^2 + k_{tz}/n_2^2}.$$

Similarly to the TE case we express both coefficients in terms of k_x:

$$R(k_x) = \frac{\dfrac{1}{n_1^2}\sqrt{k_1^2 - k_x^2} - \dfrac{1}{n_2^2}\sqrt{k_2^2 - k_x^2}}{\dfrac{1}{n_1^2}\sqrt{k_1^2 - k_x^2} + \dfrac{1}{n_2^2}\sqrt{k_2^2 - k_x^2}} \qquad (2.19a)$$

and

$$T\left(k_x\right) = \frac{\dfrac{2}{n_1^2}\sqrt{k_1^2 - k_x^2}}{\dfrac{1}{n_1^2}\sqrt{k_1^2 - k_x^2} + \dfrac{1}{n_2^2}\sqrt{k_2^2 - k_x^2}} \tag{2.19b}$$

Please note that for $\theta = 0$ the reflection and transmission coefficients given by formulas 2.18 are not equal to those given by 2.19. The reason for this discrepancy is coming from the fact that equation 2.18 are defined in terms of the electric field (E_y), whereas 2.19 for the magnetic field (H_y).

For the sake of completeness, we also define the reflection and transmission coefficients for the power. For the TE case, the component of the Poynting vector that is aligned with the direction of the wave propagation (which gives the total power flux density along the direction of the wave propagation) is equal to $\left(1/4\right)E_y^* E_y \left(1/Z^* + 1/Z\right)$, where $Z = \sqrt{\mu_0/\varepsilon}$ is the wave impedance of the medium. Therefore, the ratio of the reflected and transmitted power flux density to the incident power flux density (both measured along the direction of the wave propagation) can be expressed in terms of A_r, A_i, and A_t:

$$\bar{R} = \frac{A_r A_r^*}{A_i A_i^*} \tag{2.20a}$$

$$\bar{T} = \frac{\operatorname{Re}\left(n_1\right)}{\operatorname{Re}\left(n_2\right)} \frac{A_t A_t^*}{A_i A_i^*} = 1 - \bar{R} \tag{2.20b}$$

The coefficients $\bar{R}$ and $\bar{T}$ can be expressed in terms of the coefficients R and T by observing that multiplying both sides of $A_r = RA_i$ by A_r^* yields $A_r^* A_r = RA_i^* A_r$. However, $A_r^* = R^* A_i^*$ also holds; therefore, it follows that $A_r^* A_r = R^* RA_i^* A_i = \bar{R}A_i^* A_i$ so $\bar{R} = R^* R$. Similarly, one can obtain the expression for the transmission coefficient. First, we obtain

$$\frac{A_t^* A_t}{A_i^* A_i} = T^* T \text{ and hence: } \frac{\operatorname{Re}\left(n_1\right)}{\operatorname{Re}\left(n_2\right)} \frac{A_t^* A_t}{A_i^* A_i} = \frac{\operatorname{Re}\left(n_1\right)}{\operatorname{Re}\left(n_2\right)} T^* T = \bar{T}.$$

Similarly, the coefficients for the TM case can be derived, yielding $\bar{R} = R^* R$ and

$$\bar{T} = \frac{\operatorname{Re}\left(n_2\right)}{\operatorname{Re}\left(n_1\right)} T^* T.$$

Finally, we note that substituting $k_x = 0$ it can be verified that for the normal wave incidence the power coefficients for both polarisations are equal.

Figure 2.15 shows the dependence of the reflection and transmission coefficients R and T on k_x for the TE case when $n_1 < n_2$. In this case, the reflection coefficient is purely real and negative up to $k_x = k_1$. It becomes purely imaginary for $(n_2/n_1) k_1 > k_x > k_1$ and stays purely real for $k_x > (n_2/n_1)k_1$. When $n_1 > n_2$ (Figure 2.16), the reflection coefficient stays real up to $(n_2/n_1)k_1$, which corresponds to the angle of the total internal reflection. It becomes purely imaginary for $k_1 > k_x > (n_2/n_1)k_1$ and stays purely real for $k_x > k_1$. Figures 2.15 and 2.16 also show that for evanescent modes with sufficiently large value of k_x the interface between both regions is practically transparent. Figures 2.17 and 2.18 show the dependence of the reflection coefficient on the x-component of the propagation vector for the TM polarisation case. In general, the dependence is similar to that displayed by the TE polarisation case. The only qualitative difference is the occurrence of the zero reflectivity for one value of k_x when $n_1 > n_2$. This value of k_x corresponds to the Brewster angle.

Now, the Algorithm 2.2 can be outlined for modelling of optical beam propagation in a medium consisting of two homogenous regions.

To illustrate the implementation of Algorithm 2.2 we consider an optical beam that propagates obliquely. Figures 2.19 and 2.20 show the distribution of the absolute value of E_y calculated by Algorithm 2.2. The initial beam width (defined by formula 2.10 with $\tilde{U} = E_y$) equals 30 μm while the tilt angle $\theta = 7.5°$. The calculation window is 600 μm while the number of samples is 2^8, $L_1 = L_2 = 500$ μm. The calculation was completed using 500 2-μm long steps. For the results shown in Figure 2.19, $n_1 = 1.0$ and $n_2 = 1.5$, whereas for those from Figure 2.20, $n_1 = 1.5$ and $n_2 = 1.0$. In both figures the change of the direction of the beam propagation resulting from light refraction can be clearly seen. Furthermore, in the results shown in Figure 2.20 the transmitted wave has got a larger amplitude than the incident wave. Hence, an attempt to

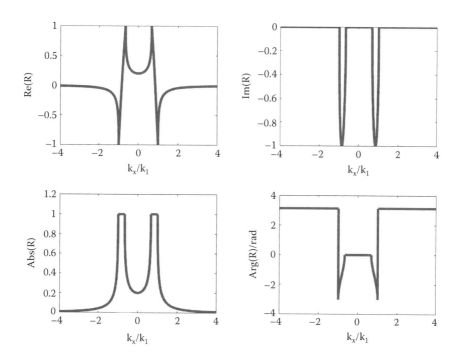

FIGURE 2.16 Dependence of the reflection coefficient R on k_x for TE polarisation when $n_1 = 1.5$ and $n_2 = 1.0$.

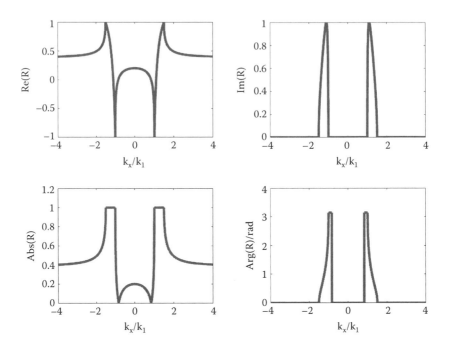

FIGURE 2.17 Dependence of the reflection coefficient R on k_x for TM polarisation when $n_1 = 1.0$ and $n_2 = 1.5$.

ALGORITHM 2.2   FOURIER METHOD FOR THE IN-PLANE PROPAGATION OF AN OPTICAL BEAM IN A MEDIUM CONSISTING OF TWO HOMOGENOUS REGIONS

1. Start
2. Set the free space wavelength λ
3. Select N for 2^N equidistant sampling points for FFT
4. Initialize: at $z = -L_1$, $\tilde{U}_i(x, -L_1)$
5. Calculate FFT of $\tilde{U}(x,z) = A\mathrm{e}^{-jk_x x}\mathrm{e}^{-jk_z z}$
6. Calculate $\tilde{V}_i(k_x,0) = \tilde{V}_i(k_x,-L_1)e^{jk_{1z}}$
7. Calculate $\tilde{V}_i(k_x,0) = \tilde{V}_i(k_x,-L_1)e^{jk_{1z}L_1}$ and
    $\tilde{V}_r(k_x,0) = R(k_x)\tilde{V}_i(k_x,0)$
8. Calculate $\tilde{V}_r(k_x,0) = R(k_x)\tilde{V}_i(k_x,0)$
9. Calculate $\tilde{V}_t(k_x,L_2) = \tilde{V}_t(k_x,0)e^{jk_{2z}L_2}$
10. Obtain $\tilde{U}_r(x,-L_1)$ by calculating inverse FFT of $\tilde{V}_r(k_x,-L_1)$
11. Obtain $\tilde{U}_t(x,L_2)$ by calculating inverse FFT of $\tilde{V}_t(k_x,-L_2)$
12. Stop

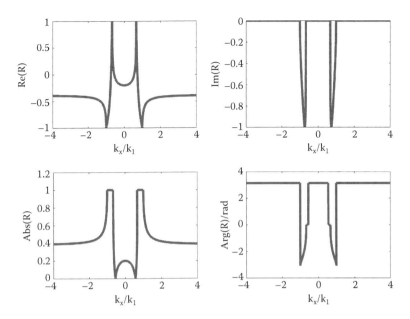

FIGURE 2.18 Dependence of the reflection coefficient R on k_x for TM polarisation when $n_1 = 1.5$ and $n_2 = 1.0$.

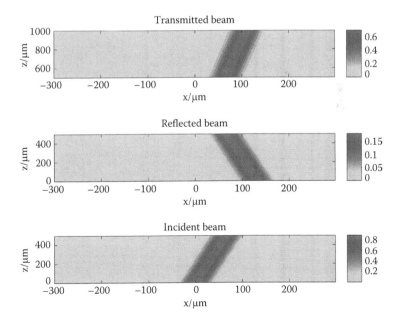

FIGURE 2.19 The distribution of the absolute value of the y electric field component calculated by Algorithm 2.2. The initial beam is TE polarised and its width equals 30 μm while the tilt angle $\theta = 7.5°$. The calculation window is 600 μm while the number of samples is 2^8, $L_1 = L_2 = 500$ μm, $n_1 = 1.0$, $n_2 = 1.5$ while the operating wavelength is 1.06 μm.

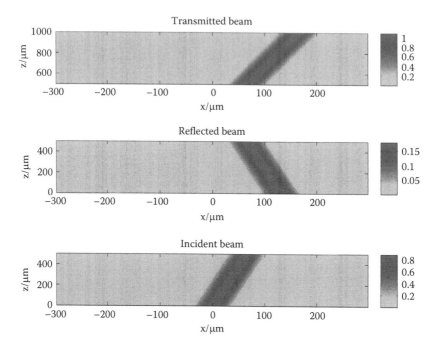

FIGURE 2.20 The distribution of the absolute value of the y electric field component calculated by Algorithm 2.2. The initial beam is TE polarised and its width equals 30 μm while the tilt angle $\theta = 7.5°$. The calculation window is 600 μm while the number of samples is 2^8, $L_1 = L_2 = 500$ μm, $n_1 = 1.5$, $n_2 = 1.0$ while the operating wavelength is 1.06 μm.

interpret the field in the traditional wave optics way would result in the violation of the energy conservation principle. Therefore, in the scalar approximation to circumvent this problem one uses the square root of the reflection coefficient for the power rather than the one for the field. Another problem resulting from a scalar approximation is illustrated by the next example. Figure 2.21 shows the absolute value of E_y calculated by Algorithm 2.2 for the transmitted wave only. The initial beam width again equals 30 μm while the tilt angle $\theta = 45°$. The calculation window is 600 μm, the number of samples is 2^{16}, $L_1 = 0$, $L_2 = 0.2$ μm, $n_1 = 3.5$, and $n_2 = 1.0$. The field distribution was calculated using 50 0.004-μm long, steps. The incidence angle is larger than the critical angle hence the wave is totally reflected. However, an evanescent wave extends into medium 2. This evanescent wave carries no net power in the z direction, which can be shown by calculating the Poynting vector.

PARAXIAL AND WIDE ANGLE APPROXIMATIONS

Paraxial and wide angle approximations allow for the significant improvement of computational efficiency in numerical simulations of optical beam propagation. Both approximations rely on extracting a quickly varying part of the wave function and solving an equation for the slowly varying envelope function. The advantage of such

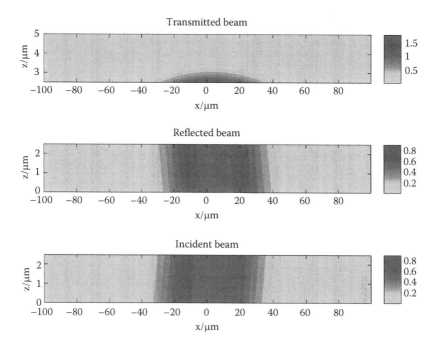

FIGURE 2.21 The distribution of the absolute value of the y electric field component calculated by Algorithm 2.2. The initial beam is TE polarised and its width equals 30 μm while the tilt angle θ = 45°. The calculation window is 200 μm while the number of samples is 2^{11}, $L_1 = L_2 = 2.5$ μm, $n_1 = 1.5$, $n_2 = 1.0$ while the operating wavelength is 1.06 μm.

an approach comes from the fact that the slowly varying envelope function requires much less sampling points than the (quickly varying) wave function. In this section, we derive the paraxial and wide angle approximations for the wave Equation 2.3 and provide illustrative examples that demonstrate the advantages and limitations of this method.

The first step in the derivation of the paraxial approximation consists of guessing the solution for 2.3 in the form of a product of a slowly varying envelope function $A(x,z)$ and a fast varying term:

$$\tilde{U}(x,z) = A(x,z)e^{-j\beta z} \tag{2.21}$$

where β is an arbitrary constant that is selected in such a way that $A(x,z)$ is a slowly varying function of the spatial variable z. Substituting 2.21 into 2.3 yields an equation for the slowly varying envelope function:

$$\frac{\partial^2 A}{\partial x^2} + \frac{\partial^2 A}{\partial z^2} + \left(k^2 - \beta^2\right)A - 2j\beta\frac{\partial A}{\partial z} = 0 \tag{2.22}$$

Now taking advantage of the fact that $A(x,z)$ is a slowly varying function of z (when β is properly selected) we can neglect the second derivative with respect to z when compared with the term containing the first derivative and obtain

$$2j\beta\frac{\partial A}{\partial z} = -\left[\frac{\partial^2 A}{\partial x^2} + \left(k^2 - \beta^2\right)A\right] \tag{2.23}$$

which is the paraxial wave equation. It will be shown in Chapter 4 that the paraxial wave equation is in many cases much easier to solve numerically than the wave equation. The main disadvantage of using 2.23 is its limited accuracy that may be insufficient for a particular application. Therefore, a wide angle approximation has been developed for the wave equation. The wide angle approximation aims at preserving the computational efficiency of the paraxial approach while providing better accuracy. To derive a wide angle approximation for the wave Equation 2.3 we first recast 2.22 in the following way:

$$\left[1 + \frac{j}{2\beta}\frac{\partial}{\partial z}\right]\frac{\partial}{\partial z}A = -\frac{j}{2\beta}\left[\frac{\partial^2}{\partial x^2} + \left(k^2 - \beta^2\right)\right]A \tag{2.24}$$

Now, a recurrence can be established, which generates wide angle approximations to Equation 2.3:

$$\left.\frac{\partial}{\partial z}\right|_n = \frac{-\dfrac{j}{2\beta}\left[\dfrac{\partial^2}{\partial x^2} + \left(k^2 - \beta^2\right)\right]}{1 + \dfrac{j}{2\beta}\dfrac{\partial}{\partial z}\bigg|_{n-1}} \tag{2.25}$$

The recurrence 2.25 was first proposed by Hadley [13] and starts with assuming that for $n = 1$ the partial derivative in the denominator of 2.25 is equal to zero. Therefore, with $n = 1$ recurrence formula 2.25 yields the paraxial approximation (cf. equation 2.23):

$$\frac{\partial}{\partial z} = -\frac{1}{2j\beta}\left[\frac{\partial^2}{\partial x^2} + \left(k^2 - \beta^2\right)\right] \tag{2.26a}$$

For $n = 2$ we obtain:

$$\frac{\partial}{\partial z} = -\frac{\dfrac{jP}{2\beta}}{1 + \dfrac{P}{4\beta^2}} \tag{2.26b}$$

for $n = 3$:

$$\frac{\partial}{\partial z} = -\frac{-\dfrac{jP}{2\beta} - \dfrac{P^2}{8\beta^3}}{1 + \dfrac{P}{2\beta^2}} \qquad (2.26c)$$

and for $n = 4$:

$$\frac{\partial}{\partial z} = -\frac{-\dfrac{jP}{2\beta} - \dfrac{jP^2}{4\beta^3}}{1 + \dfrac{3P}{4\beta^2} + \dfrac{P^2}{16\beta^4}} \qquad (2.26d)$$

where P is defined as:

$$P = \frac{\partial^2}{\partial x^2} + \left(k^2 - \beta^2\right)$$

The higher order wide angle approximations can be obtained continuing the recurrence 2.25 for larger values of n. In principle, this recursive process can be continued up to an arbitrary approximation order. However, as the approximation order increases the derivation of subsequent wide angle approximations becomes increasingly tedious. Therefore, other algorithms have been proposed for generating the wide angle approximations of 2.3. One particularly efficient technique [14] relies on formally factorising 2.3 first:

$$\left(\frac{\partial}{\partial z} + j\sqrt{\frac{\partial^2}{\partial x^2} + k^2}\right)\left(\frac{\partial}{\partial z} - j\sqrt{\frac{\partial^2}{\partial x^2} + k^2}\right)\tilde{U} = 0 \qquad (2.27)$$

So, 2.27 implicates the validity of the following two equations:

$$\left(\frac{\partial}{\partial z} + j\sqrt{\frac{\partial^2}{\partial x^2} + k^2}\right)\tilde{U} = 0 \qquad (2.28a)$$

$$\left(\frac{\partial}{\partial z} - j\sqrt{\frac{\partial^2}{\partial x^2} + k^2}\right)\tilde{U} = 0 \qquad (2.28b)$$

One of the known properties of 2.28a is that it annihilates all waves propagating in the negative z direction, whereas 2.28b annihilates all waves propagating in the positive z direction [15]. Therefore, Equation 2.28 is known in the literature under the name of "one way wave equations." If we select 2.28a and, consistently with the assumed time dependence, substitute 2.21 we obtain the following equation for the envelope A:

$$\frac{\partial A}{\partial z} = j\beta \left(1 - \sqrt{1 + \frac{P}{\beta^2}} \right) A \tag{2.29}$$

Now, we introduce a Padé expansion for the square root:

$$\sqrt{1+x} = 1 + \sum_{i=1}^{n} \frac{a_{i,n} x}{1 + b_{i,n} x} \tag{2.30}$$

whereby

$$a_{i,n} = \frac{2}{2n+1} \sin^2 \left(\frac{i\pi}{2n+1} \right) \text{ and } b_{i,n} = \cos^2 \left(\frac{i\pi}{2n+1} \right).$$

This yields the wide angle approximations for 2.3:

$$\frac{\partial A}{\partial z} = -j\beta \left(\sum_{i=1}^{n} \frac{a_{i,n} \dfrac{P}{\beta^2}}{1 + b_{i,n} \dfrac{P}{\beta^2}} \right) \tag{2.31}$$

The expansion coefficients in 2.31 are known analytically and, unlike 2.25, do not need to be derived using a recursive method. It can be shown that both 2.25 and 2.31 in fact yield wide angle approximations for 2.23 that are based on the Padé expansion of the square root operator [13]. Other techniques for deriving the wide angle approximations to 2.3 were also developed. These include the Taylor series approximation [16] and Thiele approximation [17]. We will discuss this topic in more detail in Chapter 4. Here we focus on the Padé expansion 2.31 only since this technique is most frequently used to develop software for studying optical beam propagation.

First, we observe that substituting 2.4 in 2.31 gives the approximate dependence of k_z on k_x that is implied by the wide angle approximation 2.11:

$$k_z = \beta \left(\sum_{i=1}^{n} \frac{a_{i,n} \dfrac{k^2 - \beta^2 - k_x^2}{\beta^2}}{1 + b_{i,n} \dfrac{k^2 - \beta^2 - k_x^2}{\beta^2}} \right) \qquad (2.32a)$$

Following the same procedure for the paraxial approximation we obtain:

$$k_z = \frac{1}{2\beta} \left(k^2 - \beta^2 - k_x^2 \right) \qquad (2.32b)$$

We note that since the quickly varying phase factor $\exp(-j\beta z)$ was extracted when deriving 2.31 and the paraxial approximation, the dependence of k_z on k_x given by 2.32, has to be shifted by β before it can be compared directly with 2.9.

In Figure 2.22, we compare the exact dependence of k_z on k_x given by 2.9 with an approximate one obtained using 2.32b. We assume that β equals k and plot the dependence in a normalised form with respect to k. For values of $k_x < k$ the quality of the approximation is good for small propagation angles but it gradually deteriorates

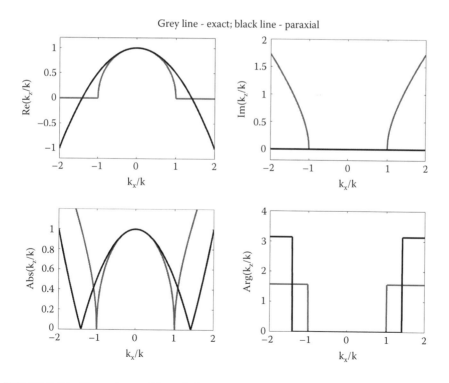

FIGURE 2.22 Dependence of k_z on k_x for paraxial approximation (formula 2.32b). The reference propagation constant is equal to the medium wavenumber $k = nk_0$.

for off-axis propagation. For $k_x > k$ the paraxial algorithm predicts the wrong value of the longitudinal component of the propagation vector. This observation can be explained by noticing that Equation 2.9 maps the negative real axis onto the imaginary one, whereas 2.32b cannot reproduce this behaviour. Such a qualitative difference between 2.9 and 2.32b means that the evanescent waves cannot be modelled accurately using the paraxial approximation. In Figure 2.23, we show the dependence of k_z on k_x for the Padé $(1,1)$ approximation. A significant improvement of accuracy is achieved for the propagating waves. Further improvement of accuracy for $k_x < k$ can be achieved using higher Padé approximation orders (Figure 2.24). However, for the evanescent modes, the Padé approximation suffers from the same problem as the paraxial one. Namely, the Padé expansion, maps the negative real axis onto the real axis and not the imaginary one. This problem can be remedied by selecting a complex value of β. Figure 2.25 shows the dependence of kz on k_x for the problem studied in the previous example. In this example however, we only keep the absolute value of β equal to the medium wave number and rotate β in the complex plane [18]. The rotation angle used to obtain the results shown in Figure 2.25 is $\pi/10$. With a complex value of β, the improvement in the quality of the wide angle approximation with the Padé expansion order is achieved for both the propagating and the evanescent waves. A noticeable improvement of the approximation can be observed primarily due to the fact that now the evanescent character of the modes is

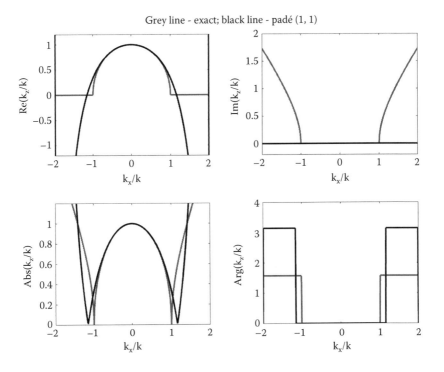

FIGURE 2.23 Dependence of k_z on k_x for Padé $(1,1)$ approximation (formula 2.32a). The reference propagation constant is equal to the medium wavenumber $k = nk_0$.

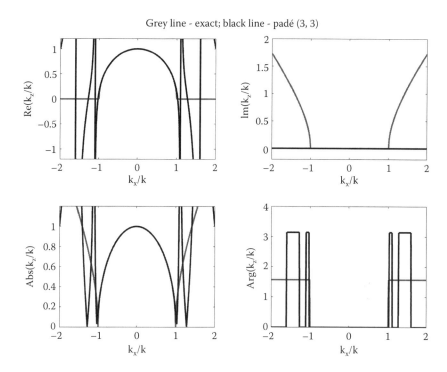

FIGURE 2.24   Dependence of k_z on k_x for Padé (3,3) approximation (formula 2.32a). The reference propagation constant is equal to the medium wavenumber $k = nk_0$.

reproduced. Furthermore, the improvement of the accuracy for $k_x > k$ is achieved by using a larger rotation angle (Figure 2.26). However, this happens at the expense of deteriorating the approximation quality for the propagating modes. In fact, an oscillatory behaviour of the Padé approximation for the imaginary part of the longitudinal component of the propagating vector may render the resulting algorithm inherently unstable. Such oscillatory behaviour can be suppressed only by increasing the wide angle approximation order (Figure 2.27). As will be shown in Chapter 4, this leads to an increased computational overhead per step of a beam propagation algorithm.

To illustrate the consequences of the errors introduced when using 2.32 instead of 2.9, we have studied the optical beam propagation from the example considered in Figure 2.21 using Algorithm 2.2 and a Padé (1,1) approximation instead of 2.9. As expected, an evanescent wave is modelled by the Padé (1,1) approximation as a propagating one when the reference propagation constant matches the medium wave number (Figure 2.28). However, when a rotation is introduced in the complex plane the correct evanescent behaviour is predicted (Figure 2.29). In practically relevant problems, one often needs to propagate both the evanescent and the propagating modes. Therefore a low Padé (1,1) approximation may not be sufficient. Because the computational overhead depends strongly on the approximation order, a careful design of the operator approximation is very important. We return to this discussion in Chapter 4.

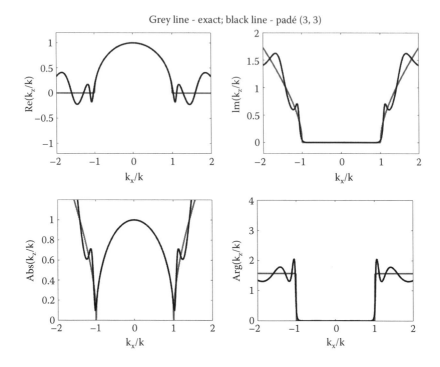

FIGURE 2.25 Dependence of k_z on k_x for Padé (3,3) approximation (formula 2.32a). The absolute value of the reference propagation constant is equal to the medium wavenumber $k = nk_0$, while its angle tilt in the complex plane equals $\pi/10$.

TRANSMISSION THROUGH THIN OPTICAL ELEMENTS

In Section 2.2 we observed that the transmission and reflection of optical beams at the interface between two homogenous media can be modelled fairly easily by considering separately all the Fourier components of the complex wave amplitude $\tilde{U}(x,z)$. A further useful extension to the Fourier method can be made under the paraxial beam approximation. It is convenient in this case to define the complex transmittance function $t(x,y) = \tilde{U}(x,z=0)/\tilde{U}(x,z=d)$. In a paraxial, scalar approximation the complex transmittance or transfer function can be relatively easily derived for several thin bulk optics components and thus the propagation through thin bulk optics elements can be included when applying the Fourier algorithm. For instance, for a transparent plate of thickness d $t(x,y) = \exp(-jnk_0 d)$ [1]. It is fairly straightforward to derive the approximate complex transmittance function for several other bulk optics elements (Table 2.1) [1].

To illustrate the usefulness of the transfer function concept we consider optical beam propagation through a biconcave lens made of N-BK7 glass. The transmission function for such lens can be obtained by observing that a biconcave lens is a concatenation of two plano-concave lenses. Hence, its transmission function can

Grey line - exact; black line - padé (3, 3)

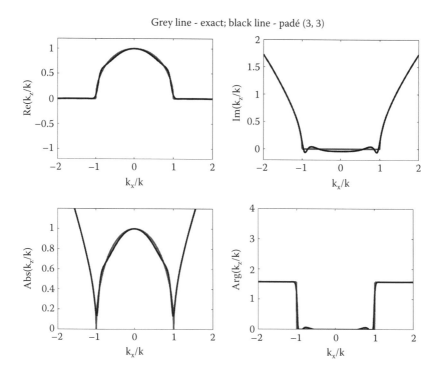

FIGURE 2.26 Dependence of k_z on k_x for Padé (3,3) approximation (formula 2.32a). The absolute value of the reference propagation constant is equal to the medium wavenumber $k = nk_0$, while its angle tilt in the complex plane equals $\pi/3$.

be obtained by squaring the transmission function of a plano-concave lens. In the calculations we assumed that each of the plano-concave lens had the following measurements: radius $R = 206.7$ mm, $d = 0.5$ mm, and $t = 3$ mm, with an operating wavelength of 1.06 μm. The refractive index at the operating wavelength is 1.50669, which was obtained using the Sellmeier dispersion formula:

$$n^2\left(\lambda\right)-1=\frac{B_1\lambda^2}{\lambda^2-C_1}+\frac{B_2\lambda^2}{\lambda^2-C_2}+\frac{B_3\lambda^2}{\lambda^2-C_3} \qquad (2.33)$$

The coefficients for N-BK7 glass are given in Table 2.2 and were obtained from SCHOTT™ optical glass collection data sheet that is available from the company's web page [19]. The width of the numerical window is 80 mm. The initial beam consists of three Gaussian lobes with beam width $w = 2$ mm (cf. formula 2.10) that are spaced at 8 mm. The calculation window is 80 mm, the number of samples is 2^{16}, and the distance from $z = 0$ plane to the lens is equal to the distance from the lens till the end of the computational window and equals 250 mm. The lens material is N-BK7 glass and ambient is air, whereas the operating wavelength is 1.06 μm. Figure 2.30 shows the light intensity distribution calculated using the FFT algorithm applying

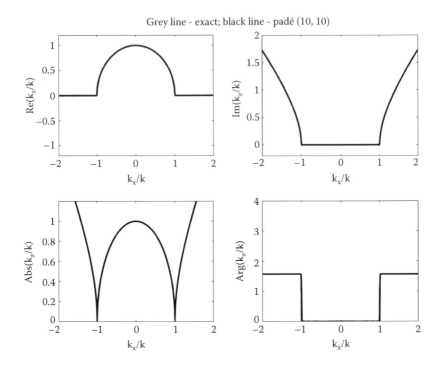

FIGURE 2.27 Dependence of k_z on k_x for Padé (10,10) approximation (formula 2.32a). The absolute value of the reference propagation constant is equal to the medium wave number $k = nk_0$, while its angle tilt in the complex plane equals $\pi/3$.

scalar approximation. The propagation through the lens was calculated using the transmittance given in the Table 2.1. Otherwise we used Algorithm 2.1. As expected, after passing the lens, all three beams diverge while the imaginary focal point is approximately 200 mm away from the lens centre.

In the last example, we have neglected the fact that the optical beam is partially reflected at each air–glass interface. This is an inherent shortcoming associated with the use of the formulae given in Table 2.1. However, within the scalar approximation the reflection from a dielectric interface is well modelled by using the square root of the power reflection and transmission coefficients for $\theta = 0$ (cf. section "Paraxial and Wide Angle Approximations"). Such an approximation is plausible because for paraxial beams there is no dependence of the reflection and transmission coefficients phase on the propagation angle (Figures 2.15 through 2.18). Multiplying a transmittance function by the transmission coefficient allows also the Fresnel loss to be taken into account. The last issue that is as yet left unresolved is how to handle the multiple reflected waves. A convenient framework for resolving this problem has been developed [20] in the form of the Bremmer series. To present the concept of the Bremmer series, let us consider a wave propagation through a set of homogenous layers (Figure 2.31). A plane wave incident from $x = -\infty$ will be partially reflected from the layers at each discontinuity. In

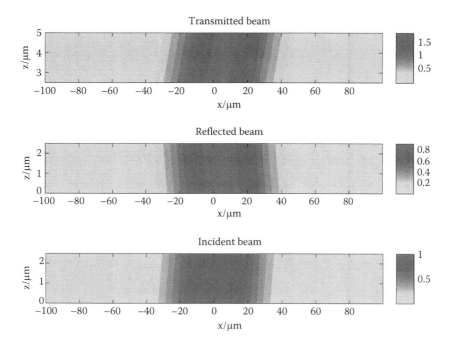

FIGURE 2.28 The distribution of the absolute value of the y electric field component calculated using Padé (1,1) approximation. The reference propagation constant is equal to wave number in each medium. The initial beam is TE polarized and its width equals 30 μm while the tilt angle $\theta = 45°$. The calculation window is 200 μm while the number of samples is 2^{11}, $L_1 = L_2 = 2.5$ μm, $n_1 = 1.5$, $n_2 = 1.0$ while the operating wavelength is 1.06 μm.

the Bremmer series approach, the wave that propagates through the stack without any reflection is termed the primary wave. The primary wave is only multiplied by the transmission coefficient at each refractive index discontinuity position. The first order correction wave is constructed by adding all waves that have been only reflected once within the dielectric stack. The second order correction wave is constructed from doubly reflected waves and so on. The convergence of the Bremmer series was studied in Atkinson [21]. When the Bremmer series converges quickly, only a few correction waves need to be considered, and an accurate result is obtained with a moderate computational effort. In Figure 2.31 the primary wave propagates along the positive z-direction and is denoted with u_0 whereas the first order correction wave propagates along the negative z direction and is denoted with u_1. The higher order correction waves propagate alternatively either along the positive or negative z direction. This intuitive description of the Bremmer series is sufficient for the rest of this section. The formal derivation of the Bremmer series can be found in Bremmer [20].

After this introductory discussion, let us consider a problem that is similar to the last example, that is, a wave propagating through a biconcave lens. This time, however, we used a beam consisting of three 4 mm wide Gaussian lobes, spaced by 8 mm, that passed through a set of three 0.250 mm wide slits that are 8 mm

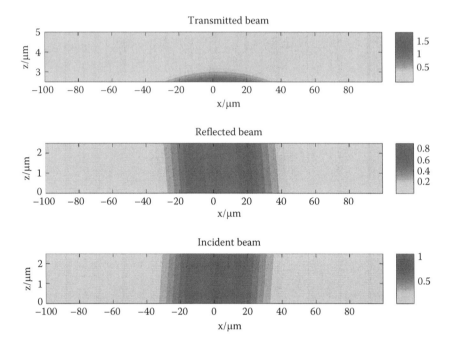

FIGURE 2.29 The distribution of the absolute value of the y electric field component calculated using Padé (1,1) approximation. The reference propagation constant is equal to wave number in the medium with refractive index n_1. In the other medium, the absolute value of the reference propagation constant matches the medium wave number while the rotation angle in the complex plane equals $\pi/3$. The initial beam is TE polarised and its width equals 30 µm while the tilt angle $\theta = 45°$. The calculation window is 200 µm while the number of samples is 2^{11}, $L_1 = L_2 = 2.5$ µm, $n_1 = 1.5$, $n_2 = 1.0$ while the operating wavelength is 1.06 µm.

apart. The width of the calculation window is 80 mm while the number of samples is 2^{16}, the distance from the slits to the lens is equal to the distance from the lens till the end of the computational window and equals 250 mm. The lens material is N-BK7 glass whereas the ambient is air. Figure 2.32 shows the calculated light intensity for the incident wave and transmitted wave (both forming the primary Bremmer series wave) and two reflected waves that form the first order correction wave. In this example, we tried to reproduce the experimentally recorded light intensity pattern (Figure 2.33). In an effort aimed at modelling an incoherent white light source that is used in the experiment, we use a wavelength comb that spans from 300 to 700 nm in steps of 5 nm and add noncoherently the light intensities calculated at each wavelength; that is, we add the light intensities and not the scalar potentials.

The calculation of the primary Bremmer series wave u_0 corresponds to recalculating the light intensity distribution as in Figure 2.30 but keeping in mind that at each air–glass interface the power transmission coefficient is not equal to 1. As mentioned earlier, the first correction wave has two components. The first one arises from the

TABLE 2.1

Transmittance of Selected Bulk Optics Elements

Element Definition	Transmittance t(x)

Thin plate

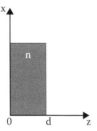

$$\exp\left(-jnk_0d\right)$$

Prism

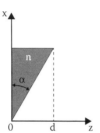

$$h_0 \exp\left(-j(n-1)k_0\alpha x\right)$$
$$h_0 = \exp\left(-jk_0d\right)$$

Plano-convex lens

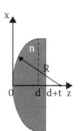

$$h_0h_1 \exp\left(jk_0\frac{x^2}{2f}\right); h_0 = \exp\left(-jnk_0d\right);$$
$$h_1 = \exp\left(-jnk_0t\right); f = \frac{R}{n-1}$$

Plano-concave lens

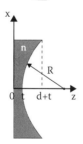

$$h_0h_1 \exp\left(jk_0\frac{x^2}{2f}\right); h_0 = \exp\left(-jnk_0d\right);$$
$$h_1 = \exp\left(-jnk_0t\right); f = -\frac{R}{n-1}$$

Sinusoidal transmission grating

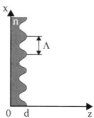

$$h_0 \exp\left(-j\frac{1}{2}(n-1)k_0d\cos\left(2\pi x / \Lambda\right)\right);$$
$$h_0 = \exp\left(-j\frac{1}{2}(n+1)k_0d\right)$$

TABLE 2.2

Sellmeier Coefficients for N-BK7 Glass

B_1	1.03961212
B_2	0.231792344
B_3	1.01046945
C_1	0.00600009867
C_2	0.0200179144
C_3	103.560653

Source: http://edit.schott.com/advanced_optics/english/abbe_data-
sheets/schott_datasheet_n-bk7.pdf

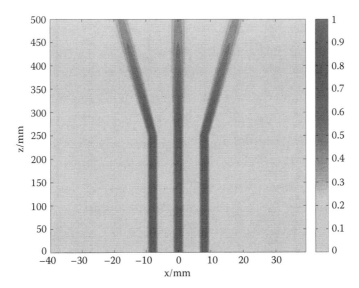

FIGURE 2.30 The distribution of the light intensity calculated using FFT algorithm in scalar approximation. The initial beam consists of three Gaussian lobes with beam width w = 2 mm (2.10) that are spaced at 8 mm. The calculation window is 80 mm while the number of samples is 2^{16}, the distance from z = 0 plane to the lens is equal to the distance from the lens till the end of the computational window and equals 250 mm. The lens material is N-BK7 glass, the ambient is air while the operating wavelength is 1.06 μm.

beam reflection at the first air–glass interface. For this wave the transmission function is given by:

$$h_0 \exp\left(-jk_0 \frac{x^2}{f}\right), \ h_0 = \exp\left(-jnk_0d\right), \text{ and } f = -\frac{R}{n-1}.$$

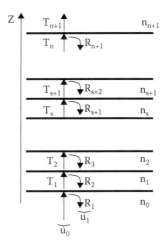

FIGURE 2.31 Schematic illustration of the Bremmer series concept.

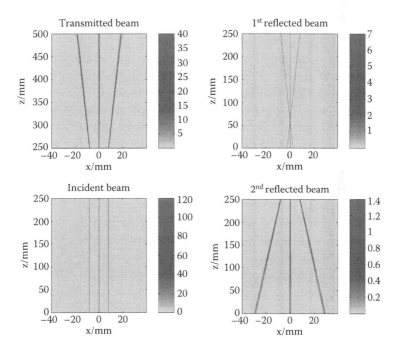

FIGURE 2.32 The distribution of the light intensity calculated using FFT algorithm in scalar approximation. The initial beam consists of three Gaussian lobes with beam width $w = 4$ mm (2.10) that are spaced at 8 mm and passed through a set of three 0.250 mm wide slits that are spaced at 8 mm. The calculation window is 80 mm while the number of samples is 2^{16}, the distance from the slits to the lens is equal to the distance from the lens till the end of the computational window and equals 250 mm. The lens material is N-BK7 glass, the ambient is air while the operating wavelength comb spans from 300 to 700 nm steps of 5 nm. The light intensities calculated at each wavelength were added (noncoherent addition).

FIGURE 2.33 An experimental setup consisting of a biconcave lens illuminated by three beams (this picture was taken from Wikipedia entry "lens"). Copyright 2000, 2001, 2002 Free Software Foundation, Inc. 51 Franklin St, Fifth Floor, Boston, MA 02110-1301 USA.

The second component of the first order correction wave u_1 comes from the reflection at the second air–glass interface. For this wave the transmission function equals to the square of the transmission function of the primary wave. The calculated light intensity distributions for all the three waves are presented in Figure 2.32. The inclusion of the higher order Bremmer series terms allows one to explain the origin of the multiple beams visible in Figure 2.33. The first component of the first correction wave is focussed at the imaginary focal point whereas the second component diverges as if from a focal point positioned at a half the distance of the actual lens focal length. Both these waves can be clearly observed in Figure 2.33. The higher order correction waves are much weaker than the primary ones and the first correction waves and hence cannot be seen in the experimental results. In the modelling results this effect is reproduced by including the correct values of the transmission and reflection coefficients when calculating the subsequent reflected and transmitted waves. This effect also explains the lowering intensity of the higher order correction waves when compared with the primary wave.

REFERENCES

1. Saleh, B.E.A. and M.C. Teich, *Fundamentals of Photonics*. 1991, New York: John Wiley & Sons Inc.
2. Poon, T.-C. and T. Kim, *Engineering Optics with MATLAB*. 2006, London: World Scientific Publishing.
3. Born, M. and E. Wolf, *Principles of Optics: Electromagnetic Theory of Propagation, Interference and Diffraction of Light*. 1999, Cambridge: Cambridge University Press.
4. Solimeno, S., B. Crosignani, and P. Di Porto, *Guiding, Diffraction and Confinement of Optical Radiation*. 1986, London: Academic Press.
5. Felsen, L.B., and N. Marcuvitz, *Radiation and Scattering of Waves*. 1994, Oxford: Oxford University Press.
6. Press, W.H., S.A. Teukolsky, W.T. Vetterling, and B.P. Flannery, *Numerical Recipes in C++: The Art of Scientific Computing*. 2002, Cambridge: Cambridge University Press.
7. Debnath, L. and P. Miskusinski, *Introduction to Hilbert Spaces with Applications*. 2005, London: Elsevier.

8. Kreyszig, E., *Introductory Functional Analysis*. 1978, London: Wiley.
9. Sujecki, S., Arbitrary truncation order three-point finite difference method for optical waveguides with stepwise refractive index discontinuities. *Optics Letters*, 2010. 35(24): p. 4115–4117.
10. Iserles, A., *A First Course in the Numerical Analysis of Differential Equations*. 2009, Cambridge: Cambridge University Press.
11. Hairer, E., C. Lubich, and G. Wanner, *Geometrical Numerical Integration: Structure-Preserving Algorithms for Ordinary Differential Equations*. 2002, Berlin: Springer.
12. Cheng, D.K., *Field and Wave Electromagnetics*. 1989, New York: Addison-Wesley.
13. Hadley, G.R., Wide angle beam propagation using Pade approximant operators. *Optics Letters*, 1992. 17(20): p. 1426–1428.
14. Anada, T., et al., Very-wide-angle beam propagation methods for integrated optical circuits. *IEICE Transactions on Electronics*, 1999. E82-C(7): p. 1154–1158.
15. Moore, T.G., et al., Theory and application of radiation boundary operators. *IEEE Transactions on Antennas and Propagation*, 1988. 36(12): p. 1797–1812.
16. Schultz, D., C. Glingener, and E. Voges, Novel generalised finite difference beam propagation method. *IEEE Journal of Quantum Electronics*, 1994. 30(4): p. 1132–1140.
17. Stralen, M.J.N.v., H. Blok, and M.V.d. Hoop, Design of sparse matrix representations for the propagator used in the BPM and directional wave field decomposition. *Optical and Quantum Electronics*, 1997. 29(2): p. 179–197.
18. Rao, H., et al., Complex propagators for evanescent waves in bidirectional beam propagation method. *Journal of Lightwave Technology*, 2000. 18(8): p. 1155–1160.
19. Data sheet of SCHOTT N-BK7 glass, http://edit. schott.com/advanced_optics/english/abbe_datasheets/schott_datasheet_n-bk7.pdf
20. Bremmer, H., The W.K.B. approximation as the first term of a geometric-optical series. *Communications on Pure and Applied Mathematics*, 1951. 4(1): p. 105–115.
21. Atkinson, F.V., Wave propagation and the Bremmer series. *Journal of Mathematical Analysis and Applications*, 1960. 1(3–4): p. 225–276.

3 Optical Waveguides

Guided wave optics, realised both in optical fibre and planar technologies, has numerous applications. Optical fibres are the backbone of modern high-speed telecommunication systems. They are also indispensable for the flexible optical power delivery in medical and manufacturing laser applications, for high-speed computer interconnects, and for decreasing the weight and improving immunity to external electromagnetic fields in modern sensor and measurement systems. Planar optical waveguides, on the other hand, are used for beam formation in edge emitting laser diodes, as sensors in modern sensor and measurement systems, to provide amplification in optical telecommunication systems, and in high-speed external laser diode modulators.

The optical fibre technology currently allows the fabrication of low loss optical waveguides providing both single mode and multimode wave guiding at a number of wavelengths and for various maximum optical power levels. The fibre optic technology is fairly advanced and provides many high-quality optical components. The weakness of fibre optic technology lays in the large dimensions of the components and in its limited suitability for low-cost mass production. Planar technology, on the other hand, has the potential to provide the low-cost large-scale production of compact optical components, which can meet the demands of the modern telecom systems and other applications.

INTRODUCTION TO OPTICAL WAVEGUIDE THEORY

Optical waveguides are characterised by very large longitudinal dimensions that easily extend over 100,000s of wavelengths and lateral dimensions ranging from nanometre-scale structural details to 100 or more wavelengths. The optical waveguides typically vary slowly along the direction of the wave propagation and are terminated by an abrupt dielectric discontinuity (Figure 3.1).

For isotropic, linear, dielectric media, the relations between the vectors of the electric and magnetic field and the electric and magnetic flux are given by Maxwell's equations:

$$\nabla \times \vec{E}(\vec{r},t) = -\frac{\partial \vec{B}(\vec{r},t)}{\partial t} \tag{3.1a}$$

$$\nabla \times \vec{H}(\vec{r},t) = \frac{\partial \vec{D}(\vec{r},t)}{\partial t} \tag{3.1b}$$

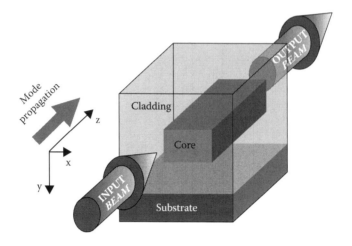

FIGURE 3.1 Optical waveguide in the rectangular coordinate system.

which for the Fourier transforms of the unknown field distributions $\vec{X}(\vec{r},\omega) = \left(\vec{E}(\vec{r},\omega),\vec{H}(\vec{r},\omega),\vec{D}(\vec{r},\omega),\vec{B}(\vec{r},\omega)\right)$ defined as:

$$\vec{X}(\vec{r},\omega) = \int_{-\infty}^{\infty} \vec{X}(\vec{r},t)\exp\left(-\mathrm{j}\omega t\right)\mathrm{d}t$$

have the form:

$$\nabla \times \vec{E}(\vec{r},\omega) = -\mathrm{j}\omega\vec{B}(\vec{r},\omega) \tag{3.1c}$$

$$\nabla \times \vec{H}(\vec{r},\omega) = \mathrm{j}\omega\vec{D}(\vec{r},\omega) \tag{3.1d}$$

where the angular frequency $\omega = 2\pi ft$. We note that the inverse Fourier transform is consequently defined as:

$$\vec{X}(\vec{r},t) = \int_{-\infty}^{\infty} \vec{X}(\vec{r},\omega)\exp\left(\mathrm{j}2\pi ft\right)\mathrm{d}f$$

Equation 3.1 is complemented by the (Gauss's law) conditions:

$$\nabla \cdot \vec{D} = 0 \tag{3.2a}$$

$$\nabla \cdot \vec{B} = 0 \tag{3.2b}$$

that hold both in the frequency and time domains, and eliminate the spurious solutions of 3.1. In addition, the dependence between the electric field and the electric flux vectors is given by:

$$\vec{D}(\vec{r},t) = \varepsilon_0 \int\limits_{-\infty}^{\infty} \left(1 + \chi^{(1)}(\vec{r},t-\tau)\right)\vec{E}(\vec{r},\tau)\,d\tau \qquad (3.3a)$$

which in the frequency domain has the form:

$$\vec{D}(\vec{r},\omega) = \varepsilon\vec{E}(\vec{r},\omega) = \varepsilon_0\varepsilon_r\vec{E}(\vec{r},\omega), \qquad (3.3b)$$

where ε_0 is the electric permittivity of the free space and ε_r is related to $\chi^{(1)}$ in the frequency domain via the equation:

$$\varepsilon_r(\vec{r},\omega) = 1 + \chi^{(1)}(\vec{r},\omega)$$

At optical frequencies for the magnetic field and the magnetic flux vectors both in the time and frequency domain the following holds:

$$\vec{B} = \mu_0\vec{H} \qquad (3.3c)$$

where μ_0 is the magnetic permeability of the free space. From 3.1 and 3.3, one can obtain the following set of wave equations in the frequency domain:

$$\nabla \times \nabla \times \vec{E} = \mu_0\varepsilon\omega^2\vec{E} \qquad (3.4a)$$

$$\varepsilon\nabla \times \left(\frac{1}{\varepsilon}\nabla \times \vec{H}\right) = \mu_0\varepsilon\omega^2\vec{H} \qquad (3.4b)$$

With the use of 3.2a, 3.3b, and the vectorial identity: $\nabla \times \nabla \times \vec{E} = \nabla(\nabla \cdot \vec{E}) - \nabla^2\vec{E}$, Equation 3.4a can be transformed to:

$$\nabla^2\vec{E} + \nabla\left(\frac{\nabla\varepsilon}{\varepsilon}\vec{E}\right) = -\mu_0\varepsilon\omega^2\vec{E} \qquad (3.5a)$$

With help of 3.2b, 3.3c, and vector identities: $\nabla \times \nabla \times \vec{H} = \nabla(\nabla \cdot \vec{H}) - \nabla^2\vec{H}$ and $\nabla \times (1/\varepsilon\nabla \times \vec{H}) = \nabla(1/\varepsilon) \times \vec{H} + (1/\varepsilon)\nabla \times \nabla \times \vec{H}$, Equation 3.4b can be recast into:

$$\nabla^2\vec{H} + \frac{\nabla\varepsilon}{\varepsilon} \times (\nabla \times \vec{H}) = -\mu_0\varepsilon\omega^2\vec{H} \qquad (3.5b)$$

The wave equations for the electric and magnetic field vectors 3.4 and 3.5, respectively, are not coupled. Consequently, in the analysis of optical waveguides, only one of the equations needs to be solved. The remaining electromagnetic field components can be obtained from 3.1. The electric field vector formulation is convenient when studying nonlinear phenomena because they are typically formulated in terms of the electric field vector (see Chapter 9). The magnetic field is generally smoother than the electric field and usually results in more efficient and accurate algorithms. It is also worth noting that the solutions of 3.4, unlike the solutions of 3.5, do not necessarily fulfil zero divergence conditions of 3.2. Consequently, Equation 3.4 admits spurious solutions as well as physically meaningful ones.

For structures with ε changing slowly on the wavelength scale, the vectorial wave equation 3.5 can be approximated by the scalar wave equation:

$$\nabla^2 \Phi = -\mu_0 \varepsilon \omega^2 \Phi \tag{3.6}$$

where Φ is a scalar potential. It is worth noting that the solutions of 3.6 obey a zero divergence condition, similarly to those of Equation 3.5, since this condition is used when deriving 3.6.

The distribution of the electromagnetic field within a section of an optical waveguide can be calculated by solving directly Equations 3.1, 3.4, 3.5, or 3.6 (in the scalar case). For this purpose finite difference, finite element, or boundary element-based numerical methods can be used in principle. These techniques can calculate the field distribution in the considered domain, subject to the distribution of excitation sources and the imposed boundary conditions. Since the dimensions of planar optical waveguides can be very large when compared with the operating wavelength, the numerical solution would usually require very large computational resources. Therefore, a different approach is used. In this approach it is assumed that the optical waveguide is perfectly straight and infinitely long. As such a (fictitious) structure is longitudinally invariant, the separation of variables can be performed. Consequently, if we align the waveguide with the z-axis (Figure 3.1), ε does not depend on z and the solution of 3.5a in the frequency domain can be expressed in the form of a product of two functions:

$$\vec{E}(\vec{r},\omega) = E(x,y,\omega)\exp(-j\beta z) \tag{3.7}$$

Substituting 3.7 in Equations 3.5a and if ε does not depend on z, then

$$\nabla \varepsilon = \vec{i}_x \frac{\partial \varepsilon}{\partial x} + \vec{i}_y \frac{\partial \varepsilon}{\partial y}$$

we obtain the following equations for the components of the vector $E(x,y,\omega) = \vec{i}_x E_x(x,y,\omega) + \vec{i}_y E_y(x,y,\omega) + \vec{i}_z E_z(x,y,\omega)$:

$$\frac{\partial^2 E_x}{\partial y^2} + \frac{\partial}{\partial x}\left(\frac{1}{\varepsilon}\frac{\partial(\varepsilon E_x)}{\partial x}\right) + \left(\omega^2 \mu_0 \varepsilon - \beta^2\right)E_x + \frac{\partial}{\partial x}\left(\frac{1}{\varepsilon}\frac{\partial(\varepsilon E_y)}{\partial y}\right) - \frac{\partial^2 E_y}{\partial x \partial y} = 0 \tag{3.8a}$$

$$\frac{\partial^2 E_y}{\partial x^2} + \frac{\partial}{\partial y}\left(\frac{1}{\varepsilon}\frac{\partial\left(\varepsilon E_y\right)}{\partial y}\right) + \left(\omega^2\mu_0\varepsilon - \beta^2\right)E_y + \frac{\partial}{\partial y}\left(\frac{1}{\varepsilon}\frac{\partial\left(\varepsilon E_x\right)}{\partial x}\right) - \frac{\partial^2 E_x}{\partial y\partial x} = 0 \quad (3.8b)$$

$$\frac{\partial^2 E_z}{\partial x^2} + \frac{\partial^2 E_z}{\partial y^2} + \left(\omega^2\mu_0\varepsilon - \beta^2\right)E_z - j\beta\left(\frac{1}{\varepsilon}\frac{\partial\varepsilon}{\partial x}E_x + \frac{1}{\varepsilon}\frac{\partial\varepsilon}{\partial y}E_y\right) = 0 \quad (3.8c)$$

Similarly, substituting $\vec{H}(\vec{r},\omega) = H(x,y,\omega)\exp(-j\beta z)$ in 3.5b, we obtain the following equations for the components of the magnetic field vector $H(x,y,\omega) = \vec{i}_x H_x(x,y,\omega) + \vec{i}_y H_y(x,y,\omega) + \vec{i}_z H_z(x,y,\omega)$:

$$\frac{\partial^2 H_x}{\partial x^2} + \varepsilon\frac{\partial}{\partial y}\left(\frac{1}{\varepsilon}\frac{\partial H_x}{\partial y}\right) + \left(\omega^2\mu_0\varepsilon - \beta^2\right)H_x - \varepsilon\frac{\partial}{\partial y}\left(\frac{1}{\varepsilon}\frac{\partial H_y}{\partial x}\right) + \frac{\partial^2 H_y}{\partial y\partial x} = 0 \quad (3.8d)$$

$$\frac{\partial^2 H_y}{\partial y^2} + \varepsilon\frac{\partial}{\partial x}\left(\frac{1}{\varepsilon}\frac{\partial H_y}{\partial x}\right) + \left(\omega^2\mu_0\varepsilon - \beta^2\right)H_y - \varepsilon\frac{\partial}{\partial x}\left(\frac{1}{\varepsilon}\frac{\partial H_x}{\partial y}\right) + \frac{\partial^2 H_x}{\partial x\partial y} = 0 \quad (3.8e)$$

$$\varepsilon\frac{\partial}{\partial x}\left(\frac{1}{\varepsilon}\frac{\partial H_z}{\partial x}\right) + \varepsilon\frac{\partial}{\partial y}\left(\frac{1}{\varepsilon}\frac{\partial H_z}{\partial y}\right) + \left(\omega^2\mu_0\varepsilon - \beta^2\right)H_z - j\beta\frac{1}{\varepsilon}\left(\frac{\partial\varepsilon}{\partial x}H_x + \frac{\partial\varepsilon}{\partial y}H_y\right) = 0 \quad (3.8f)$$

Equations 3.6 and 3.7 can be expressed compactly using matrix notation:

$$\left(\begin{bmatrix} a_{11} & a_{12} & a_{13} \\ a_{21} & a_{22} & a_{23} \\ a_{31} & a_{32} & a_{33} \end{bmatrix} + \omega^2\mu_0\varepsilon\right)\begin{bmatrix} E_x \\ E_y \\ E_z \end{bmatrix} = \beta^2\begin{bmatrix} E_x \\ E_y \\ E_z \end{bmatrix} \quad (3.9a)$$

$$\left(\begin{bmatrix} b_{11} & b_{12} & b_{13} \\ b_{21} & b_{22} & b_{23} \\ b_{31} & b_{32} & b_{33} \end{bmatrix} + \omega^2\mu_0\varepsilon\right)\begin{bmatrix} H_x \\ H_y \\ H_z \end{bmatrix} = \beta^2\begin{bmatrix} H_x \\ H_y \\ H_z \end{bmatrix} \quad (3.9b)$$

where the elements of the matrices a_{ij} and b_{ij} are given in Table 3.1 and Table 3.2, respectively, where $\circ$ is a "placeholder" symbol that holds the place for a function that is acted upon by the preceding operator.

TABLE 3.1
Elements a_{ji}

$a_{j,i}$	$i = 1$	$i = 2$	$i = 3$
$j = 1$	$\dfrac{\partial^2 \circ}{\partial y^2} + \dfrac{\partial}{\partial x}\left(\dfrac{1}{\varepsilon}\dfrac{\partial(\varepsilon\circ)}{\partial x}\right)$	$\dfrac{\partial}{\partial x}\left(\dfrac{1}{\varepsilon}\dfrac{\partial(\varepsilon\circ)}{\partial y}\right) - \dfrac{\partial^2 \circ}{\partial x \partial y}$	0
$j = 2$	$\dfrac{\partial}{\partial y}\left(\dfrac{1}{\varepsilon}\dfrac{\partial(\varepsilon\circ)}{\partial x}\right) - \dfrac{\partial^2 \circ}{\partial y \partial x}$	$\dfrac{\partial^2 \circ}{\partial x^2} + \dfrac{\partial}{\partial y}\left(\dfrac{1}{\varepsilon}\dfrac{\partial(\varepsilon\circ)}{\partial y}\right)$	0
$j = 3$	$-j\beta\dfrac{1}{\varepsilon}\dfrac{\partial\varepsilon}{\partial x}\circ$	$-j\beta\dfrac{1}{\varepsilon}\dfrac{\partial\varepsilon}{\partial y}\circ$	$\dfrac{\partial^2}{\partial x^2} + \dfrac{\partial^2}{\partial y^2}$

TABLE 3.2
Elements b_{ji}

$b_{j,i}$	$i = 1$	$i = 2$	$i = 3$
$j = 1$	$\dfrac{\partial^2 \circ}{\partial x^2} + \varepsilon\dfrac{\partial}{\partial y}\left(\dfrac{1}{\varepsilon}\dfrac{\partial\circ}{\partial y}\right)$	$-\varepsilon\dfrac{\partial}{\partial y}\left(\dfrac{1}{\varepsilon}\dfrac{\partial\circ}{\partial x}\right) + \dfrac{\partial^2 \circ}{\partial y\partial x}$	0
$j = 2$	$-\varepsilon\dfrac{\partial}{\partial x}\left(\dfrac{1}{\varepsilon}\dfrac{\partial\circ}{\partial y}\right) + \dfrac{\partial^2 \circ}{\partial x\partial y}$	$\dfrac{\partial^2 \circ}{\partial y^2} + \varepsilon\dfrac{\partial}{\partial x}\left(\dfrac{1}{\varepsilon}\dfrac{\partial\circ}{\partial x}\right)$	0
$j = 3$	$-j\beta\dfrac{1}{\varepsilon}\dfrac{\partial\varepsilon}{\partial x}\circ$	$-j\beta\dfrac{1}{\varepsilon}\dfrac{\partial\varepsilon}{\partial y}\circ$	$\varepsilon\dfrac{\partial}{\partial x}\left(\dfrac{1}{\varepsilon}\dfrac{\partial\circ}{\partial x}\right) + \varepsilon\dfrac{\partial}{\partial y}\left(\dfrac{1}{\varepsilon}\dfrac{\partial\circ}{\partial y}\right)$

In 3.9, a_{13}, a_{23}, b_{13}, and b_{23} are equal to zero. Hence, the transverse field components do not couple with the longitudinal ones and it is sufficient to solve the wave equations for one pair of the transverse components only:

$$\left(\begin{bmatrix} a_{11} & a_{12} \\ a_{21} & a_{22} \end{bmatrix} + \omega^2\mu_0\varepsilon\right)\begin{bmatrix} E_x \\ E_y \end{bmatrix} = \beta^2\begin{bmatrix} E_x \\ E_y \end{bmatrix} \tag{3.10a}$$

$$\left(\begin{bmatrix} b_{11} & b_{12} \\ b_{21} & b_{22} \end{bmatrix} + \omega^2\mu_0\varepsilon\right)\begin{bmatrix} H_x \\ H_y \end{bmatrix} = \beta^2\begin{bmatrix} H_x \\ H_y \end{bmatrix} \tag{3.10b}$$

If Equation 3.10a is solved, then the E_z component can be calculated from 3.2a:

$$E_z = -\frac{1}{j\beta\varepsilon}\left(\frac{\partial\left(\varepsilon E_x\right)}{\partial x} + \frac{\partial\left(\varepsilon E_y\right)}{\partial y}\right) \tag{3.11a}$$

The magnetic field components can then be obtained from:

$$\frac{\partial E_z}{\partial y} + j\beta E_y = -j\omega\mu_0 H_x \tag{3.11b}$$

$$-\frac{\partial E_z}{\partial x} - j\beta E_x = -j\omega\mu_0 H_y \tag{3.11c}$$

$$\frac{\partial E_y}{\partial x} - \frac{\partial E_x}{\partial y} = -j\omega\mu_0 H_z \tag{3.11d}$$

Alternatively, if Equation 3.10b is solved, then the H_z component can be calculated from 3.2b:

$$H_z = -\frac{1}{j\beta}\left(\frac{\partial H_x}{\partial x} + \frac{\partial H_y}{\partial y}\right) \tag{3.11e}$$

The electric field components can be obtained from:

$$\frac{\partial H_z}{\partial y} + j\beta H_y = j\omega\varepsilon E_x \tag{3.11f}$$

$$-\frac{\partial H_z}{\partial x} - j\beta H_x = j\omega\varepsilon E_y \tag{3.11g}$$

$$\frac{\partial H_y}{\partial x} - \frac{\partial H_x}{\partial y} = j\omega\varepsilon E_z \tag{3.11h}$$

Again using the matrix notation, Equation 3.10 can be recast into a more compact form for use in later derivations:

$$\left(A + \omega^2\mu_0\varepsilon\right)E_t = \beta^2 E_t \tag{3.12a}$$

$$\left(B + \omega^2\mu_0\varepsilon\right)H_t = \beta^2 H_t \tag{3.12b}$$

where E_t and H_t stand for the transverse (w.r.t. z) components of the electric and magnetic field, respectively. The matrices A and B are given by:

$$A = \begin{bmatrix} a_{11} & a_{12} \\ a_{21} & a_{22} \end{bmatrix}, \ B = \begin{bmatrix} b_{11} & b_{12} \\ b_{21} & b_{22} \end{bmatrix} \tag{3.13}$$

with the elements a_{ji} and b_{ji} given in Table 3.1 and Table 3.2.

The vectorial equations 3.12 are relatively difficult to solve. However, they can be significantly simplified for waveguiding structures, which support polarised modes. For such structures two types of modes are guided, that is, modes that are polarised either along the x direction (x-polarised, quasi-TE modes) or the y direction (y-polarised, quasi-TM modes). The main transverse components of electromagnetic field for the x-polarised modes are E_x and H_y while for the y-polarised ones: E_y and H_x. The polarised approximation proved very successful in modelling planar optical waveguides and is therefore of large practical importance [1]. For the x-polarised modes, while neglecting the minor field component, Equation 3.12 reduces to:

$$\left(a_{11} + \omega^2 \mu_0 \varepsilon\right) E_x = \beta^2 E_x \tag{3.14a}$$

and

$$\left(b_{22} + \omega^2 \mu_0 \varepsilon\right) H_y = \beta^2 H_y \tag{3.14b}$$

whereas for y-polarised modes one obtains the following equations:

$$\left(a_{22} + \omega^2 \mu_0 \varepsilon\right) E_y = \beta^2 E_y \tag{3.14c}$$

and

$$\left(b_{11} + \omega^2 \mu_0 \varepsilon\right) H_x = \beta^2 H_x \tag{3.14d}$$

Unlike 3.12, the equations for the transverse components of the electric and magnetic fields are not coupled under the polarised approximation. Hence, they can be solved separately, which reduces the required computational effort. Numerical results published in the literature have shown that both the electric and magnetic fields approach yielded the same values of the propagation constants for the corresponding modes [1]. Theoretically, this can be demonstrated by showing that the operators present in Equations 3.14a,c and 3.14b,d are mutually adjoint.

If ε varies slowly in the transverse plane, one can use the scalar approximation. Substituting $\Phi(\vec{r}, \omega) = \phi(x, y, \omega) \exp(-j\beta z)$ in 3.6 yields:

$$\left(s + \omega^2 \mu_0 \varepsilon\right)\phi = \beta^2 \phi \tag{3.15}$$

where s is given by:

$$s = \frac{\partial^2}{\partial x^2} + \frac{\partial^2}{\partial y^2} \tag{3.16}$$

Equations 3.12 through 3.15 form the basis for the theory of the optical waveguides. With the exception of several simple waveguide structures, these equations cannot be solved analytically. When considering the light propagation in optical fibres, it is convenient to recast Equations 3.12 and 3.16 in the cylindrical coordinate system. This will be discussed in detail in the section "Optical Fibres."

Finally, we consider the orthogonality of the modes. Let's consider two solutions of Maxwell's equations denoted with indexes "1" and "2." The mode orthogonality condition can be derived from the Maxwell's equations in the form akin to the Lorenz reciprocity theorem [2]:

$$\nabla \cdot \left(\vec{E}_1 \times \vec{H}_2^* + \vec{E}_2^* \times \vec{H}_1\right) = 0 \tag{3.17}$$

From 3.17, using the field distributions corresponding to two eigenmodes of 3.12:

$$\vec{E}_1 = \left(\vec{E}_{tm} + \vec{E}_{zm}\right)e^{j(\omega t - \beta_m z)}$$

$$\vec{H}_1 = \left(\vec{H}_{tm} + \vec{H}_{zm}\right)e^{j(\omega t - \beta_m z)}$$

$$\vec{E}_2 = \left(\vec{E}_{tp} + \vec{E}_{zp}\right)e^{j(\omega t - \beta_p z)}$$

$$\vec{H}_2 = \left(\vec{H}_{tp} + \vec{H}_{zp}\right)e^{j(\omega t - \beta_p z)}$$

where the index "t" denotes the transverse part of the vector, the index "z" the longitudinal one while the indices m and p designate the modes, and integrating over the entire transverse plane S upon the application of the divergence theorem yields [2,3]:

$$\left(\beta_m - \beta_p\right)\iint_S \left(\vec{E}_{tm} \times \vec{H}_{tp}^* + \vec{E}_{tp}^* \times \vec{H}_{tm}\right)dxdy = 0 \tag{3.18}$$

Therefore, if $\beta_m \neq \beta_p$ it must hold that:

$$\iint\limits_S \left(\vec{E}_{tm} \times \vec{H}_{tp}^* + \vec{E}_{tp}^* \times \vec{H}_{tm} \right) dxdy = 0 \tag{3.19}$$

which is the orthogonality condition for two eigenmodes of 3.12 that correspond to two distinct eigenvalues β_m and β_p.

One can observe that the square of the propagation constant is an eigenvalue of 3.12. This means that for each eigenvalue of 3.12 there are two possible values of the propagation constant—one with a positive sign and one with the negative sign. The positive sign corresponds to modes that propagate in the positive z-direction while the negative sign to the modes propagating in the opposite direction. It is obvious that both modes have the same distribution of either the transverse electric or magnetic field, depending on whether we solve 3.12a or 3.12b. However, the components of the other electromagnetic field vector that are obtained from 3.11 differ in sign depending on the direction of the mode propagation. In the electromagnetic field, distribution consists of modes that propagate only in one direction, 3.19 reduces to [2]:

$$\iint\limits_S \left(\vec{E}_{tm} \times \vec{H}_{tp}^* \right) dxdy = 0 \tag{3.20}$$

One objection that could be raised when deriving 3.19 is that that the divergence theorem applied to functions do not necessarily form continuously differentiable vector fields. However, it should be noted that an orthogonality condition is needed to form an inner product space. The definition of an inner product in general does not need to be derived. It is sufficient to guess one and verify that it fulfils the conditions given by the definition of the inner product [4,5]. When $\beta_m = \beta_p = \beta_{mp}$, then from the Poynting theorem it can be shown that:

$$\iint\limits_S \left(\vec{E}_{tmp} \times \vec{H}_{tmp}^* + \vec{E}_{tmp}^* \times \vec{H}_{tmp} \right) dxdy = 4 \iint\limits_S S_z dxdy \tag{3.21}$$

where S_z stands for the longitudinal component of the time averaged Poynting vector. Hence, if the coordinate system is right-handed, then 3.21 is positive, and thus 3.19 is positively defined. The linearity and conjugate symmetry property can be shown similarly. Hence 3.19 defines an inner product.

Finally, for later use we write explicitly the longitudinal component of the Poynting vector in terms of the electromagnetic field components of the waveguide modes:

$$S_z = \frac{1}{4} \left(E_x H_y^* - E_y H_x^* + E_x^* H_y - E_y^* H_x \right) \tag{3.22}$$

Once the E_x and E_y or H_x and H_y are calculated from 3.12 the other two transverse field components can be obtained self-consistently from 3.11.

TABLE 3.3
Classical and Quantum Theory Measures of Power Flow

Classical		Quantum Theory
Vectorial and Polarised Approximation	**Scalar**	
Power density:	Light intensity:	Photon flux density:
$S_z = \dfrac{1}{4}\left(E_x H_y^* - E_y H_x^* + E_x^* H_y - E_y^* H_x\right)$	$I = \Phi^* \Phi$	$\dfrac{I}{h\nu}; \dfrac{S_z}{h\nu}$
Power:	Power:	Photon flux:
$P = \displaystyle\iint_S S_z \,\mathrm{d}x\mathrm{d}y$	$P = \displaystyle\iint_S I \,\mathrm{d}x\mathrm{d}y$	$\dfrac{P}{h\nu}$

Formulae 3.19 and 3.22 are also applicable in the case of the polarised modes obtained from 3.14. In the scalar case the orthogonality of the modes follows from the fact that the operator $s+\omega^2\mu\varepsilon$ is self-adjoint with respect to the scalar product defined as:

$$\langle \varphi_m, \varphi_p \rangle = \iint_S \varphi_m^* \varphi_p \,\mathrm{d}x\mathrm{d}y \tag{3.23}$$

If $m = p$, then 3.23 yields the definition of the light intensity for the mode mp in terms of the scalar potential φ obtained by solving 3.15:

$$I_{mp} = \varphi_{mp}^* \varphi_{mp} \tag{3.24}$$

In Table 3.3 we have put together the measures of power and power flux for the vectorial solution 3.12, polarised approximation 3.14, and the scalar approximation 3.15 and added formulas for extracting the quantum theory equivalents, which are needed in Chapters 7 through 9.

The rest of the material presented in this chapter is split into two sections. The first section is devoted to the analysis and modelling of optical waveguides that have been fabricated using a planar technology while the latter one presents the theory and modelling techniques that are applicable to optical fibres.

PLANAR OPTICAL WAVEGUIDES

In this section, we first discuss various optical waveguides that are used in integrated optics and planar photonic devices. Then we present the theory and the numerical methods used for the calculation of the mode field distributions and of the propagation constants. Examples of simple programs written using MATLAB are presented and used to obtain results that illustrate the application of the theory.

WAVEGUIDING IN PLANAR OPTICAL WAVEGUIDES

All the functional elements of the integrated optical circuits use optical waveguides to achieve the optical field confinement in the transverse plane. The design of a photonic device therefore requires the selection of a particular optical waveguide that meets the design specifications within the parameter space imposed by a selected fabrication technology and device application. There are three waveguiding mechanisms that are used for achieving wave guiding in the planar optical waveguides, they are:

- Index guiding
- Low loss leaky wave guiding
- Gain guiding

The index-guided structures rely on the phenomenon of the total internal reflection. As a result, the core of these structures must have a larger refractive index than the cladding and substrate layers. This puts a restriction on the selection of the possible materials and fabrication technology. This limitation can be often relaxed by using low loss leaky wave guiding. The gain guiding is used exclusively in laser diodes.

Index Guiding Planar Optical Waveguides

There are several basic index guiding planar waveguide structures, which are of interest in integrated optics:

- Slab waveguide (Figure 3.2)
- Rectangular waveguide (Figure 3.3)
- Rib waveguide (Figure 3.4)
- Ridge waveguide (Figure 3.5, which is just a special case of a rib waveguide with $H = 0$)
- Rib-loaded waveguide (Figure 3.6)
- In-diffused waveguide (Figure 3.7)

The simplest waveguiding structure, which is easily realisable using planar technology, is the slab waveguide (Figure 3.2). The slab waveguide consists of a core, with thickness t and refractive index n_f, and cladding and substrate regions with refractive indices n_c and n_s, respectively. In order to provide index guiding, the refractive index of the core must be larger than that of the substrate and the cladding, that is, $n_f > \max(n_s, n_c)$. A slab waveguide supports two types of modes: transverse electric (TE) and transverse magnetic (TM). Both these types of modes have three components of the electromagnetic field, only. If the slab waveguide is symmetrical, that is, $n_s = n_c$, it guides at least one mode at a given wavelength. Otherwise, if $n_s > n_c$, no modes are guided if

$$t \frac{2\pi}{\lambda} \sqrt{n_f^2 - n_s^2} < \sqrt{A} \text{ , where } A = \frac{n_s^2 - n_c^2}{n_f^2 - n_s^2} \text{ for TE and } A = \frac{n_f^4}{n_c^4} \frac{n_s^2 - n_c^2}{n_f^2 - n_s^2}$$

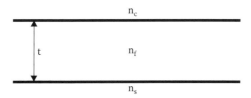

FIGURE 3.2   Slab waveguide.

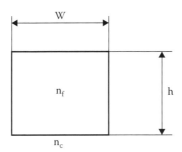

FIGURE 3.3   Rectangular waveguide.

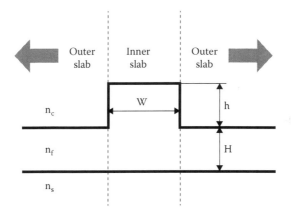

FIGURE 3.4   Rib waveguide.

for TM modes [3]. There is one major difficulty with applying slab waveguide in integrated optical circuits: The slab waveguide does not provide lateral modal confinement. A slab waveguide can have more than one layer between the substrate and the cladding. Such a structure is typically referred to as a multilayered slab waveguide. The slab waveguide is important mainly from the theoretical point of view because it is the only planar optical waveguide structure that can be analysed.

The rectangular dielectric waveguide (Figure 3.3) is the simplest structure used as a planar waveguide, which provides both lateral and vertical confinement. The light beam is guided by the core, which has the refractive index n_f, which is larger

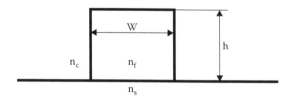

FIGURE 3.5   Ridge waveguide.

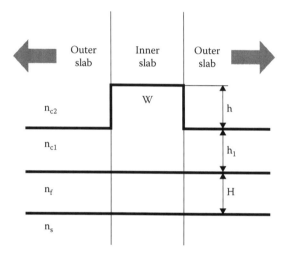

FIGURE 3.6   Rib-loaded waveguide.

FIGURE 3.7   In-diffused waveguide.

than that of the cladding n_c. For a square waveguide of side w, the single mode condition is given approximately by $w*(4\pi/\lambda)*\sqrt{n_f^2 - n_c^2} < 2$ [6], where λ is the operating wavelength. The single mode condition changes with the aspect ratio w/h. For large aspect ratios w/h, the single-mode condition for rectangular waveguide can be approximated by the single mode condition of a slab waveguide with thickness h, that is, $h*(2\pi/\lambda)*\sqrt{n_f^2 - n_c^2} < \pi/2$ [7]. As the single mode condition contains the product of waveguide dimension and refractive index contrast, it is not possible to design a single-mode rectangular waveguide, which has both a large cross-section and large refractive index contrast between the core and the cladding. Consequently, it is difficult to efficiently couple a rectangular optical waveguide of a large refractive index contrast with a single mode optical fibre. The detailed study of the general properties of the rectangular waveguide modes can be found in Majewski and Sujecki [6,9] and Goell [8].

Single mode planar waveguides with both a large cross-section and using a dielectric material with a large refractive index contrast between the core and the cladding can be realised by the use of the rib waveguide structure. The application of rib waveguides in integrated optics was originally suggested by Goell [10] to ease fabrication tolerances for directional couplers. Figure 3.4 shows the cross-section of an optical rib waveguide. The rib waveguide with refractive index n_f is situated on the substrate of refractive index n_s and covered with cladding layer having the refractive index n_c. H is the outer slab thickness, h the rib height, and w the rib width. The areas between the vertical dashed lines and outside of the vertical dashed lines are referred to as the inner and outer slab, respectively. The refractive index of the core is larger than that of the substrate and cladding, that is, $n_f > $ max (n_s, n_c), in order to provide vertical index guiding. The horizontal index guiding results from the height of the inner slab being larger than that of the outer slab (see, for example, Ebeling [3]).

One particular type of rib waveguides has been identified in the literature, that is, the large rib waveguide. A large rib waveguide fulfils the condition [11]:

$$\left((H+h) \middle/ \lambda \right) * \sqrt{n_f^2 - n_s^2} > 1$$

Large rib waveguides allow single mode structures with both large cross-section and large refractive index contrast between the core and the cladding to be designed. This results from the fact that the number of guided modes for this structure depends only on the ratios $h/(h+H)$ and $w/(h+H)$ [11,12]. The corrections to simple analytical design formulas for single-mode large rib waveguides are given in Soref et al. [11] and Marcatili [12], determined using more accurate numerical and experimental techniques are given in Dagli and Fonstad [13], Pogossian et al. [14], Powell [15], and Lousteau et al. [16]. The detailed studies of general properties of optical rib waveguides, including modal polarisation, can be found in Majewski and Sujecki [17,18] and Chiang and Wong [19].

The ridge waveguide (Figure 3.5) is a special case of a rib waveguide, although it may be also understood as a generalised rectangular waveguide. The ridge waveguide has similar single mode properties to a rectangular waveguide. From the effective index analysis, it follows that slightly larger single mode waveguide cross sections, than in the case of rectangular waveguides, can be achieved [3]. However, the cut-off wavelength of the fundamental mode is not equal to zero.

The rib-loaded waveguide (Figure 3.6) provides a way to fabricate waveguides without etching through the core guiding layer (important, for example, in the case of semiconductor laser diodes). In this case the core layer of thickness H and refractive index n_f is deposited on a substrate with refractive index n_s. Then the core layer is covered with a rib waveguide with cladding of refractive index n_{c1}, which provides lateral guiding and covered with another layer with refractive index n_{c2} (typically air or metal contact). One disadvantage of the rib-loaded waveguides is the large aspect ratio of the fundamental mode beam shape, which makes efficient coupling with a single-mode fibre difficult.

Another structure often used in integrated optics is an in-diffused waveguide (Figure 3.7). It is fabricated by diffusion of dopants, which increase locally the refractive index in the substrate of refractive index n_s. The single-mode properties of this waveguide depend on the distribution of the refractive index, which is a function of the distribution of the dopants. In order to design theoretically the in-diffused waveguides, the doping profile and the resulting refractive index profile have to be measured. An experimental study of the single mode properties of in-diffused optical waveguides is given in Petermann [20].

Often, in the case of in-diffused, rib, ridge, rib-loaded, and ridge-loaded waveguides the cladding is air. However, it is often advantageous to cover these structures with another material to protect the waveguide surface from contamination by dust; and to decrease the refractive index contrast between core and cladding in order to minimise the roughness scattering losses. Such waveguides are referred to in the literature as "buried waveguides," for example, a buried rib or ridge waveguide.

Low Loss Leaky and Gain Guided Planar Optical Waveguides

Optical waveguides can also be designed to guide low loss leaky waves. In particular, antireflection resonant optical waveguides (ARROWs) use this guidance mechanism. The ARROW was originally suggested in Duguay et al. [21] to design an integrated optical polariser. Usually ARROWs rely typically on index guiding in one direction and on antiresonant mirrors in the other, see, for instance, Gerces et al. [22] and Zmudzinski et al. [23]. ARROWs generally are more sensitive to fabrication tolerances than the index-guided structures. For the sake of completeness it should be also mentioned that there are other waveguide structures using leaky wave guiding mechanism, for example, the waisted-rib waveguide [24].

Gain guiding structures are exclusively used in semiconductor lasers. The current flowing through the active region of a laser diode creates an area of positive gain, which guides the light laterally while vertically index guiding is used [25]. The modes of gain guided waveguides are known to have curved phase fronts, unlike index guided waveguide modes [25].

EXAMPLES OF PLANAR OPTICAL WAVEGUIDES

A number of materials are used for fabrication of optical integrated structures, these include: III–V semiconductors, silicon, silica glass, lithium niobate, and polymers. The applicability of one material and technology for a particular optical planar device depends on a number of issues:

- Wafer size, cost, and availability
- Epitaxial deposition quality, thickness, and cost
- Etching quality, depth, and cost
- Transparency at the operating wavelength
- Possibility of performing switching operations
- Possibility of fabricating active devices: sources and amplifiers
- Possibility of fabricating detectors
- Possibility of cointegrating with integrated electronic circuits.

III–V semiconductors offer good quality wafers and well-developed processing technology. The presence of electro-optic and electro-absorption effects allow for the realisation of fast electro-absorption and electro-optic switches. The III–V semiconductor based laser diode sources cover wavelengths from 700 to 1600 nm, that is, all relevant telecom wavelengths. III–V semiconductors are also used for the realisation of near infrared light detectors. The cointegration of passive and active components is still difficult, though there is a constant progress in this area, for example, the quantum well intermixing technology [26].

Silicon on insulator (SOI) technology is based on a silicon wafer and uses standard silicon processing steps developed for integrated electronic circuits. This technology is very well developed and offers at the moment very good quality passive waveguide integrated optical circuits. A major advantage of Si technology is the possibility of cointegration with integrated electronic circuits. However, silicon is not suitable for the realisation of fully functional photonic circuits. One disadvantage of applying silicon is that this material is not well suited for the realisation of high-quality photodiodes operating at the main optical telecom wavelengths that is 1.3 and 1.55 μm; even though, silicon-based photodetectors for 1.55 μm range can, in principle, be realised [27]. The nonexistence of the electro-optic effect due to crystal symmetry renders silicon unsuitable for the realisation of fast electro-optic modulators. So far a silicon-based modulator with a bandwidth of 2.5 GHz [28] has been demonstrated. Furthermore, due to the indirect bandgap, efficient electrically pumped coherent light sources based on Si have not been realised despite many attempts, for example, Raman effect-based silicon coherent light sources [29].

Lithium niobate has a large electro-optic coefficient and is used primarily to fabricate high-speed modulators [30]. There are no light sources and detectors offered by this technology. However, optical amplifiers at 1.55 μm in lithium niobate have been demonstrated by implanting Er ions [31].

Silica glass based technology [32] offers waveguides with record low losses at 1.55 μm. This technology also provides the first commercially available erbium waveguide amplifiers (EDWAs) with gain of 20 dB. Silica glass is a competitive low-cost technology but it is not well suited for the realisation of fast optical modulators, optical detectors, and electrically pumped coherent light sources.

Another group of materials, which have been recently developed, are polymers. Their main advantage is low cost and, in theory, simple fabrication technology, using either imprinting or spin coating. The polymers indicate large electro-optic constant and hence can be used for fast modulators [33]. The amplification in rear earth doped polymer waveguides has also been demonstrated [34].

More detail on various technological issues in planar waveguide technology can be found in standard books on semiconductor lasers and integrated optics [2,3,35,36].

Figure 3.8 shows some examples of planar optical waveguide structures. A typical technology using a rectangular waveguide is silica on silicon. A rectangular optical waveguide (Figure 3.8a) can be fabricated using flame hydrolysis to deposit a germanium-doped silica glass layer on a thermally oxidised silicon wafer. The deposited layer is then etched to form a rectangular waveguide structure and covered with another layer of silica glass, which is deposited again by the flame hydrolysis process [32]. In Figure 3.8b a typical rib waveguide fabricated by SOI technology is shown.

A rib waveguide structure is etched in Si layer that is optically insulated from the substrate by a sufficiently thick SiO_2 layer [37]. A polymer waveguide (Figure 3.8c) can be fabricated by etching first a groove in the silicon substrate. Then a silica layer is deposited and spin coated with a polymer [38]. Figure 3.8d shows a typical example of a rib-loaded waveguide structure. This rib-loaded waveguide has been applied to realise an Er-doped integrated optical amplifier [39]. The use of a rib-loaded structure allows roughness scattering loss to be reduced. The disadvantage of using a rib-loaded waveguide is the large beam astigmatism and consequently a poor overlap with the fundamental mode field distribution of a standard single mode optical fibre. Figure 3.8e shows another typical rib waveguide-loaded structure, which is used in semiconductor lasers [3]. In this case, etching through the active layer is avoided, because it affects adversely the reliability of the device. The last example is a Ge in-diffused silica waveguide (Figure 3.8f). The advantage of this technology is its simplicity. However, it does not offer much design flexibility. Diffusion is also used to fabricate waveguides in lithium niobate substrates.

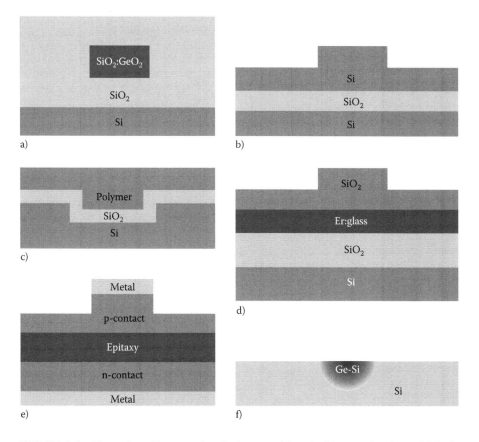

FIGURE 3.8 Examples of integrated optical waveguides: a) silicon on insulator (SOI) rib waveguide, b) rib waveguide, c) polymer rib waveguide, d) erbium ion–doped rib-loaded waveguide, e) semiconductor laser rib-loaded waveguide, and f) in-diffused waveguide.

SLAB OPTICAL WAVEGUIDE

In this section we present numerical methods for the calculation of the field distributions and the corresponding propagation constants for a slab waveguide. First, we consider Equation 3.1 and split them into a set of six equations, one for each of the three components of the vector fields that stand on both sides of 3.1a and 3.1b. Further, we assume that the modes of the slab waveguide propagate along the z-axis and that the refractive index distribution depends on the x-coordinate only (Figure 3.9) and also that the refractive index of the core is larger than that of the substrate and cladding. Since the structure is shift invariant along the y- and z-axis, while z is the assumed direction of propagation, each component of the electromagnetic field distribution of a waveguide mode can be expressed in the frequency domain as a product of two functions:

$$X(x,z,\omega) = F(x,\omega)\exp(-j\beta z) \tag{3.25}$$

Substituting 3.25 in 3.1 and using 3.2 yields two sets of equations for complex field envelopes $F(x,\omega)$ of each of the three components of the electric and magnetic field vector:

$$\left|\begin{array}{l} j\beta E_y = -j\omega\mu_0 H_x \\[2mm] \dfrac{dE_y}{dx} = -j\omega\mu_0 H_z \\[2mm] -j\beta H_x - \dfrac{dH_z}{dx} = j\omega\varepsilon E_y \end{array}\right. \tag{3.26a}$$

$$\left\{\begin{array}{l} j\beta H_y = j\omega\varepsilon E_x \\[2mm] \dfrac{dH_y}{dx} = j\omega\varepsilon E_z \\[2mm] -j\beta E_x - \dfrac{dE_z}{dx} = -j\omega\mu_0 H_y \end{array}\right. \tag{3.26b}$$

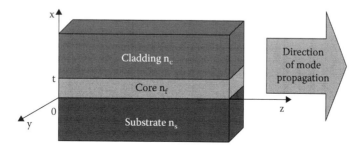

FIGURE 3.9   Three-layer slab waveguide in a rectangular coordinate system.

Therefore, a slab waveguide guides two types of modes, which have three components of the electromagnetic field only. The three components of the field for the modes that are the solution to the set of Equation 3.26 are E_y, H_x, and H_z. Since the only longitudinal (i.e., directed along the direction of propagation) component of the electromagnetic field vector is H_z, this mode is denoted as "transverse electric" (TE). By analogical arguments, the modes described by Equation 3.9b are denoted as "transverse magnetic" (TM).

Eliminating H_x and H_z from 3.26 yields:

$$\frac{d^2 E_y}{dx^2} + \left(n^2 k_0^2 - \beta^2\right) E_y = 0 \tag{3.27a}$$

Similarly, eliminating H_y and E_z from 3.9b gives:

$$\frac{d}{dx}\left(\frac{1}{n^2}\frac{d\left(n^2 E_x\right)}{dx}\right) + \left(n_2 k_0^2 - \beta^2\right) E_x = 0 \tag{3.27b}$$

For the transverse magnetic field component, the following equations are obtained from 3.26 for the TE and TM case, respectively:

$$\frac{d^2 H_x}{dx^2} + \left(n^2 k_0^2 - \beta^2\right) H_x = 0 \tag{3.28a}$$

$$n^2 \frac{d}{dx}\left(\frac{1}{n^2}\frac{dH_y}{dx}\right) + \left(n_2 k_0^2 - \beta^2\right) H_y = 0 \tag{3.28b}$$

Thus, we have reduced the solution of a set of partial differential equations 3.1 to the solution of two decoupled second order ordinary differential equations. It is important to observe that 3.27 and 3.28 are valid for any waveguide whose refractive index distribution is dependent on the x-coordinate only if the direction of propagation is selected along the z-direction (in fact the same equations follow if y-axis is assumed as the propagation direction).

When a slab waveguide has a piece-wise homogenous refractive index distribution, an analytical solution of the Maxwell's equations can be obtained. As an illustrating example, we derive an analytical solution to 3.27 and 3.28 in the case of a slab waveguide consisting of three homogenous layers (Figure 3.9). Since the derivation of the dispersion equation for the calculation of the propagation constants of the eigenmodes can be found in a number of textbooks [2,3,40–42], here we only retrace the main steps of this process.

In each homogenous layer, the solution of Equations 3.27 and 3.28 can be expressed analytically. The form of the solution depends on the sign of the $n^2 k_0^2 - \beta^2$ factor. We therefore define three constants for the three layers of the slab waveguide (Figure 3.9):

$$p^2 = \beta^2 - n_c^2 k_0^2$$
$$q^2 = \beta^2 - n_s^2 k_0^2$$
$$h^2 = n_f^2 k_0^2 - \beta^2 \tag{3.29}$$

It is obvious that if β is larger than $n_c k_0$ and $n_s k_0$, then the solution is a combination of the sine and cosine functions in the core. Otherwise, the dependence of the solution on x is exponential, that is, the field decays away from the centre of the waveguide (Figure 3.10). In order to specify exactly the field distribution within the waveguide structure, it is necessary to calculate the constants A, B, C, D (Figure 3.10), and, the corresponding propagation constant, β. Imposing continuity conditions on the tangential field components at $x = 0$ and $x = t$ (Figure 3.9) yields four homogenous algebraic equations for the TE case:

$$\begin{bmatrix} 1 & 0 & 0 & -1 \\ \cos(ht) & \sin(ht) & -1 & 0 \\ 0 & h & 0 & -q \\ -h\sin(ht) & h\cos(ht) & p & 0 \end{bmatrix} \begin{bmatrix} A \\ B \\ C \\ D \end{bmatrix} = 0 \tag{3.30a}$$

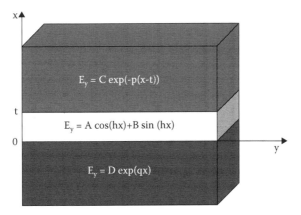

FIGURE 3.10 The functions describing the dependence of E_y on x for TE mode of a slab waveguide.

and for the TM one:

$$
\begin{bmatrix}
1 & 0 & 0 & -1 \\
\cos(ht) & \sin(ht) & -1 & 0 \\
0 & \dfrac{h}{n_f^2} & 0 & -\dfrac{q}{n_s^2} \\
-\dfrac{h}{n_f^2}\sin(ht) & \dfrac{h}{n_f^2}\cos(ht) & \dfrac{p}{n_c^2} & 0
\end{bmatrix}
\begin{bmatrix} A \\ B \\ C \\ D \end{bmatrix} = 0
\tag{3.30b}
$$

From linear algebra it is known that the homogenous algebraic equations have a nontrivial solution only if the determinant is equal to zero. Calculating the determinant of the coefficient matrix in 3.30 and equating it to zero yields the following equation for the TE case:

$$
tg(ht) = \frac{\dfrac{q}{h} + \dfrac{p}{h}}{1 - \dfrac{q}{h}\dfrac{p}{h}}
\tag{3.31a}
$$

Analogically, from 3.30b one obtains the following equation for the TM case:

$$
tg(ht) = \frac{\dfrac{q}{h}\dfrac{n_f^2}{n_s^2} + \dfrac{p}{h}\dfrac{n_f^2}{n_c^2}}{1 - \dfrac{q}{h}\dfrac{n_f^2}{n_s^2}\dfrac{p}{h}\dfrac{n_f^2}{n_c^2}}
\tag{3.31b}
$$

Since, $tg(\alpha+\beta) = (tg\alpha+tg\beta)/(1-tg\alpha tg\beta)$ we can recast 3.31a as:

$$
ht = \arctg\frac{q}{h} + \arctg\frac{p}{h} + m\pi
\tag{3.32a}
$$

and 3.31b as:

$$
ht = \arctg\left(\frac{q}{h}\frac{n_f^2}{n_s^2}\right) + \arctg\left(\frac{p}{h}\frac{n_f^2}{n_c^2}\right) + m\pi
\tag{3.32b}
$$

where m is a parameter that sets the mode order. Introducing the normalised frequency V, relative propagation constant B, and an asymmetry parameter A_E:

$$V = k_0 t \sqrt{n_f^2 - n_s^2}; \quad B = \frac{\beta^2 - n_s^2 k_0^2}{n_f^2 k_0^2 - n_s^2 k_0^2}; \quad A_E = \frac{n_s^2 - n_c^2}{n_f^2 - n_s^2}$$

and recasting 3.32, further, into a more convenient normalised form:

$$V\sqrt{1-B} - \text{arctg}\sqrt{\frac{B + A_E}{1-B}} - \text{arctg}\sqrt{\frac{B}{1-B}} - m\pi = 0 \qquad (3.33a)$$

for the TE mode and

$$V\sqrt{1-B} - \text{arctg}\left(\frac{n_f^2}{n_c^2}\sqrt{\frac{B + A_E}{1-B}}\right) - \text{arctg}\left(\frac{n_f^2}{n_s^2}\sqrt{\frac{B}{1-B}}\right) - m\pi = 0 \qquad (3.33b)$$

for the TM mode.

From 3.33 one can easily obtain the explicit dependence of V on B. However, in practice one typically needs to know the propagation constant for a given wavelength (and the slab waveguide parameters: n_f, n_s, n_c, and t), which when expressed using normalised parameters means that we need to calculate B for a given value of V. For this purpose, the numerical solution of Equation 3.33 is necessary.

Equation 3.33 can be solved using standard numerical methods used for calculating zeros of nonlinear algebraic equations, for example, the bisection method. Algorithm 3.1 summarises the main steps required for calculating the propagation constants of a three layer slab waveguide.

A MATLAB script based on the Algorithm 3.1 is given as follows.

ALGORITHM 3.1   CALCULATION OF THE TM PROPAGATION CONSTANTS OF A THREE LAYER SLAB WAVEGUIDE BY FINDING ZEROS OF 3.33B

1. Start
2. Set the values of, wavelength, refractive indices, waveguide thickness and mode order
3. Calculate the endpoints of the interval that contains the value of the propagation constant by bracketing the corresponding zero of (3.33b)
4. Refine the value of the propagation constant using the bisection method
5. Stop

```
% program calculates propagation constants for slab waveguide
clear  % clears variables
clear global  % clears global variables
format long e
% initial constants
i=sqrt(-1); % remember not to overwrite pi or i !
pi=3.141592653589793e+000;

lam=1.55; %wavelength [micrometers]
nc=3.2874; % refractive index in cladding
nf=3.3704; % refractive index in core
ns=3.2874; % refractive index in substrate
g=2.; % waveguide thickness [micrometers]
m=0; % mode order

aE=(ns*ns-nc*nc)/(nf*nf-ns*ns); % parameter AE for TE modes
nH(1)=nf*nf/(ns*ns);
nH(2)=nf*nf/(nc*nc);

% calculation parameters
k=2.0*pi/lam; % wavenumber
V=g*k*sqrt(nf^2-ns^2); % normalised frequency V

%%%%%%%%%%%%%%%%%%%%%%%%%%%%%%%%%%%%%%%%%%%%%%%%%%%%%%%%%%%%%%%%%%
% Calculation of initial interval locating B
% for bisection method
b=(0.0199999:0.01:0.9999999);
% Dispersion equation
DE=m*pi+atan(nH(1)*sqrt(b./(1-b)))+...
atan(nH(2)*sqrt((b+aE)./(1-b)))-V*sqrt(1-b);

% precise initial location of B
length=size(b);
for j=length(2):-1:2
   if DE(j)/DE(j-1)<0
      break
   end
end

if j==2
   disp('no zero found, change initial interval')
   disp('the results are incorrect')
end

btop = b(j);
bbot = b(j-1);

%%%%%%%%%%%%%%%%%%%%%%%%%%%%%%%%%%%%%%%%%%%%%%%%%%%%%%%%%%%%%%%%%%
% bisection method for root finding starting
% from the initial interval
DEtop=m*pi+atan(nH(1)*sqrt(btop/(1-btop)))+...
```

```
atan(nH(2)*sqrt((btop+aE)/(1-btop)))-V*sqrt(1-btop);
DEbot=m*pi+atan(nH(1)*sqrt(bbot/(1-bbot)))+...
atan(nH(2)*sqrt((bbot+aE)/(1-bbot)))-V*sqrt(1-bbot);
jmax=100; % maximum number of bisections
tolerance=0.00000000000001; % tolerance for finding B

for j=1:jmax
    bcen=(btop-bbot)/2+bbot;
    DEcen=m*pi+atan(nH(1)*sqrt(bcen/(1-bcen)))+...
    atan(nH(2)*sqrt((bcen+aE)/(1-bcen)))-V*sqrt(1-bcen);
    if DEtop/DEcen > 0
        btop=bcen;
        DEtop=DEcen;
    else
        bbot=bcen;
        DEbot=DEcen;
    end

    delta=btop-bbot;
    if delta < tolerance
        break
    end
end
%end of bisection method loop
%%%%%%%%%%%%%%%%%%%%%%%%%%%%%%%%%%%%%%%%%%%%%%%%%%%%%%%%%%%%%%%%%
%%%%%%%%%%%%%%%%%%%%%%%%%%%%%%%%%%%%%%%%%%%%%%%%%%%%%%%%%%%%%%%%%
% checks for tolerance
if j==jmax
    disp('maximum number of bisections exceeded before reaching
    tolerance')
else
    out=bcen;
end
% variable 'out' stores the relative propagation constant B

nin=abs(sqrt(out*(nf*nf-ns*ns)+ns*ns));
% calculation of effective index

[V out nin] % plotting results in one line
```

The TE mode propagation constant can be obtained by setting the two element vector "nH" to 1. The aforementioned program first brackets the value of B in the interval $0.0199999 < B < 0.9999999$ taking the advantage of the fact that for a guided mode B is positive and less than 1. This initial guess should be sufficient for most situations, but it might need a modification if a mode is either very near to or very far from cut-off. In the second step, the initial guess is refined and finally the bisection method is used to obtain the solution with the accuracy set by the variable "tolerance." Once the propagation constant is calculated, the constants A, B, C, D, and the field distribution can be calculated by substituting β and solving 3.30. Since the algebraic equations 3.30 are linearly dependent, one needs to set one constant to a

**ALGORITHM 3.2   CALCULATION OF THE TE
PROPAGATION CONSTANTS OF A SLAB WAVEGUIDE
USING THE FINITE DIFFERENCE METHOD**

1. Start
2. Set the finite difference mesh
3. Set the values of, wavelength, refractive indices and waveguide thickness and mode order
4. Assemble the finite difference coefficient matrix
5. Solve the eigenvalue problem
6. Select the eigenvalue and eigenvector that corresponds to the required mode
7. Plot the field distribution of the mode
8. Stop

number, for example, set $A = 1$, discard one equation, and calculate B, C, and D from the remaining three equations.

The analytical solution can also be obtained in the case of a multilayered slab waveguide if the refractive index distribution is homogenous within each layer of an isotropic [43] or uniaxial anisotropic medium [44], and also for several cases of optical waveguides with a graded index profile [2]. For finding more information on this topic, including the most recent developments, one can consider reading Ding and Chan [45], Anemogiannis et al. [46,47], and Sujecki [48] and the references included therein.

When considering waveguides with an arbitrary refractive index distribution $n(x)$, it is particularly easy to solve numerically 3.26a and 3.26c using the finite difference method. Since the unknown function is continuous, the standard finite difference approximation of the second derivative at $x = x_0$ can be used:

$$\left. \frac{d^2 H_x}{dx^2} \right|_{x=x_0} \approx \frac{1}{\Delta x^2} \left(H_x \left(x_0 - \Delta x \right) - 2 H_x \left(x_0 \right) + H_x \left(x_0 + \Delta x \right) \right) \qquad (3.34)$$

Substituting 3.34 into 3.26c and following the standard procedure for the finite difference method [49] yields an algebraic eigenvalue problem that can be solved using standard numerical methods [50]. The algorithm 3.2 summarises the main steps required for the finite difference solution of the wave equation for a slab waveguide.

A simple example of the MATLAB implementation of finite difference method to the solution of 3.26c is shown next.

```
% calculation of slab waveguide modes using FDs
pi = 3.141592653589793e+000;
N = 800;% total number of mesh points
A = zeros(N);% initialisation of matrix A
Nw = 100;% number of mesh points in waveguide
```

```
t = 2;% slab thickness in [micrometers]
dx = t/Nw;% dx in [micrometers]
lam = 1.55;% wavelength in [micrometers]
k0 = 2*pi/lam;% wave number k0
nf = 3.3704;% refractive index in core
ns = 3.2874;% refractive index in cladding
nc = 3.2874;% refractive index in substrate
x = dx*(1:N);% vector with the positions of sampling points

c = 1/(dx*dx);
% setting off diagonals
for j = 1:N-1
A(j,j+1) = c;
A(j+1,j) = c;
end

% setting up vector storing the values of refractive index

n = ns*ones(N,1);
for j = N/2:N
n(j) = nc;
end

for j = (N/2-Nw/2+1):(N/2-Nw/2+Nw)
n(j) = nf;
end

% setting main diagonal
A = A+diag(n.*n*k0*k0-2*c);

[V,Ba] = eig(A);

[max_eig,max_pos] = max((real(diag(Ba))));
% the largest eigenvalue corresponds to fundamental
% propagating mode:
beta = sqrt((((Ba(max_pos,max_pos))))))
neff = beta/k0%effective index
```

The performance of this software in the MATLAB environment can be enhanced by using sparse matrices. It should be noted, however, that using standard higher order finite difference approximations for the second derivative of H_x will not improve the convergence rate, because higher order derivatives of H_x are not continuous [51]. In the TM case the standard finite difference approximation 3.34 cannot be used for the second derivative, but a slight modification yields a correct finite difference stencil [52]. More information on the finite difference discretisation techniques that can be used instead can be found in Sujecki [48] and the references listed therein.

In Table 3.4, we present a set of reference propagation constant values calculated using an analytical method [43]. These results can be useful when developing software for modal analysis of optical slab waveguides. Structure 1 is a GaAs-based symmetrical slab waveguide. The refractive index is equal to 3.3704 and 3.2874

TABLE 3.4

Selected Reference Values of the Effective Refractive Index

Effective Refractive Index	Mode	
	TE$_0$	TM$_0$
Structure 1	3.35798693676	3.35770896615
Structure 2	3.28735394812	3.28551251240
Structure 3	2.35113096409	1.82906299792

in the core and cladding, respectively. The width of the waveguide core is 2 µm. Structures 2 and 3 are multilayered waveguides corresponding to a multiquantum well structure [53]. Both waveguides consist of 56 barriers of width 0.0012 µm and 55 wells of width 0.007 µm. The refractive index in the well is equal to 3.3704. For the second structure the refractive index of the barrier is 3.2874, while that of the cladding is 3.2224. For the third structure the refractive indices of barrier and cladding are 1.6 and 1.5, respectively. For all structures the operating wavelength is 1.55 µm. The analytical software presented in this section can be used to reproduce the reference results for the slab waveguide "Structure 1" on all quoted decimal places. The finite difference method can achieve agreement only on a limited number of digits on a standard PC, but can be applied to all three test structures, cf. [54].

A slab waveguide, apart from the guided modes, supports the substrate modes and free space radiation modes. The derivation of the relevant formulae that describes the field distribution of these modes can be found in many textbooks [3,40,41]. The guided modes together with the substrate modes and the free space radiation modes form a complete function basis set that is used in many numerical methods when studying the light propagation in planar optical waveguides. These techniques include the mode matching method which we discuss in the section "Propagation Constant Calculation Techniques for Planar Optical Waveguides" and the beam propagation method (Chapter 4).

Even if the refractive index distribution is purely real, Equations 3.31 through 3.33 admit solutions for the complex values of the propagation constant β. In the literature these modes are referred to as "leaky modes." A characteristic feature of the leaky modes is an exponential growth of the field away from the core. A simple explanation of this field behaviour is provided in Rozzi and Mongiardo [42]. Leaky modes are very important when studying the optical properties of waveguide modes below cut-off [41], and when calculating propagation constants of the ARROW modes. Leaky modes also provide a convenient way of handling the continuous spectra of the radiation modes [55,56]. A review of leaky mode theory can be found in Hu and Menyuk [57].

Finally, Equations 3.27 and 3.28 also admit solutions that correspond to pairs of modes with complex conjugate propagation constants, again, even if the refractive index distribution is purely real. In the literature these modes are referred to as "complex modes" and should not be confused with the leaky modes. The distinction between these two types of modes is discussed in Jablonski [58]. The presence of the complex modes has significant implications for the numerical stability of the beam propagation algorithms [59–61] (see Chapter 4).

EFFECTIVE INDEX METHOD

If there is no variation of the refractive index in the x direction (Figure 3.1), Equation 3.10 reduces to Equations 3.27 and 3.28, which can be compactly expressed as:

$$\left(a_{\mathrm{TE}} + \omega^2 \mu_0 \varepsilon\right) E_x = \beta^2 E_x \tag{3.35a}$$

$$\left(b_{\mathrm{TE}} + \omega^2 \mu_0 \varepsilon\right) H_y = \beta^2 H_y \tag{3.35b}$$

$$\left(a_{\mathrm{TM}} + \omega^2 \mu_0 \varepsilon\right) E_y = \beta^2 E_y \tag{3.35c}$$

$$\left(b_{\mathrm{TM}} + \omega^2 \mu_0 \varepsilon\right) H_x = \beta^2 H_x \tag{3.35d}$$

The operators a_{TE}, a_{TM}, b_{TE}, and b_{TM} are given in Table 3.5.

Similarly, in the case of the scalar approximation from 3.15 we obtain:

$$\left(s_{1D} + \omega^2 \mu_0 \varepsilon\right) \phi = \beta^2 \phi \tag{3.36}$$

where

$$s_{1D} = \frac{\partial^2}{\partial x^2}.$$

Hence in the 1D case the same values of the propagation constants are obtained for the TE-polarised mode and when applying the scalar approximation.

For a number of planar waveguiding structures, it is possible to approximate a 2D structure by an equivalent 1D structure. The standard way of obtaining such approximation consistently is the calculation of effective indices, that is, the use of the effective index method (EIM). The EIM allows reducing the solution of Equation 3.10 to the

TABLE 3.5

The Operators a_{TE}, a_{TM}, b_{TE}, and b_{TM}

	a	b
TE	$\dfrac{\partial^2}{\partial x^2}$	$\dfrac{\partial^2}{\partial x^2}$
TM	$\dfrac{\partial}{\partial x}\left(\dfrac{1}{\varepsilon}\dfrac{\partial(\varepsilon\circ)}{\partial x}\right)$	$\varepsilon\dfrac{\partial}{\partial x}\left(\dfrac{1}{\varepsilon}\dfrac{\partial(\circ)}{\partial x}\right)$

solution of 3.34, which is easier and was discussed in the previous section. The EIM can be used to obtain approximate solutions of both semivectorial and scalar 2D equations. The EIM is based on the separation of variables, that is, it relies on the assumption that a 2D field distribution can be represented by a product of two 1D field distributions. As an example, we derive an EIM solution in the case of an x-polarised fundamental mode. Setting the solution in the form $E_x(x,y,\omega) = X(x,\omega)*Y(y,\omega)$ into 3.14a gives:

$$X\frac{\partial^2 Y}{\partial y^2} + Y\frac{\partial}{\partial x}\left(\frac{1}{\varepsilon}\frac{\partial(\varepsilon X)}{\partial x}\right) + \omega^2 \mu_0 \varepsilon XY = \beta^2 XY \qquad (3.37)$$

Dividing both sides of 3.37 by XY and moving the constants to the right hand side (RHS) results in:

$$\frac{1}{Y}\frac{\partial^2 Y}{\partial y^2} + \frac{1}{X}\frac{\partial}{\partial x}\left(\frac{1}{\varepsilon}\frac{\partial(\varepsilon X)}{\partial x}\right) = -\left(\omega^2 \mu_0 \varepsilon - \beta^2\right) \qquad (3.38)$$

which implies that a separation constant can be introduced to obtain two coupled 1D equations:

$$\frac{\partial^2 Y}{\partial y^2} + \omega^2 \mu_0 \varepsilon Y = \beta_{\text{eff}}^2 Y \qquad (3.39a)$$

$$\frac{\partial}{\partial x}\left(\frac{1}{\varepsilon}\frac{\partial(\varepsilon X)}{\partial x}\right) + \omega^2 \mu_0 \varepsilon_0 n_{\text{eff}}^2 X = \beta^2 X \qquad (3.39b)$$

where the effective index of the slab waveguide fundamental mode is $n_{eff} = \beta_{\text{eff}}/k_0$ and $k_0 = 2\pi/\lambda$. Once the dependence of n_{eff} on x is calculated using 3.39a, Equation 3.39b is used to obtain the propagation constant for the fundamental mode of the "effective" slab waveguide, which gives the approximate value of the propagation constant of the fundamental mode of the original 2D waveguide. Thus the EIM allows the solution of the original 2D problem to be reduced to a solution of a 1D problem, which can be handled analytically. As an illustrating example, Figure 3.11 shows schematically the application of the EIM in the case of a rib-loaded waveguide. Analogically, the effective index approximation can be derived for y-polarised and scalar cases.

The EIM was originally proposed by Ramaswamy [62] and is closely related to an approximate method for the analysis of the rectangular waveguides proposed by Marcatili [7]. Due to its simplicity, robustness, and versatility the EIM is often used in photonics. The main shortcoming of the EIM is its limited accuracy for many optical waveguide structures of practical importance. Therefore other approximate methods that allow higher accuracy to be achieved while preserving the computational efficiency have been developed [63–67]. For air clad optical waveguides that have a small refractive index contrast between the core and the substrate particularly

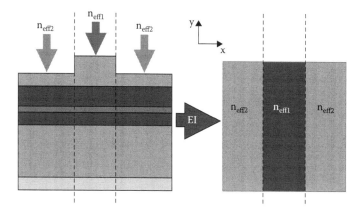

FIGURE 3.11   Schematic illustration of the effective index method in the case of a rib-loaded optical waveguide.

effective is the spectral index method [68,69]. A fairly comprehensive description of the approximate numerical techniques that are applicable to planar optical waveguides can be seen in the works of Robson and Kendall [70].

Propagation Constant Calculation Techniques for Planar Optical Waveguides

The modelling methods for planar optical waveguides are divided into three classes depending on which equation is being attempted to be solved. The vectorial methods attempt to solve the vectorial wave equation 3.12. The semivectorial methods attempt to solve the wave equation under the polarised approximation 3.14 while the scalar ones solve the scalar wave equation 3.15. There are four main numerical methods, which are used for the solution of 3.12, 3.14, and 3.15 in the case of a planar optical waveguide:

- Mode matching method (MMM)
- Method of lines (MOL)
- Finite difference method (FDM)
- Finite element method (FEM).

Each of the numerical techniques used for planar waveguide analysis has its unique advantages. In general, the two former techniques, that is, MMM and MOL, require small to moderate computer resources, while being less versatile. In particular, the MMM and the MOL may prove numerically ill-conditioned for some quite common structures [71]. On the other hand, the two latter techniques, that is, the FEM and the FDM require large computer resources but are comparatively robust.

The MMM is an example of an analytical method. Analytical methods exploit the analytical techniques that have been developed for the solution of partial differential equations. In the case of planar optical waveguides, analytical methods rely on dividing the waveguide cross-section, considered, into subregions within which an analytical solution of the partial differential equations can be easily obtained. These analytical local solutions are then used to construct the global analytical solution, which is valid in the entire domain considered, that is, it satisfies the imposed boundary conditions. The solution of the problem also yields the dispersion equation for the calculation of the propagation constants. There are two types of analytical methods used in the analysis of the planar optical waveguides. One of them (the MMM) splits the waveguide cross-section into a number of subregions with a refractive index distribution dependent on only one of the transverse spatial variables [20,71–83]. The other one uses a set of subregions with a constant value of the refractive index [8,84]. As a consequence, these analytical methods are efficient only if the waveguide cross-section can be divided into a moderate number of subregions with a constant value of the refractive index.

The analytical technique, which divides the entire waveguide cross-section into subregions with a constant value of the refractive index, provides a very efficient way of solving the vectorial wave equation, especially for rectangular waveguides [8,84]. This technique relies on matching the field expansions in different regions along the region boundaries at a number of points. Hence this technique is often referred to as point matching method (PMM). A PMM can be used in principle for various waveguide structures with piecewise continuous refractive index distribution. However, so far it has been only applied to simple core-clad optical waveguide structures [8,84].

The most often used analytical method for planar optical waveguides is the MMM. As previously noted, the MMM is based on dividing the waveguide cross-section into a number of regions within which the refractive index depends on one spatial variable only. The field distribution within each region is represented by the sum of the modal field distributions of the local slab waveguides that are formed within each region. The continuity along the longitudinal direction is imposed by selecting the same spatial dependence along the longitudinal direction for all modes used in the local expansion. This operation is usually, conveniently represented by "rotating" the functional basis [85]. In the transverse plane at the boundaries between the regions, the continuity conditions are imposed for both transverse field components: the electric field if 3.12a is solved or magnetic field if 3.12b is solved. This procedure yields a set of equations that can be converted into a set of linear algebraic equations, whereby the unknowns correspond to the local mode expansion coefficients. This latter step is usually accomplished by calculating the overlap integrals with the local mode distributions, cf., in particular, [74]. The propagation constants are calculated by finding the determinant zeros of the coefficient matrix for the set of linear algebraic equations.

If the local mode expansions are used to obtain the solution of Equation 3.14 or 3.15, then a semivectorial and scalar version of the MMM is respectively obtained. In principle the MMM follows the same steps as the EIM, which was discussed in the previous subsection. The only difference is that in the EIM only the first term of the modal expansion is retained [13]. The MMM is most efficient in the case of loss-less structures whose cross-section can be divided into a number of rectangles with constant refractive index. The MMM typically requires a relatively large effort to derive and evaluate analytical expressions when developing the software. A transmission line formalism [72,76,77] can be used to simplify the derivations. A number of helpful guidelines on a development of a stable MMM code can be found in Sudbo [71,73]. A version of the MMM tailored for the analysis of optical waveguide structures with a complex refractive index distribution can be found in Bienstman [85]. The MMM has also proved very helpful when deriving approximate analytical techniques for the analysis of planar optical waveguides [20,74,75,77,80,81].

Another class of analytical methods form techniques that use Fourier transform. There are two types of methods that have been developed, so far. In one type of these techniques, Fourier series expansion is used to represent the solution within the entire cross-section [86,87]. These methods are simple to implement and result in a linear eigenvalue problem with a dense matrix, which is very easy to handle with standard linear algebra packages. The other class of techniques divides the cross-section into regions were the refractive index depends only on one coordinate (in the same manner as the MMM). The solution is expressed in the Fourier domain in each region and matched along the boundaries between the regions to form the global solution. This procedure yields a transcendental equation, which is solved for the propagation constants of the guided modes [81,88–90]. When the free space radiation modes of slab waveguides are approximated by the free space radiation modes of a homogenous medium with a suitably chosen value of the refractive index, one obtains the free space radiation method (FSRM) that is very efficient and accurate for optical waveguide structures with a low refractive index contrast [79]. For a long time, the application of Fourier method was limited to the solution of the wave equation in the scalar and polarised approximation and only fairly recently, a fully vectorial version of the Fourier series-based method was developed [91].

Unlike the MMM, the FDM is a very robust and versatile technique. In general, it is as efficient in the case of a rectangular waveguide as it is for an in-diffused one. Therefore, the FDM is the preferred method for structures with arbitrary refractive index distribution, for example, laser diodes with refractive index perturbed by carrier, stress, and temperature distributions. The FDM is also relatively straightforward and requires a small effort for analytical derivations, hence, making it relatively easy to program, when compared with the MMM. On the other hand, the FDM requires much larger computational resources than the MMM. Hence, a large research effort was invested into an improvement of the numerical efficiency of the FDM. Such improvement, however, usually comes at the cost of reduced robustness and an increased effort put into the analytical derivations and software development. One of the ways proposed to improve the efficiency of the finite difference method is the application of nonuniform meshes either directly [92,93], using similarity transformations [94], or by applying subgridding techniques [95]. As the FDM typically results in a standard sparse matrix eigenvalue problem (examples of the FDM formulation that lead to a generalised eigenvalue problem can be found in Sujecki [48] and Chiou et al. [96]), another way to improve the efficiency of the finite difference method is by using modern numerical methods to handle the sparse eigenvalue problem [97–99]. In the case of optical waveguide structures with a large refractive index contrast, the use of effective penetrative depth was demonstrated to improve the computational efficiency [100]. Another way of making a vectorial FDM more efficient is the calculation of the solution using either scalar or a semivectorial FDM and then applying a vectorial correction to obtain a more accurate value of the propagation constant [101].

The improvement of the efficiency follows from the fact that typically the vectorial solution requires matrices four times larger than in the scalar or polarised case.

The dielectric discontinuities, which are present in most planar optical wave-guides, result in field discontinuities. Since the standard finite difference approximations are only valid for continuous functions, modified FD stencils had to be developed that directly incorporate function discontinuities. In general, there are two ways of discretising the wave equation using finite differences. One places the dielectric discontinuities between the mesh nodes [51,52,102,103] while in the other technique the mesh nodes coincide with the dielectric interfaces [104–109]. The latter method has been applied for techniques solving the wave equation using the magnetic field or the magnetic vector potential only [110]. More recently FD stencils were developed that allow for an arbitrary position of the dielectric discontinuity at the expense of larger complexity of the formulae describing the FD coefficients, cf. [48,53,54,111,112] and references listed therein.

The FDM has also been applied successfully to optical waveguide structures with oblique boundaries and high refractive index contrast between the core and the cladding [111,113–115]. The FDM is also a robust technique for analysis of waveguides with gain [116] and anisotropy [109]. Finally, we note that the FDM usually relies on the strong formulation of the boundary value problem but a weak (variational) formulation of the FDM, mimicking the approach typically followed when applying a FEM, is also possible [117–119].

The MOL is a technique that reduces the solution of the partial differential equations 3.13 through 3.15 to a solution of a set of ordinary differential equations. This is usually achieved by the application of either the finite element method (FEM) or the finite difference method along one of the transverse waveguide dimensions. In the context of the optical waveguide analysis, a FDM has so far been the preferred approach. In this latter version the MOL attempts to combine the advantages of the MMM and the FDM. If a three-point finite difference approximation is used, this approach results in a set of ordinary differential equations of the form:

$$\frac{d^2 \phi_i}{dx^2} + M \phi_i = \beta^2 \phi_i$$

where M is a tri-diagonal matrix and ϕ_i is the field distribution along the line "i." Typically, the set of ordinary differential equations is solved by diagonalising the matrix M first, which can be achieved fairly easily by using finite differences to calculate the modal spectrum of a slab waveguide and by applying the transverse resonance formalism to obtain a set of linear algebraic equations. Calculating the determinant of the set of the linear algebraic equations yields a nonlinear eigenvalue problem in scalar, polarised, and vectorial cases [120,121]. In this formulation, the MOL has similar characteristics to the FMM. When compared with the FMM, however, it benefits from a much simpler way in which the slab waveguide modes are calculated. It may, however, prove less accurate than the FMM for the same order of the determinant. It should be noted that the diagonalisation of the matrix M is in principle not necessary. Neither does the implementation of the MOL have to result in a nonlinear eigenvalue problem. For instance, a linear eigenvalue problem can be obtained using the MOL if instead of the transverse resonance method, a sine function basis is used to approximate the field distribution along the lines [122]. An in depth study on the application of the MOL to optical waveguides can be found in Pregla [123].

The finite element method (FEM) has similar general characteristics to the FDM. It is robust but not as numerically efficient as the MMM and the MOL. The FEM, unlike the FDM, uses a weak (variational) formulation associated with Equation 3.4b. In the vectorial case, a standard FEM, which is based on the variational expression [124], is the result:

$$\omega^2 = \frac{\int (\nabla \times H)^* \varepsilon^{-1} (\nabla \times H) d\Omega}{\int H^* \mu_0 H d\Omega} \quad (3.40)$$

suffers from spurious modes. This results from the fact that the variational expression 3.40 is derived directly from 3.4a, which does not automatically force the zero divergence condition. By taking the rotation of 3.1b, it can be shown that the solutions obtained by finding the stationary points of 3.40 fulfil constant divergence condition only. In the scalar case on the other hand, the spurious modes do not appear [125]. A number of methods have been suggested to eliminate or reduce the number of spurious modes in the vectorial FEM. One of such techniques consists of adding a penalty function [124]. This technique allows the number of spurious modes to be reduced at the expense of decreased accuracy. An effective technique eliminating the spurious solutions is the application of tangential or edge elements [126,127]. Other ways to eliminate spurious modes were suggested by Hayata [128], Koshiba et al. [129], and Silveira and Gophinat [130]. Similarly as for FDM, the computational efficiency of FEM depends critically on the way in which the matrix eigenvalue problem is treated. However, unlike FDM the application of FEM results in a generalised eigenvalue problem [131], which can be efficiently solved using modern projection methods that are specially designed for sparse matrices [132]. Until fairly recently no special treatment of the field discontinuities in finite element formulation was given. This reduces the accuracy of FEM and limits its application to the solution of the wave equation for the magnetic field only. Li et al. [133] included the direct treatment of discontinuities in the finite element representation of the wave equation for waveguide analysis. They used nine-point inhomogenous finite elements for the solution of the wave equation under a polarised approximation.

The FEM proved also to be an efficient method for the analysis of optical waveguides with gain [134] and anisotropy [135]. An advanced and detailed description of the application of FEM to optical waveguide analysis can be found in Wang [136], Fernandez et al. [137], Koshiba [138], and Rahman and Agrawal [139].

Unlike the heat transfer modelling, integral equations methods have not yet found major applications in the analysis of the modal properties of planar optical waveguides. More recently however, several efficient techniques implementing the boundary element and integral equations methods to the calculation of the propagation constants of planar optical waveguides were proposed [140–144]. In general, these techniques have similar properties to the analytical methods, that is, they offer high computational efficiency at the expense of robustness.

Finally, we note that in the case of planar optical waveguides the presence of dielectric corners at high refractive index contrast interfaces results in field singularities. These field singularities impede the convergence of any numerical method to the exact solution since the rapid variations of the field near the corner cannot be accurately represented using a small number of terms taken from the set of standard basis functions [145]. An improvement of the convergence can be only achieved if around the corner a local field expansion is applied that accurately represents the field behaviour with a small number of terms [107,141,146–148].

COMPARISON OF POLARISED, SCALAR, AND EFFECTIVE INDEX APPROXIMATIONS

The accuracy of the calculations carried out under scalar, polarised, or effective index approximations depends on the waveguide structure. In this subsection, a comparison between results obtained using 1D and 2D scalar, polarised, and vectorial methods is given for a set of reference planar optical waveguide structures that are typically used in the literature.

Table 3.6 presents the values of the normalised propagation constant

$$B = \frac{\beta^2 - n_s^2 k_0^2}{n_f^2 k_0^2 - n_s^2 k_0^2},$$

which have been obtained for a rectangular waveguide structure with aspect ratio 2:1. The refractive index values are 1.5 and 1.45 in the core and cladding regions, respectively. The operating wavelength is 1.15 μm. All results have been obtained using FDM [1] and are believed to be calculated with an absolute error of 5×10^{-5}. The results presented in Table 3.6 show that the difference between the polarised and vectorial analysis results is no more than one on the fourth decimal place. However, the results obtained by the scalar method differ even in the first decimal place (e.g., y-polarised mode for $V = 1.0$,

TABLE 3.6

Normalised Propagation Constants B of Rectangular Dielectric Waveguide ($a/b = 2$, $n_f = 1.5$, $n_s = 1.45$, $\lambda = 1.15$ μm, cf. Figure 3.3)

	SFD	Polarised	Vectorial	Polarised	Vectorial
V	Scalar	x-Polarised		y-Polarised	
0.4	0.0361	0.0334	0.0334	0.0305	0.0306
0.5	0.1118	0.1069	0.1068	0.1004	0.1005
0.6	0.2048	0.1991	0.1990	0.1900	0.1901
0.7	0.2962	0.2908	0.2907	0.2806	0.2806
0.8	0.3789	0.3740	0.3739	0.3635	0.3636
0.9	0.4508	0.4466	0.4465	0.4365	0.4366
1.0	0.5125	0.5090	0.5089	0.4996	0.4997

$$V = \frac{2b}{\lambda} \sqrt{n_c^2 - n_a^2},$$

that is, shorter side full length b = 1.497172476279475 μm) for this relatively low refractive index contrast structure.

Table 3.7 shows the values of the normalised propagation constant obtained using the EIM for the same structure as in Table 3.6. The difference between 1D and the corresponding 2D results decreases as the normalised frequency V increases. This is caused by the fact that the assumption of the separability of the variables used by effective index method becomes less plausible near to mode cut-off [7]. The results from Table 3.7 also show that the 1D effective index analysis can be more accurate than the much more numerically intensive 2D scalar analysis (e.g., y-polarised mode for V = 1.0). Furthermore, the values of propagation constants calculated with EIM and the scalar method provide an upper bound for the results obtained using more accurate and numerically intensive polarised and vectorial methods.

Table 3.8 presents the normalised propagation constant values, which have been obtained from a rectangular waveguide structure with an aspect ratio of 2:1, with the refractive index value of 1.5 in the core. The refractive index in the cladding region was varied between 1.2 and 1.45. The operating wavelength was 1.15 μm.

TABLE 3.7

Normalised Propagation Constants B of Rectangular Dielectric Waveguide Calculated by Effective Index Method (a/b = 2, n_f = 1.5, n_s = 1.45, λ = 1.15 μm, cf. Figure 3.3)

	Normalised Propagation Constant		
	Scalar	x-Polarised	y-Polarised
0.4	0.0720	0.0706	0.0633
0.5	0.1476	0.1451	0.1344
0.6	0.2334	0.2301	0.2178
0.7	0.3175	0.3140	0.3014
0.8	0.3942	0.3909	0.3788
0.9	0.4619	0.4588	0.4476
1.0	0.5207	0.5179	0.5077

TABLE 3.8

Normalised Propagation Constants B of Rectangular Dielectric Waveguide (V = 1, a/b = 2, n_f = 1.5, λ = 1.15 μm, cf. Figure 3.3)

	SFD	Polarised	Vectorial	Polarised	Vectorial
n_s	Scalar	x-Polarised (Quasi-TE Mode)		y-Polarised (Quasi-TM Mode)	
1.45	0.5125	0.5090	0.5090	0.4996	0.4996
1.4	0. 5125	0.5052	0.5051	0.4860	0.4859
1.3	0. 5125	0.4976	0.4970	0.4571	0.4567
1.2	0. 5125	0.4897	0.4881	0.4258	0.4245

Again all results have been obtained using FDM [1] and are believed to be calculated with an absolute error of 5×10^{-5}. The results from Table 3.8 show that the difference between the outcomes of a polarised and a vectorial method increases with the increase of the refractive index contrast but does not exceed two on the third decimal place even for large refractive index contrast structures. In fact, for most of the practically realisable rectangular optical waveguides, the refractive index contrast $\Delta_n = (n_f - n_s)/n_f$ does not exceed 7%, because of the side wall scattering loss [32]. For such structures the difference between the vectorial and polarised results does not exceed one in the fourth decimal place, which again confirms high accuracy of the polarised approximation when applied to planar optical waveguides. In contrast, the results obtained by the scalar method differ even up to one on the first decimal place (e.g., y-polarised mode for $n_s = 1.2$).

Table 3.9 shows the values of the normalised propagation constant obtained using EIM for the same structure as in Table 3.8. These results show that 1D effective index results are more accurate than the 2D scalar for structures with high refractive index contrast, for both x- and y-polarised modes. For the y-polarisation the 1D effective index results are more accurate than the 2D scalar results even for small refractive index contrast structures. Again, the results from Tables 3.8 and 3.9 confirm that the scalar method and EIM provide an upper bound for the propagation constants calculated with vectorial and semivectorial methods. It is also observed that the difference between the 1D effective index results and the polarised and vectorial results decreases with a decrease in the refractive index contrast between the core and cladding.

Table 3.10 presents normalised propagation constant values, which have been obtained from a rib waveguide structure with $H = 1$ μm, $W = 1.5$ μm, $n_f = 3.44$, $n_s = 3.4$, $n_c = 1.0$ (Figure 3.5). The outer slab thickness is between 0.1 and 0.9 μm. Operating wavelength is 1.15 μm. The scalar and polarised results have been obtained using FDM [1] while the vectorial ones were taken from Hadley [108] and Vassallo [149] for x- and y-polarised modes, respectively. All the results are believed to be calculated with an absolute error of 5×10^{-5}. The difference between the polarised and vectorial analysis results does not exceed three on the fourth decimal place, which confirms the generally observed good accuracy of the polarised approximation in the case of rib waveguides. The results obtained by the scalar method, however, differ

---

TABLE 3.9

Normalised Propagation Constants B of Rectangular Dielectric Waveguide Calculated by Effective Index Method ($V = 1$, $a/b = 2$, $n_f = 1.5$, $\lambda = 1.15$ μm, cf. Figure 3.3)

| | Normalised Propagation Constants | | |
n_c	Scalar	x-Polarised	y-Polarised
1.45	0.5207	0.5179	0.5077
1.4	0.5207	0.5150	0.4942
1.3	0.5207	0.5087	0.4654
1.2	0.5207	0.5016	0.4340

TABLE 3.10

Normalised Propagation Constants *B* of Rib Dielectric Waveguide (*H+h* = 1 μm, *W* = 1.5 μm, *n*$_f$ = 3.44, *n*$_s$ = 3.4, *n*$_c$ = 1.0, λ = 1.15 μm, cf. Figure 3.4)

	SFD	Polarised	Vectorial [108]	Polarised	Vectorial [149]
h (μm)	Scalar	*x*-Polarised		*y*-Polarised	
0.1	0.3095	0.3018	0.3019	0.2674	0.2674
0.2	0.3129	0.3055	0.3057	0.2706	—
0.3	0.3176	0.3108	0.3110	0.2750	0.2751
0.4	0.3239	0.3178	0.3181	0.2810	—
0.5	0.3320	0.3267	0.3270	0.2888	0.2890
0.6	0.3420	0.3377	0.3380	0.2985	—
0.7	0.3542	0.3509	0.3512	0.3104	0.3107
0.8	0.3693	0.3672	0.3675	0.3253	—
0.9	0.3893	0.3884	0.3886	0.3453	0.3455

TABLE 3.11

Normalised Propagation Constants *B* of Rib Dielectric Waveguide (*H+h* = 1 μm, *W* = 1.5 μm, *n*$_f$ = 3.44, *n*$_s$ = 3.4, *n*$_c$ = 1.0, λ = 1.15 μm, cf. Figure 3.4)

	Normalised Propagation Constant		
h [μm]	Scalar	*x*-Polarised	*y*-Polarised
0.1	0.3555	0.3552	0.3122
0.2	0.3555	0.3552	0.3122
0.3	0.3555	0.3552	0.3122
0.4	0.3555	0.3552	0.3122
0.5	0.3555	0.3552	0.3122
0.6	0.3586	0.3583	0.3122
0.7	0.3652	0.3649	0.3185
0.8	0.3751	0.3750	0.3279
0.9	0.3909	0.3909	0.3436

up to eight on the third decimal place for *x*-polarised modes and, even, in the first decimal place for y polarised modes.

Table 3.11 shows the values of normalised propagation constants calculated by the EIM for the same structure as in Table 3.10. For small values of outer slab height, the difference between 1D and 2D results is large because the outer slab is below cut-off, and the substrate refractive index has been used instead of the outer slab effective index. The inability to obtain an effective index when the outer slab is below cut-off is one of the major limitations of the EIM. The difference between the 2D and 1D results decreases as the outer slab height increases. This is because a shallow etch rib waveguide differs only slightly from the slab waveguide, for which the separation of variables can be carried out without incurring an error. Again, the results shown in Tables 3.10 and 3.11 confirm that scalar and effective index results provide an upper bound for propagation constants calculated with semivectorial and vectorial methods. Also, similar to the case of the rectangular waveguide, the results calculated by the 1D polarised mode solver can be more accurate than the results obtained by a 2D scalar technique (e.g., the y-polarised mode at $h = 0.7$ μm).

In summary, the results obtained confirm that for many planar optical waveguides of practical importance the polarised approximation offers a very high accuracy when compared with the full vectorial analysis. The 2D scalar analysis is only as accurate as the polarised and vectorial methods in the case of very low refractive index contrast optical waveguides. The results obtained show that the 1D effective index analysis provides a fairly accurate approximation of the propagation constant in the case of shallow etch optical rib waveguides and low refractive index contrast rectangular optical waveguides. However, the accuracy of the EIM rapidly decreases near the modal cut-off. In the case of the y-polarised modes, the 1D effective index calculations are often more accurate than the more numerically intensive 2D scalar calculations. The results obtained confirm also that the propagation constants obtained with the 2D scalar and effective index methods provide an upper bound for the propagation constant values obtained with the semivectorial and vectorial techniques.

OPTICAL FIBRES

Optical fibres have characteristic geometric features that distinguish them from the planar optical waveguides. This is primarily the consequence of a different technology of fabrication. Even though there are examples of optical waveguides that can be realised using both the planar and fibre optic technology, the large majority of structures are characteristic only to the specific technology. One feature that distinguishes an optical fibre from a planar optical waveguide is the presence of curved boundaries between the core and the cladding. Numerical methods discussed in the previous section are in general not well suited for handling such structures with curved boundaries. Therefore it was necessary to develop numerical methods that are specifically tailored for optical fibres. We discuss these methods in this section.

WAVEGUIDING IN OPTICAL FIBRES

There are two main types of optical fibres: classical fibres and microstructured optical fibres (MOFs). The classical fibres have been developed from the early 1960s as a transmission medium for long-distance optical telecoms systems. The work on MOFs was initiated in the 1990s to address the shortcomings of the classical core-clad fibres in a number of usually nontelecom related applications.

The classical fibre can use either the index guiding or other guiding mechanisms to achieve low transmission loss. The most simple classical fibre structure is the step index circular fibre (Figure 3.12). In this fibre structure, the core of the refractive index n_c and radius a is surrounded by an infinitely extending cladding with refractive index n_a, whereby $n_c > n_a$. (In practically realised structures, the cladding has a large but limited radius.) For a step index circular fibre, Maxwell's equations can be solved analytically [150]. A step-index circular fibre guides three types of modes: hybrid modes, transverse electric (TE), and TM modes. Hybrid modes have all six components of the

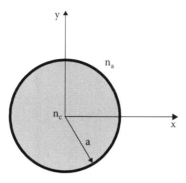

FIGURE 3.12   Step index circular optical fibre.

electromagnetic field and are divided into two groups: HE_{mn} and EH_{mn} modes. The TE modes have only E_θ, H_r, and H_z components while TM modes have H_θ, E_r, and E_z ones. For structures with a moderate refractive index contrast, the fundamental mode of the step-index circular fibre is the HE_{11} mode. The HE_{11} mode is nearly linearly polarised and is guided at any wavelength. Due to the circular symmetry of the fibre structure, the orientation of the linear polarisation is arbitrary. The first higher order mode is the TE_{01} mode, which belongs to a group of three modes, that is, TE_{01}, TM_{01}, and HE_{21}. The single-mode region cut-off condition is set by the first zero of the equation $J_0(V) = 0$, where J_0 is the zero order Bessel function of the first kind, $V = \dfrac{2\pi}{\lambda} a \sqrt{n_c^2 - n_a^2}$ and λ is the operating wavelength. When the refractive index contrast between the core and cladding is small all the modes of the circular fibre become linearly polarised. The detailed description of the optical properties of step-index circular fibres can be found in a number of text books [55,151–154].

In principle the step index fibre core can have any shape. In the literature, fibres with elliptical, triangular, rectangular (cf. subsection "Waveguiding in Planar Optical Waveguides"), and other core shapes were discussed. A structure that found particular interest is the elliptical core fibre. Similarly, to the case of the circular fibre Maxwell's equations can be solved analytically for the elliptical core fibre (Figure 3.13) [155]. All modes of an elliptical fibre are hybrid (i.e., have all six components of the electromagnetic field) and they divide into four groups: $_oHE_{mn}$, $_eHE_{mn}$, $_oEH_{mn}$, and $_eEH_{mn}$ whereby index "o" stands for odd while "e" for even. There are two fundamental modes: $_oHE_{11}$ and $_eHE_{11}$ that have zero cut-off frequency. Both fundamental modes have orthogonal polarisations. The transverse electric field component of the even mode is parallel to the minor axis while the odd one to the major axis of the core ellipsis. Both fundamental modes have different values of the propagation constants. The elliptical fibre can therefore preserve the polarisation of the guided light. More details on elliptical fibres can be found in Dyott [156].

In order to improve the chromatic dispersion characteristics of step index circular fibres, multistep circular fibres were introduced. In a multistep circular fibre, the core is surrounded by a number of rings of varying width and refractive index (Figure 3.14). If the core of the multistep circular fibre is made of air such a structure

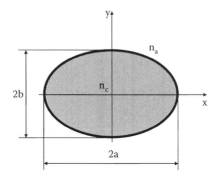

FIGURE 3.13   Step index elliptical optical fibre.

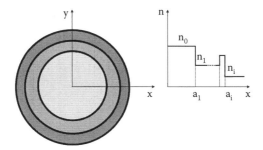

FIGURE 3.14   Multistep circular optical fibre.

is referred to as a hollow core fibre. A Bragg fibre is formed when the refractive index of the rings changes periodically between a low and high value. A more general structure that allows for more flexibility in the refractive index distribution, and that is akin to the Bragg fibre, is the 1D photonic crystal fibre. The optical propagation properties of multistep circular fibres can vary within a wide range depending on the refractive index distribution, which allows optimisation of this structure for a wide range of applications. For instance, a multistep circular fibre can be designed to achieve flat dispersion characteristics within a wide range of wavelengths [157] while the concept of the hollow core fibre can be used to design fibres with low loss at long wavelengths [158]. Fairly recently, a multistep circular fibre structure has been suggested for realizing an all dielectric coaxial optical waveguide [159]. More details on the multistep index fibres can be found in the references [152–154].

Lastly, we also mention also anisotropic optical fibres that are currently used in most applications that require a polarisation maintaining property. The best known fibres of this genre are the "panda" and "bow tie" fibres. In these fibres a specially selected glass introduces a mechanical stress that results in optical anisotropy of the material. The optical anisotropy of the material results in modal birefringence, which is defined as $(\beta_x - \beta_y)/k$ where β_x and β_y are the propagation constants of an x- and y-polarised mode, respectively, while k is the wavenumber. A fairly exhaustive review of anisotropic fibres is given in Noda et al. [160].

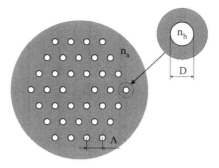

FIGURE 3.15   Photonic crystal fibre.

Without any question the MOFs offer the most flexibility in the design of the optical characteristics. This statement is confirmed by a large number of publications that report on the design and practical realisation of optical fibres with unique properties using the MOF concept. Probably the most spectacular example is the endlessly single-mode fibre proposed by Birks et al. [161], which has been cited nearly 2,000 times at the time of publication of this book. The fibre proposed by Birks et al. is single mode for any wavelength, which is a property that was not achievable using a standard step index circular fibre.

The simplest example of a MOF is a solid core photonic crystal fibre (Figure 3.15). This fibre consists usually of set of air holes that surround the solid core. The air holes make typically a triangular lattice pattern. However, other arrangements of holes are also possible. Typically the holes are also round, but fibres with other hole shapes have also been proposed. The properties of a photonic crystal fibre depend strongly on the hole diameter and spacing. A fairly comprehensive review on MOFs can be found in recently published books [162–164].

EXAMPLES OF OPTICAL FIBRES

The growing number of optical fibre applications drives an intensive research in the field of material science. With the exception of the silver halide, the materials that are used for the fabrication of optical fibre can be divided into two groups: glasses and polymers. There are several factors that decide about the usefulness of a particular material for fibre optics:

- Low cost
- Attenuation at the operating wavelength
- Mechanical robustness
- Structural robustness (resistance to crystallisation)
- Resistance to environmental factors (day light, air, humidity)
- Optical nonlinear properties
- Optical bending loss
- Ability to be drawn into a fibre
- Ability to dissolve lanthanide ions

Silica glass is an example of an oxide glass, and, currently, it is the most often used material for the fabrication of optical fibres. This material is very robust, low cost, can be fairly easily drawn into a fibre, has low bending loss, comparatively low optical nonlinearity, and dissolves the lanthanide ions. The silica glass fibres are primarily applied in long-haul telecom systems as the transmission medium and also for the realisation of fibre lasers and amplifiers. The main limiting factor for silica glass is its large attenuation for long wavelengths due to high phonon energy (~1100 cm^{-1}). Its application is therefore practically limited to operating wavelengths not exceeding 2 μm. However, other oxide glasses can be used for longer wavelengths. For instance, germanium oxide glass can be used for optical beam delivery up to 3 μm. Even longer wavelengths can be reached using tellurites [165].

The nonoxide glasses have been developed particularly to address the applications in the mid-infrared wavelength rage. The absence of oxygen allows reducing the phonon energy. An example of a nonoxide glass is fluoride glass, of which a particular example is ZBLAN. Fluoride glass was initially intended as a replacement for the silica glass as the transmission medium in long haul telecom links. However, this idea was abandoned because fluoride glass is significantly less robust than silica glass and also has a tendency to crystallise, which eventually increases the glass attenuation. At the moment the main area of application of fluoride glass is the Er^{3+} doped lasers operating at 3 μm and optical beam delivery for wavelengths of up to 4 μm.

The phonon energy of ZBLAN is ~650 cm^{-1}. This effectively prevents the application of this glass beyond 4 μm. Longer wavelengths can be achieved using a chalcogenide glass. The phonon energy of the chalcogenide glass can be as low as 350 cm^{-1}. Chalcogenide glass has a much larger nonlinear optical coefficients than the silica glass and has also been demonstrated to dissolve the lanthanide ions [166].

Another material that has been developed for long wavelengths is silver halide, which is a polycrystalline material. Silver halide fibres are nonhygroscopic and flexible. However, their optical properties gradually deteriorate when exposed to daylight. Silver halide fibres are suitable for the optical beam delivery in CO$_2$ laser systems [167].

Polymers on the other hand are used mainly for the transmission of light in the visible part of the optical spectrum. A polymer that is of particular interest for photonics is PMMA. The polymer fibres are easy to handle and fairly robust therefore they find an increasing application in optical local area networks [168].

A fairly recent survey of glass materials for optical fibres can be found in Richardson et al. [169] and Harrington [170], for the infrared wavelength region.

Figure 3.16 shows cross-section examples of selected optical fibres. The standard core-clad silica glass fibre is shown in Figure 3.16a [151]. It consists of a silica glass cladding and GeO$_2$-doped silica glass core. Other dopant materials can also be used for modifying the value of the silica glass in order to achieve the desired light propagation characteristics. Figure 3.16b shows an example of a silica glass-based photonic crystal fibre [161]. A set of air holes confines the light in the core. The optical waveguide characteristics can be modified by varying the hole pattern and hole diameter. Figure 3.16c shows a panda fibre [160]. When compared

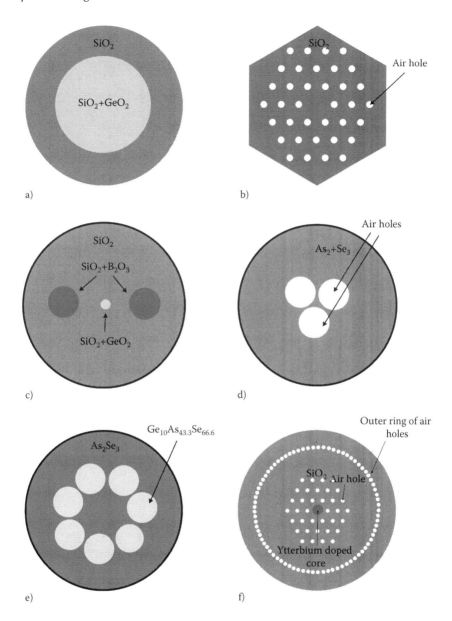

FIGURE 3.16 Examples of optical fibres: a) core-clad silica glass fibre, b) silica glass photonic crystal fibre, c) PANDA fibre, d) suspended core fibre, e) arsenide selenide microstructured fibre, and f) ytterbium-doped microstructured fibre.

with a standard silica glass fibre, panda fibre has two additional rods made of B_2O_3. These two rods introduce a mechanical stress that results in a significant modal birefringence. Figure 3.16d shows a suspended core fibre made of arsenic selenide glass [171]. The large refractive index difference between the glass core and air cladding allows for achieving large light intensity within the core. Another example of an arsenide selenide glass fibre is the microstructured fibre shown in Figure 3.16e [172]. In this case the light confinement in the core is achieved by a set of seven lower refractive $Ge_{10}As_{43.3}Se_{66.6}$ glass rods. Finally, Figure 3.16f shows an extended mode area microstructured fibre doped with ytterbium in the core [173]. Such fibres are used for the realisation of fibre lasers. The outer ring of air holes provides optical confinement for the pump while the signal light is guided in the ytterbium-doped core.

The large variety of cross-sectional shapes of optical fibres necessitates the use of a wide range of modelling techniques that is required for the calculation of the propagation constants and modal field distributions. In subsequent subsections, we discuss these techniques starting from the classical fibres and moving on to MOFs.

STEP INDEX CIRCULAR OPTICAL FIBRE

The field distributions and propagation constants of the modes of a step index optical fibre can be calculated by an analytical solution of the Maxwell equations. In order to derive this solution we follow the same procedure as in the case of the slab waveguide. We obtain the solution in the core and cladding first and use the boundary conditions to obtain the solution that is valid in the entire considered domain.

First, we observe that in homogenous medium Equations 3.8c and 3.8f take the form of the Helmholtz equation:

$$\frac{\partial^2 E_z}{\partial x^2} + \frac{\partial^2 E_z}{\partial y^2} + \left(\omega^2 \mu_0 \varepsilon - \beta^2\right) E_z = 0 \tag{3.41a}$$

$$\frac{\partial^2 H_z}{\partial x^2} + \frac{\partial^2 H_z}{\partial y^2} + \left(\omega^2 \mu_0 \varepsilon - \beta^2\right) H_z = 0 \tag{3.41b}$$

We can recast 3.41 into the cylindrical coordinate system and, taking the advantage of the circular symmetry, we express the solution in the form of the following product:

$$U(r, \theta, z, \omega) = U(r, \omega) e^{j(-\beta z + m\theta)} \tag{3.42}$$

where U stands for either E_z or H_z, and $m \geq 0$ is the azimuthal mode number. Substituting 3.42 into 3.41, expressed in the cylindrical coordinate system, yields:

$$\frac{\partial^2 U_z}{\partial r^2} + \frac{\partial U_z}{r \partial r} + \left(\kappa^2 - \left(\frac{m}{r}\right)^2\right) U_z = 0 \tag{3.43a}$$

for the core area and

$$\frac{\partial^2 U_z}{\partial r^2} + \frac{\partial U_z}{r \partial r} - \left(\gamma^2 + \left(\frac{m}{r} \right)^2 \right) U_z = 0 \qquad (3.43b)$$

for the cladding, whereby $\kappa^2 = \omega^2 n_c^2 \varepsilon_0 \mu_0 - \beta^2$, $\gamma^2 = \beta^2 - \omega^2 n_a^2 \varepsilon_0 \mu_0$, and n_c and n_a is the refractive index in the core and cladding, respectively.

The solutions of Equation 3.43 can be expressed as:

$$U_z = A_1 J_m(\kappa r) + A_2 Y_m(\kappa r)$$

in the core and

$$U_z = A_3 I_m(\gamma r) + A_4 K_m(\gamma r)$$

in the cladding. J_m and Y_m are the Bessel functions of the first and second kind, respectively, while I_m and K_m denote the modified Bessel functions of the first and second kind, and m is the function order. Since the unbound growth of the field in the core and cladding is not physically justified, the axial components of the electromagnetic field in the core can be expressed as:

$$E_z \left(r, \theta, z, \omega \right) = A_E J_m \left(\kappa r \right) e^{j(-\beta z + m\theta)} \qquad (3.44a)$$

$$H_z \left(r, \theta, z, \omega \right) = A_H J_m \left(\kappa r \right) e^{j(-\beta z + m\theta)} \qquad (3.44b)$$

while in the cladding as:

$$E_z \left(r, \theta, z, \omega \right) = B_E K_m \left(\gamma r \right) e^{j(-\beta z + m\theta)} \qquad (3.44c)$$

$$H_z \left(r, \theta, z, \omega \right) = B_H K_m \left(\gamma r \right) e^{j(-\beta z + m\theta)} \qquad (3.44d)$$

After some algebra the transverse electromagnetic field components can be obtained by substituting 3.44 into 3.1, expressed in the cylindrical coordinate system:

$$E_r = -\frac{j}{\kappa^2} \left[\beta \kappa A_E J_m' \left(\kappa r \right) + j \omega \mu_0 \frac{m}{r} A_H J_m \left(\kappa r \right) \right] e^{j(-\beta z + m\theta)} \qquad (3.45a)$$

$$E_\theta = -\frac{j}{\kappa^2} \left[j \beta \frac{m}{r} A_E J_m \left(\kappa r \right) - \omega \mu_0 \kappa A_H J_m' \left(\kappa r \right) \right] e^{j(-\beta z + m\theta)} \qquad (3.45b)$$

$$H_r = -\frac{j}{\kappa^2} \left[\beta \kappa A_H J_m' \left(\kappa r \right) - j \omega \varepsilon_0 n_c^2 \frac{m}{r} A_E J_m \left(\kappa r \right) \right] e^{j(-\beta z + m\theta)} \qquad (3.45c)$$

$$H_\theta = -\frac{j}{\kappa^2}\left[j\beta\frac{m}{r}A_H J_m(\kappa r)+\omega\varepsilon_0 n_c^2\kappa A_E J'_m(\kappa r)\right]e^{j(-\beta z+m\theta)} \qquad (3.45d)$$

in the core and

$$E_r = \frac{j}{\gamma^2}\left[\beta\gamma B_E K'_m(\gamma r)+j\omega\mu_0\frac{m}{r}B_H K_m(\gamma r)\right]e^{j(-\beta z+m\theta)} \qquad (3.45e)$$

$$E_\theta = \frac{j}{\gamma^2}\left[j\beta\frac{m}{r}B_E K_m(\gamma r)-\omega\mu_0\gamma B_H K'_m(\gamma r)\right]e^{j(-\beta z+m\theta)} \qquad (3.45f)$$

$$H_r = \frac{j}{\gamma^2}\left[\beta\gamma B_H K'_m(\gamma r)-j\omega\varepsilon_0 n_a^2\frac{m}{r}B_E K_m(\gamma r)\right]e^{j(-\beta z+m\theta)} \qquad (3.45g)$$

$$H_\theta = \frac{j}{\gamma^2}\left[j\beta\frac{m}{r}B_H K_m(\gamma r)+\omega\varepsilon_0 n_a^2\gamma B_E K'_m(\gamma r)\right]e^{j(-\beta z+m\theta)} \qquad (3.45h)$$

in the cladding, where the "prime" stands for the differentiation with respect to the function argument. Imposing the continuity of the tangential electromagnetic field components at the boundary between the core and the cladding, $r = a$, yields a set of four algebraic equations:

$$\begin{bmatrix} J_m(u) & 0 & -K_m(w) & 0 \\ 0 & J_m(u) & 0 & -K_m(w) \\ j\frac{\beta m}{a\kappa^2}J_m(u) & +\frac{\omega\mu_0}{\kappa}J'_m(u) & j\beta\frac{m}{a\gamma^2}K_m(w) & +\frac{\omega\mu_0}{\gamma}K'_m(w) \\ \frac{\omega\varepsilon_0}{\kappa}n_c^2 J'_m(u) & -j\frac{\beta m}{a\kappa^2}J_m(u) & +\frac{\omega\varepsilon_0}{\gamma}n_a^2 K_m(w) & -j\beta\frac{m}{a\gamma^2}K_m(w) \end{bmatrix}\begin{bmatrix} A_E \\ A_H \\ B_E \\ B_H \end{bmatrix} = 0$$

$$(3.46)$$

where $u = \kappa a$ and $w = \gamma a$. Since Equation 3.46 is homogenous, we proceed in the same way as in the case of the slab waveguide. Calculating the determinant and equating it to zero yields the Hondros-Debye equation:

$$(X_m + Y_m)(n_c^2 Y_m + n_a^2 X_m) - \frac{m^2 N^2}{(u^2 B)^2} = 0 \qquad (3.47)$$

where

$$X_m = \frac{K'_m(w)}{wK_m(w)}, \; Y_m = \frac{J'_m(u)}{uJ_m(u)}, \; B = \frac{N^2 - n_a^2}{n_c^2 - n_a^2}$$

is the relative propagation constant, $N = \beta/k$ is the effective refractive index, and k is the wave number. For $m = 0$, Equation 3.47 splits into two equations, the solutions of which give the propagation constants of the TE modes:

$$\frac{J_1(u)}{uJ_0(u)} + \frac{K_1(w)}{wK_0(w)} = 0 \tag{3.48a}$$

and TM modes:

$$n_c^2 \frac{J_1(u)}{uJ_0(u)} + n_a^2 \frac{K_1(w)}{wK_0(w)} = 0 \tag{3.48b}$$

Once the propagation constant of a mode is calculated, the field distribution can be obtained from 3.44 and 3.45 after calculating the unknown constants A_E, A_H, B_E, and B_H from 3.46. Algorithm 3.3 summarises all the steps required to calculate the propagation constant of a step index circular fibre by finding the zeros of the Hondros-Debye equation.

The following MATLAB script performs the calculation of the propagation constants using the bisection method for solving numerically the Hondros-Debye equation:

```
% program calculates normalised propagation constant for step
% index
% optical fibres
format long e
% fibre definition
```

ALGORITHM 3.3   CALCULATION OF THE PROPAGATION CONSTANTS OF A STEP INDEX CIRCULAR FIBRE BY FINDING ZEROS OF 3.47

1. Start
2. Set the values of, wavelength, refractive indices and core radius
3. Calculate the interval that contains the value of the propagation constant by bracketing the corresponding zero of 3.46
4. Refine the value of the propagation constant using the bisection method
5. Stop

```
lambda = 1.55;
radius = 7.5;
na = 1.46;
nc = 1.47;
m = 1;

% calculation parameters
k = 2.0*pi/lambda;
V = radius*k*sqrt(nc^2-na^2);
B = (0.05:0.00001:0.975);

% Hondros-Debye equation
HD = hdf(m,nc,na,V,B);
zero = 0.*B;
plot(B,HD,B,zero)

% precise initial location of B
length = size(B);
for j = length(2):-1:2
if HD(j)/HD(j-1)<0
break
end
end
if j = =2
disp('no zero found, change initial interval')
disp('the results are incorrect')
end

btop = B(j);
bbot = B(j-1);
%%%%%%%%%%%%%%%%%%%%%%%%%%%%%%%%%%%%%%%%%%%%%%%%%%%%%%%%%%%%%%%%%%%%%%%%%
%%%%%%%

% bisection method for root finding starting from the initial
% interval
DEtop = hdf(m,nc,na,V,btop);
DEbot = hdf(m,nc,na,V,bbot);
jmax = 100;    % maximum number of bisections
tolerance = 0.000000000001;       % tolerance for finding B

for j = 1:jmax
bcen = (btop-bbot)/2+bbot;
DEcen = hdf(m,nc,na,V,bcen);
if DEtop/DEcen > 0
btop = bcen;
DEtop = DEcen;
else
bbot = bcen;
DEbot = DEcen;
end
```

```
delta = btop-bbot;
if delta < tolerance
break
end
end
%end of bisection method loop
%%%%%%%%%%%%%%%%%%%%%%%%%%%%%%%%%%%%%%%%%%%%%%%%%%%%%%%%%%%%%%%%%%%%%%%%%%%%
%%%%%%
%%%%%%%%%%%%%%%%%%%%%%%%%%%%%%%%%%%%%%%%%%%%%%%%%%%%%%%%%%%%%%%%%%%%%%%%%%%%
%%%%%%
% checks for tolerance
if j = =jmax
disp('maximum number of bisections exceeded')
else
out = bcen;
end
% variable 'out' stores the relative propagation constant B

% calculation of effective index
nin = abs(sqrt(out*(nc*nc-na*na)+na*na))

function HDebye = hdf(m,nc,na,V,B)
% Hondros Debye equation evaluation
w = sqrt(B)*V;
u = sqrt(1-B)*V;
N2 = B*(nc^2-na^2)+na^2;

Jminus = besselj(m-1,u)./(u.*besselj(m,u));
Jplus = besselj(m+1,u)./(u.*besselj(m,u));
Kminus = besselk(m-1,w)./(w.*besselk(m,w));
Kplus = besselk(m+1,w)./(w.*besselk(m,w));
X1 = -0.5*(Kminus+Kplus);
Y1 = 0.5*(Jminus-Jplus);
HDebye = (X1+Y1).*(nc^2*Y1+na^2*X1)-m*m*N2./((u.^2.*B).^2);
```

Following the standard programming rules in MATLAB, the function needs to be saved in a file hdf.m. In order to evaluate 3.46 for given β, we used the following formulae:

$$X_m = -\frac{1}{2}\left(K^- + K^+\right), \quad Y_m = \frac{1}{2}\left(J^- - J^+\right) \quad \text{while} \quad J^\pm = \frac{J_{m\pm1}(u)}{uJ_m(u)} \quad \text{and} \quad K^\pm = \frac{K_{m\pm1}(w)}{wK_m(w)}.$$

Since, unlike in the slab waveguide case, Equation 3.46 can have more than one zero, a plot of the function is helpful when locating the propagation constants. As a reference we used this software to calculate the propagation constant for the fundamental mode HE_{11} and TE_{01} mode of a fibre with the radius $a = 7.5$ μm, $n_c = 1.47$, and $n_a = 1.46$ at the wavelength of 1.55 μm, and obtained respectively 5.9528599309 μm^{-1} and 5.9438095599 μm^{-1}, which agrees on all digital places with the result quoted in Chiang et al. [111]. The extension of the analytical method of the solution of Equation 3.41 for double clad and multilayered optical fibres can be found in Majewski et al. [9]. Alternatively, standard numerical techniques like the FEM or FDM can be applied. High-order finite difference approximations for optical fibres can be found in Du and Chiou [174] and Lu et al. [175].

Now we will present the scalar approximation for the step index fibre. For this purpose, we transform 3.15 into the cylindrical coordinate system and substitute the solution in the product form:

$$U(r,\theta,z,\omega) = \tilde{U}(r)e^{j(-\beta z + l\theta)} \tag{3.49}$$

This procedure yields:

$$\frac{\partial^2 \tilde{U}}{\partial r^2} + \frac{\partial \tilde{U}}{r \partial r} + \left(\kappa^2 - \left(\frac{m}{r}\right)^2\right)\tilde{U} = 0 \tag{3.50a}$$

in the core and

$$\frac{\partial^2 \tilde{U}}{\partial r^2} + \frac{\partial \tilde{U}}{r \partial r} - \left(\gamma^2 + \left(\frac{m}{r}\right)^2\right)\tilde{U} = 0 \tag{3.50b}$$

in the cladding, whereby $\kappa^2 = \omega^2 n_c^2 \varepsilon_0 \mu_0 - \beta^2$, $\gamma^2 = \beta^2 - \omega^2 n_a^2 \varepsilon_0 \mu_0$, and n_c and n_a are the refractive indexes in the core and cladding, respectively. Similarly, as in the case of 3.43, the scalar potential in the core can be expressed as:

$$\tilde{U}(r,\theta,z,\omega) = AJ_l(\kappa r)e^{j(-\beta z + l\theta)} \tag{3.51a}$$

And that in the cladding as:

$$\tilde{U}(r,\theta,z,\omega) = BK_l(\gamma r)e^{j(-\beta z + l\theta)} \tag{3.51b}$$

Imposing the continuity of the scalar potential and its derivative at the boundary between the core and the cladding yields a set of two algebraic equations:

$$\begin{bmatrix} J_l(u) & -K_l(w) \\ uJ_l'(u) & -wK_l'(w) \end{bmatrix}\begin{bmatrix} A \\ B \end{bmatrix} = 0 \tag{3.52}$$

The calculation of the determinant yields the dispersion equation:

$$\frac{u J_{l+1}(u)}{J_l(u)} = \frac{w K_{l+1}(w)}{K_l(w)} \qquad (3.53)$$

The solution of 3.53 give the propagation constants of the modes in the scalar approximation. The corresponding field distributions are calculated from 3.51 so that 3.52 is satisfied.

The propagation constant of the fundamental mode in the scalar approximation is obtained with $l = 0$. Tables 3.12 and 3.13 show the comparison for the normalised propagation constant

$$B = \frac{\beta^2 - n_a^2 k^2}{n_c^2 k^2 - n_a^2 k^2}$$

of the fundamental mode obtained solving the vectorial dispersion equation 3.47 and the scalar dispersion equation 3.53. These results are believed to be calculated with an absolute error of 5×10^{-7}. The results collected in the Table 2.12 were obtained for a fibre with $n_a = 1.46$, $n_c = 1.47$, and $\lambda = 1.55$ μm while varying the core radius. The relative error Δ is defined as:

$$\Delta = \frac{\left| B_{sc} - B_{vec} \right|}{B_{vec}},$$

where B_{sc} and B_{vec} are the relative propagation constants obtained with the scalar and vectorial methods, respectively. Table 3.12 shows that the difference between the scalar and vectorial results diminishes as the core radius increases, that is, when the mode is further away from the cut-off. Table 3.13 shows the relative propagation constant values that have been obtained for the fibre with $n_c = 1.47$, $\lambda = 1.55$ μm, and $V = ak\sqrt{n_c^2 - n_a^2}$ while varying the cladding refractive index. These results show that the difference between the vectorial and scalar results grows with the increase in the refractive index contrast between the core and the cladding.

It is important to note that Equation 3.53 is the same as the dispersion equation of the linearly polarised (LP) modes [176]. However, one should not confuse the scalar approximation with the LP approximation. Although both are only valid for a small refractive index contrast between the core and the cladding, and use the same dispersion relation, they differ since the scalar approximation is obtained through solving the scalar wave equation while the LP modes are obtained as an approximate solution of the Maxwell's equations. The field distributions of the scalar modes are given by scalar potentials while the field distributions of the LP modes are given by the distributions of the appropriate components of the electromagnetic field, cf. [176].

The LP modes should also not be confused with the modes obtained under polarised approximation derived in the section "Introduction to Optical Waveguide

TABLE 3.12

Normalised Propagation Constants of the Fundamental Mode of a Circular Single Step Optical Fibre with n_c = 1.47, n_a = 1.46, λ = 1.55 µm, cf. Figure 3.12

a (µm)	B Vectorial 3.46	B Scalar 3.52	Relative Difference
1.5	0.05115	0.05229	2.2e-002
2.5	0.32061	0.32273	6.6e-003
3.5	0.53550	0.53702	2.8e-003
4.5	0.66974	0.67074	1.5e-003
5.5	0.75495	0.75562	8.8e-004
6.5	0.81153	0.81199	5.7e-004

TABLE 3.13

Normalised Propagation Constant of the Fundamental Mode of a Circular Single Step Optical Fibre with V = 2.4, n_c = 1.47, λ = 1.55 µm, cf. Figure 3.12

n_a	B Vectorial 3.46	B Scalar 3.52	Relative Difference
1.45	0.52691	0.53003	5.9e-003
1.4	0.51882	0.53003	2.2e-002
1.3	0.50126	0.53003	5.7e-002
1.2	0.48171	0.53003	1.0e-001
1.1	0.45996	0.53003	1.5e-001
1.0	0.43582	0.53003	2.3e-001

Theory" and given by the solutions of Equation 3.14. In fact, the polarised approximation is quite accurate when applied to calculate the propagation constants of the fundamental HE_{11} mode. However, it is not generally applicable for the calculation of the propagation constants of the higher order modes [102].

To show the clear difference between LP and scalar modes, we derive the field distribution and dispersion equation for the LP_{01} mode. The field distribution of LP_{01} mode can be obtained from the field distribution of the HE_{11} mode under the weakly guiding approximation, that is, $n_c \approx n_a$. Substituting $B_H = 1$ Equation 3.46 yields:

$$B_E = \frac{1}{j} \frac{\omega \mu_0}{\beta m} Bu^2 \left(X_m + Y_m \right)$$ (3.54a)

$$A_E = \frac{1}{j} \frac{\omega \mu_0}{\beta m} Bu^2 \left(X_m + Y_m \right) A_H$$ (3.54b)

$$A_H = \frac{K_m(w)}{J_m(u)} \tag{3.54c}$$

Substituting 3.54 in 3.46 and using Bessel function identities [177] yields:

$$E_r = \frac{A_H}{\kappa}\omega\mu_0 \left[\frac{1}{2}\left(\frac{C}{m}+1\right)J_{m+1}(\kappa r) + \frac{1}{2}\left(-\frac{C}{m}+1\right)J_{m-1}(\kappa r)\right]e^{j(-\beta z + m\theta)} \tag{3.55a}$$

$$E_\theta = -\frac{jA_H}{\kappa}\omega\mu_0 \left[\frac{1}{2}\left(\frac{C}{m}+1\right)J_{m+1}(\kappa r) + \frac{1}{2}\left(\frac{C}{m}-1\right)J_{m-1}(\kappa r)\right]e^{j(-\beta z + m\theta)} \tag{3.55b}$$

$$H_r = -\frac{jA_H}{\kappa}\beta \left[\frac{1}{2}\left(-\left(\frac{n_c}{N}\right)^2\frac{C}{m}-1\right)J_{m+1}(\kappa r) + \frac{1}{2}\left(-\left(\frac{n_c}{N}\right)^2\frac{C}{m}+1\right)J_{m-1}(\kappa r)\right]e^{j(-\beta z + m\theta)}$$
$$\tag{3.55c}$$

$$H_\theta = \frac{A_H}{\kappa}\beta \left[\frac{1}{2}\left(\left(\frac{n_c}{N}\right)^2\frac{C}{m}+1\right)J_{m+1}(\kappa r) + \frac{1}{2}\left(-\left(\frac{n_c}{N}\right)^2\frac{C}{m}+1\right)J_{m-1}(\kappa r)\right] \tag{3.55d}$$

in the core and for the cladding:

$$E_r = -\frac{\omega\mu_0}{\gamma} \left[\frac{1}{2}\left(\frac{C}{m}+1\right)K_{m+1}(\gamma r) + \frac{1}{2}\left(\frac{C}{m}-1\right)K_{m-1}(\gamma r)\right]e^{j(-\beta z + m\theta)} \tag{3.55e}$$

$$E_\theta = \frac{j\omega\mu_0}{\gamma} \left[\frac{1}{2}\left(\frac{C}{m}+1\right)K_{m+1}(\gamma r) + \frac{1}{2}\left(-\frac{C}{m}+1\right)K_{m-1}(\gamma r)\right]e^{j(-\beta z + m\theta)} \tag{3.55f}$$

$$H_r = \frac{j\beta}{\gamma} \left[\frac{1}{2}\left(-\left(\frac{n_a}{N}\right)^2\frac{C}{m}-1\right)K_{m+1}(\gamma r) + \frac{1}{2}\left(\left(\frac{n_a}{N}\right)^2\frac{C}{m}-1\right)K_{m-1}(\gamma r)\right]e^{j(-\beta z + m\theta)}$$
$$\tag{3.55g}$$

$$H_\theta = -\frac{\beta}{\gamma}\left[\frac{1}{2}\left(\left(\frac{n_a}{N}\right)^2\frac{C}{m}+1\right)K_{m+1}(\gamma r)+\frac{1}{2}\left(\left(\frac{n_a}{N}\right)^2\frac{C}{m}-1\right)K_{m-1}(\gamma r)\right]e^{j(-\beta z+m\theta)}$$

(3.55h)

where $C = Bu^2(X_m + Y_m)$. Substituting $m = 1$ into 3.55 and calculating the transverse field components x and y under the assumption that $n_a \approx n_c \approx N$ yields at $z = 0$:

$$E_x = -j\frac{A_H}{\kappa}\omega\mu_0 J_0(\kappa r)$$

(3.56a)

$$E_y = 0$$

(3.56b)

$$H_x = 0$$

(3.56c)

$$H_y = -j\frac{A_H}{\kappa}\beta J_0(\kappa r)$$

(3.56d)

in the core and for the cladding:

$$E_x = -j\frac{\omega\mu_0}{\gamma}K_0(\gamma r)$$

(3.56e)

$$E_y = 0$$

(3.56f)

$$H_x = 0$$

(3.56g)

$$H_y = -j\frac{\beta}{\gamma}K_0(\gamma r)$$

(3.56h)

The formulae 3.56 show that under the weakly guiding approximation the fundamental mode distribution becomes linearly polarised. It can also be noted that the ratio of the transverse electric field amplitude to the transverse magnetic field vector is

$$Z = \frac{E_x}{H_y} = \frac{\omega\mu_0}{\beta} \approx \frac{1}{n_c}\sqrt{\frac{\mu_0}{\varepsilon_0}} \approx \frac{1}{n_a}\sqrt{\frac{\mu_0}{\varepsilon_0}}$$

(3.57)

The weakly guiding approximation was originally proposed by Gloge [178] and then proved very successful in explaining the experimentally observed mode patterns of silica glass optical fibres. The derivations of the field distribution of higher order LP modes can be found for instance in Majewski and Sujecki [9] and Yariv [176].

As in the case of the planar slab waveguide, analytical expressions can be derived for the radiation modes of step index optical fibres [40]. More information on leaky waves and complex modes can be found in Rozzi and Mongiardo [41], Snyder and Love [55], and Jablonski [58].

A "Poor Man's Approach" to Modelling MOFs

A particularly simple numerical technique for modelling MOFs has been proposed [179]. It relies on the scalar approximation and the finite difference methods. In the scalar approximation, modes of an optical waveguide can be calculated by solving 3.17. If the standard three-point finite difference approximation [49] is used for both second order partial derivatives, one obtains from 3.17 for the point (x_i, y_j) the following equation that relates five samples of the scalar potential $\varphi_{i,j}$:

$$\frac{\phi_{i-1,j} - 2\phi_{i,j} + \phi_{i+1,j}}{h_x^2} + \frac{\phi_{i,j-1} - 2\phi_{i,j} + \phi_{i,j+1}}{h_y^2} + n_i^2 k^2 \phi_{i,j} = \beta^2 \phi_{i,j}$$

Writing analogous equations for each point in the considered domain, including the boundary conditions and collating the equations together, yields an algebraic eigenvalue problem in β^2:

$$A\overline{\phi} = \beta^2 \overline{\phi} \qquad (3.58)$$

where $\overline{\phi}$ is a vector that collates all scalar potential samples and A is a real penta-diagonal symmetric matrix, the eigenvalues of which can be easily calculated using standard numerical methods.

When considering a typical microstructured fibre, one can observe that the refractive index distribution of such fibre has a number of symmetry planes that can be used in order to reduce the size of the computational domain. For instance, a microstructured fibre based on an equilateral triangular mesh has six symmetry planes and a sixfold rotation axis. Such symmetry properties cannot be fully exploited in the rectangular coordinate system. However, in the cylindrical coordinate system, or by using a hexagonal finite difference mesh, such symmetry properties can be easily taken advantage of whilst allowing truncating the computational window. For instance, in the rectangular coordinate system the smallest possible computational window for an equilateral mesh based microstructured fibre consists of one quadrant (Figure 3.17a) while in the cylindrical coordinate system a 30° sector can be considered (Figure 3.17b).

In a cylindrical coordinate system, the scalar wave equation 3.15 has the following form:

$$\frac{\partial^2 \phi}{\partial r^2} + \frac{\partial \phi}{r \partial r} + \frac{1}{r^2} \frac{\partial^2 \phi}{\partial \theta^2} + \left(k^2 - \beta^2\right)\phi = 0 \qquad (3.59)$$

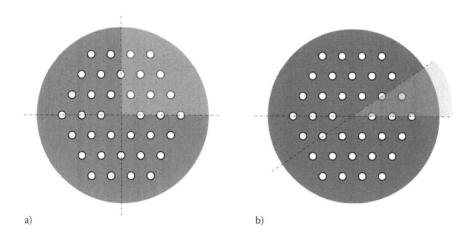

a)   b)

FIGURE 3.17 Computational window for an equilateral triangular mesh microstructured fibre in a) rectangular and b) cylindrical coordinate system.

ALGORITHM 3.4   CALCULATION OF THE PROPAGATION CONSTANTS OF A MICROSTRUCURED FIBRE USING THE SCALAR FINITE DIFFERENCE METHOD IN THE CYLINDRICAL COORDINATE SYSTEM

1.   Start
2.   Set the finite difference mesh
3.   Set the value of the wavelength
4.   Set the refractive index distribution
5.   Assemble the finite difference coefficient matrix
6.   Solve the eigenvalue problem
7.   Select the eigenvalue and eigenvector that corresponds to the required mode
8.   Plot the field distribution of the mode
9.   Stop

Substituting the standard three-point finite difference approximation for second order partial derivatives and the two-point central difference approximation for the first derivative in 3.59 yields at a point (x_i, y_j) the following equation that relates five samples of the scalar potential $\varphi_{i,j}$ in the cylindrical coordinate system:

$$\frac{\phi_{i-1,j} - 2\phi_{i,j} + \phi_{i+1,j}}{h_r^2} + \frac{\phi_{i,j-1} - \phi_{i,j}}{2r_{i,j}h_r} + \frac{\phi_{i,j-1} - 2\phi_{i,j} + \phi_{i,j+1}}{r_{i,j}^2 h_\theta^2} + n_i^2 k^2 \phi_{i,j} = \beta^2 \phi_{i,j}$$

Similarly, as in the rectangular coordinate system, if one collects together finite difference equations at all sampling points in the considered computational domain, an eigenvalue problem is obtained in β^2. The handling of a nodal point for $r = 0$ is described in Smith [49]. Algorithm 3.4 summarises the main steps required for obtaining the finite difference solution of the scalar wave equation for a microstructured fibre.

The following MATLAB script performs the calculation of the propagation constants of a MOF according to Algorithm 3.4.

```
% program calculates propagation constants of MOF,
% cylindrical c.s.
% boundary conditions: zero field on external border and
% zero normal derivative on symmetry plane
clear % clears variables
clear global % clears global variables
format long e
% initial constants
I = sqrt(-1);% remember not to overwrite pi or i !
pi = 3.141592653589793e+000;

wavelength = 1.5;% wavelength [mi]
k0 = 2*pi/wavelength;
ref_in_ba = 1.45;% background ref index
ref_in_ho = 1.;% hole ref index
hole_rad = 2;% hole radius [mi]
LAM = 10.;% lattice constant [mi]
PC_first_lay = 2;% first layer index
PC_no_lay = 9;
% number of PC hole layers (so if first = 2, no_lay = 4 then
% the last PC hole layer is no 5)

boundary_condition = 1;
%boundary condition = 0 - zero field on x axis;
%boundary condition = 1-zero derivative on x axis
shift = k0*1.4488;

% mesh params
window_radius = 30;%window radius [mi]
dr =.2;% radial mesh size [mi]
window_angle = pi/6;%window angle [rad]
dt = window_angle/150;
% angular mesh size [rad]-denominator gives the
%number of nodes on the arc
no_nodes_r = round(window_radius/dr);
% number of nodes in x direction
%(the x waveguide dimension = dx*n_nodes_x*2)
no_nodes_t = round(window_angle/dt);
% number of nodes in y direction
%(the y waveguide dimension = dx*n_nodes_x*sqrt(3)*2)
% - in y direction %we do not count shifted nodes
no_nodes = no_nodes_r*no_nodes_t;%total number of nodes
```

```
%creating a complex matrix with mesh node positions
%x-real part, y-imaginary part
%as the origin of the coordinate system is placed in
%the centre of symmetry and we consider only one quater
%the refernece point is 0+I*0
FD_radius = 0.5*dr:dr:window_radius;
%vector storing radial positions
FD_angle = 0.5*dt:dt:window_angle;
%vector storing angular positions
FD_mesh = ones(no_nodes,1);
%matrix storing node positions row wise

for j1 = 1:no_nodes_r
    for j2 = 1:no_nodes_t
        FD_mesh((j1-1)*no_nodes_r+j2) = FD_radius(j1)*…
        exp(I*FD_angle(j2));
    end
end

% calculating the lattice nodes positions (first part)
loop_count = 1;
for j1 = PC_first_lay:PC_no_lay+PC_first_lay-1
    for j2 = 1:j1
        PC_mesh(loop_count) = (j1-1)*LAM-(j2-1)*LAM/2+…
        I*(j2-1)*LAM*sqrt(3)/2;
        loop_count = loop_count+1;
    end
end


% calculating the lattice nodes positions
% (rotated by 60 deg part)
for j1 = PC_first_lay:PC_no_lay+PC_first_lay-1
    for j2 = 1:j1
        PC_mesh(loop_count) = ((j1-1)*LAM-(j2-1)*LAM/2+…
        I*(j2-1)*LAM*sqrt(3)/2)*exp(I*pi/3);
        loop_count = loop_count+1;
    end
end
PC_no_nodes = loop_count-1;
'finished calculating lattice positions'

% creating refractive index map in the mesh nodes
ref_in = ref_in_ba*ones(no_nodes,1);
%matrix storing refractive indices%in nodes.
%It is initiated to background ref index
for j1 = 1:no_nodes
    test_con = 1;
    j2 = 1;
    while test_con
        if abs(FD_mesh(j1)-PC_mesh(j2)) < hole_rad
```

```
                    ref_in(j1) = ref_in_ho;
                    test_con = 0;
                end
            j2 = j2+1;
            if j2 > PC_no_nodes
                test_con = 0;
            end
        end
    end
end
'set ref index values'

%setting up the matrix
FD_matrix1 = zeros(no_nodes,5);
%reserving space for complex matrix with
%FD coefficients

%setting the initial pattern
loop_count = 1;
for j1 = 1:no_nodes_r
    for j2 = 1:no_nodes_t
        FD_matrix1(loop_count,2) =…
        1/(dt*dt*FD_radius(j1)*FD_radius(j1));
        FD_matrix1(loop_count,3) = -2/(dr*dr)-…
        2/(dt*dt*FD_radius(j1)*FD_radius(j1))+…
        k0*k0*ref_in(loop_count)*ref_in(loop_count)-…
        shift*shift;
        FD_matrix1(loop_count,4) =…
        1/(dt*dt*FD_radius(j1)*FD_radius(j1));
        FD_matrix1(loop_count,1) =…
        1/(dr*dr)-1/(2*dr*FD_radius(j1));
        FD_matrix1(loop_count,5) =…
        1/(dr*dr)+1/(2*dr*FD_radius(j1));
        loop_count = loop_count+1;
    end
end

% adding periodic features
%%%%%%%%%%%%%%%%%%%%%%%%%%%%%%%%%%%%%%%%%%%%%%%%%%%%%%%%%%
loop_count = 1;
for j1 = 1:no_nodes_r
    FD_matrix1(loop_count,3)=FD_matrix1(loop_count,3)+…
    FD_matrix1(loop_count,2);
    FD_matrix1(loop_count+no_nodes_t-1,3)=…
    FD_matrix1(loop_count+no_nodes_t-1,3)+…
    FD_matrix1(loop_count+no_nodes_t-1,4);
    loop_count = loop_count+no_nodes_t;
end

%including zero field option
if abs(boundary_condition)<1.d-16
    'boundary zero field'
```

```
    loop_count = 1;
    for j1 = 1:no_nodes_r
        FD_matrix1(loop_count,3) =…
        FD_matrix1(loop_count,3)-…
        2*FD_matrix1(loop_count,2);
        %FD_matrix1(loop_count,1) = 0 so omitted
        loop_count = loop_count+no_nodes_t;
    end
end

% setting zeros in off diagonals
loop_count = 1;
for j1 = 1:no_nodes_r-1
    FD_matrix1(loop_count+no_nodes_t,2) = 0;
    FD_matrix1(loop_count+no_nodes_t-1,4) = 0;
    loop_count = loop_count+no_nodes_t;
end


%initial matrix created
%%%%%%%%%%%%%%%%%%%%%%%%%%%%%%%%%%%%%%%%%%%%%%%%%%%%%%%%%%%%%%%%%%%%

%shifting columns for fitting spdiag format
FD_matrix1(:,1) = circshift(FD_matrix1(:,1),-no_nodes_t);
FD_matrix1(:,2) = circshift(FD_matrix1(:,2),-1);
FD_matrix1(:,4) = circshift(FD_matrix1(:,4),1);
FD_matrix1(:,5) = circshift(FD_matrix1(:,5),no_nodes_t);
%
FD_matrix2 = spdiags(FD_matrix1, …
[-no_nodes_t -1 0 1 no_nodes_t], no_nodes, no_nodes);

'started calculating eigenvalues'
%calculating eigenvalues and plotting
aug = ones(no_nodes,1);%augmented vector
augp = ones(no_nodes,1);
%augmented vector after forward substitution

tol =.05e-1;
maxit = 150;
droptol = 0.0001;
[L,U] = luinc(FD_matrix2,droptol);
'finished ilu'
for i = 1:8
    x0 = aug;
    augp = aug;
    aug = bicgstab(FD_matrix2,aug,tol,maxit,L,U,x0);
    eigenval = (aug'*augp)/(aug'*aug);
    [i,eigenval,sqrt(eigenval+shift*shift)/k0]
end
```

The eigenvalues are calculated using the inverse shifted power method [50]. The value of the shift is given in terms of the effective index and needs to be selected near to the value of the effective index of the mode in order to achieve fast convergence of the power method. However, a value too close to the effective index of the mode will make the coefficient matrix singular and will result in a poor convergence of the BICGSTAB method that is used for matrix inversion. Therefore, an optimum value needs to be selected by "trial and error." It should be noted that no convergence test is performed for the inverse power method. The user should inspect if the results converge to a solution by comparing the values of the effective index from subsequent iterations. As can be seen in the code, we used the shift equal to 1.4488 for the fibre structure considered. The structure considered is a MOF consisting of a triangular lattice of holes within a silica glass host. The lattice constant is 10 μm while the hole radius is 2 μm. The operating wavelength is 1.5 μm. The variables "window_radius" and "window_angle" set the dimensions of the computational window. In the presented code these parameters have been set to 30 μm and $\pi/6$, respectively. At the edge of the computational window the field is assumed equal to zero. In the case considered, 30 μm radius was found sufficiently large for this assumption to have a negligible influence on the value of the effective index. Figure 3.18 shows the dependence of the effective refractive index of the fundamental mode on the radial FD mesh constant h_r. The number of mesh points along the azimuthal direction has been selected equal to the number of mesh points along the radial direction. The results obtained are compared with the results obtained using the scalar finite difference method on a hexagonal grid [180]. The results of the latter method are plotted as a function of the mesh constant h_h. Both methods converge to the same result to at least five decimal places.

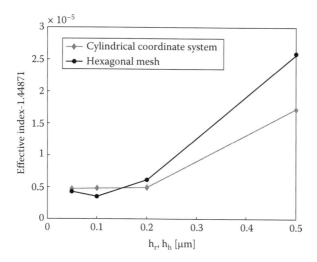

FIGURE 3.18 Dependence of the effective index on the finite difference mesh size for a microstructured optical fibres (MOF) with a triangular mesh of cylindrical air holes with a mesh constant of 10 μm and a hole diameter of 2 μm. The operating wavelength is 1.5 μm. One hole in the centre is missing, forming a defect.

The scalar finite difference method, although comparatively very simple to implement and use, suffers from the limitations of the scalar approximation. This problem is particularly severe because of the presence of abrupt refractive index discontinuities in many practically relevant MOFs, for example, between silica glass and air. Some extension of the applicability of the scalar finite difference method can be achieved by applying vectorial corrections [181]. However, in a general case vectorial methods have to be used.

PROPAGATION CONSTANT CALCULATION TECHNIQUES FOR MOFs

In principle, most techniques that are used for the numerical analysis of propagation characteristics of the planar optical waveguides can be also used for MOFs. Therefore quite a number of techniques were directly imported from the field of integrated optics. This in particular applies to the FEM [182–184] and the FDM [114,185–188]. The main problem that makes such adaptation not straightforward is the presence of the curved boundaries between the domains with distinctly different value of the refractive index, for example, air silica glass. The mode solvers that are used for planar devices use a staircase approximation, which results in a nonphysical behaviour of the field near strong refractive index discontinuities. Therefore some adaptation was necessary when using the finite element and finite difference codes for MOFs. This was either achieved by using a more flexible treatment of the curved interfaces, for example, curvilinear elements in FEM [189] or refractive index averaging in FD [190]. In general, FEM and FDM are very robust and can be applied to practically any type of MOF. Both techniques suffer, however, from low computational efficiency.

Due to the shortcomings of FEM and FDM, a number of analytical techniques were developed to speed up the MOF design process. As for the planar optical waveguides, an effective index method (EIM) has found wide spread use. There are in principle two types of EIM that are used for MOFs. We discuss the difference between the two techniques using the MOF structure from Figure 3.16 as an example. The first technique relies on identifying two distinct regions for the structure from Figure 3.16. One of the regions contains the central part (where the hole is missing) and is referred to as a core in analogy to a standard step index optical fibre. The other region consists of the photonic crystal structure that extends outside of the core. The effective index of the core area is assumed equal to the core material refractive index while the effective index of the photonic crystal area is equal to the effective index of the fundamental mode of the infinitely extending photonic crystal, that is, neglecting the presence of the core (defect) [191,192]. The calculation of this latter effective index is not trivial and requires an application of vectorial numerical methods. Therefore in Li et al. [193] a simplified technique was applied that approximates the elementary cell of a photonic crystal by a circular cell and uses a simple analytical expression to obtain the effective index. The main difficulty with applying the EIM, whereby the fibre is divided into the core and photonic crystal cladding region, is a lack of the ability to define clearly the boundary between the core and the cladding. A number of authors tried to find optimal solution for this problem but, as one would expect, the applicability of these solutions only holds for a small subset of all MOFs. The other EIM approach is not suffering from this difficulty. It divides the cross-section of MOF (Figure 3.16) in the core region and a set of concentric rings [194]. For each ring, an effective index is calculated along the azimuthal direction. In this manner an equivalent 1D structure is formed with an effective index distribution depending on the radial dimension only. This problem can be solved using the standard methods for classical multiclad fibres. It can be noticed that this latter approach closely resembles the EIM used for planar optical waveguides. The only difference is that the effective indices are calculated along the azimuthal and radial directions, instead of the orthogonal directions in the Cartesian coordinate system. Unlike its planar counterpart, however, the EIM for MOFs can be, and usually is, a "vectorial" technique. This follows from the fact that once a MOF is approximated by a classical fibre, a well-established vectorial analysis can be applied (cf. subsection "Step Index Circular Optical Fibre"). The "vectorial" character of EIM in the case of MOFs should not, however, be taken as a guarantee of accuracy. In fact the EIM is an approximate technique and can only be used for a qualitative analysis. For quantitative analysis, other analytical techniques should be used for obtaining an accurate rather than an approximate solution of the wave equation.

One of the very widely used techniques in the analysis of MOFs is the localised function method. It is particularly designed to be used for holey fibres. Similarly, to the first EIM approach it identifies two regions in the cross-section of a holey fibre: the core region and the photonic crystal cladding region. The transverse components of the magnetic [195] or electric field [196,197] are represented globally by the sum of Hermite-Gaussian and cosine functions. The Hermite-Gaussian function set is selected to approximate efficiently the localised field within the core region, while the cosine functions approximate the field distribution in the cladding. In order to simplify the calculation of the overlap integrals, the refractive index distribution is similarly approximated by Hermite-Gaussian and cosine functions. One of the advantages of this technique is that it leads to a linear eigenvalue problem, which can be solved using standard numerical routines. More detail on the localised function method and its extensions can be found in Knudsen and Bjarklev [198] and Zhi et al. [199].

A widely used technique that can be used for more general MOF structures is the multipole method (MPM) [200]. In the MPM, the cross-section is divided into regions of constant refractive index, for example, for the structure from Figure 3.16d these would be the As_2Se_3 glass region and the three air hole regions. Within each of the regions the longitudinal components of the electric and magnetic field are expanded using the Hankel and Bessel functions. Then at the boundaries between the regions with different values of the refractive index the continuity of the tangential field components is imposed, which leads to a nonlinear eigenvalue problem. The values of the propagation constant, for which the coefficient matrix is singular, yield the values of the modal propagation constants. Originally the MPM was formulated for circular subregions only; however, it was later extended to structures with non-circular subregions. A technique that is fairly, closely related to the MPM is the source-model technique (SMT) [201]. As in the MPM the cross-section is divided into regions with constant refractive index. Within each region the longitudinal field components of the electric and magnetic field are approximated by the Hankel functions that give the fundamental solution to the Helmholtz equation in each region.

The Hankel functions are centred at a set of suitably distributed fictitious source points and then used to represent the tangential field distribution along the boundaries of the regions. As in the MPM, imposing the continuity of the tangential field components leads to a nonlinear eigenvalue problem. The solution of this eigenvalue problem yields the propagation constant. The third technique that is related closely to both MPM and SMT is the boundary element method (BEM) [202] or more generally integral equation method (IEM) [203,204]. The BEM is a technique that is a blend of a purely numerical method and an analytical one and can be viewed as both. It takes advantage of a very efficient quadrature formulae to perform the calculation of the line integrals appearing in the integral equations that are solved. Therefore, the BEM is much more computationally efficient than either the FEM or FDM. Another advantage of the BEM when compared with the FEM and the FDM is that it reduces a 2D problem to a 1D one by converting the surface integrals to line integrals using Green's identity. This latter characteristic, however, is as much of an advantage for the BEM as it is its weak point since it is not straightforward to find the fundamental solution for a general nonhomogeneous medium. It is therefore necessary to divide the fibre cross-section into regions with a constant value of the refractive index (as in both MPM and SMT). An integral equation is then derived that is fulfilled by the unknown components of the electromagnetic field along the region boundaries. It should be noted that the BEM is particularly efficient if the boundaries between the regions are smooth curves, that is, without any corners. By imposing the continuity conditions across boundaries and representing numerically the line integrals, BEM yields a nonlinear eigenvalue problem, which is solved for the propagation constants of the guided modes. Several variants of this technique have been proposed so far. The most distinct difference between the BEM and MPM and SMT is that both the latter techniques avoid handling the singularity of the Hankel function at the origin, while in BEM, handling the singularity is a part of the standard procedure. The BEM takes advantage of the fact that the singularities that appear in the equations are integrable and can be handled with appropriate quadratures. One of the main limitations of MPM, SMT, and BEM is that neither of these techniques can be efficiently applied if there is a continuous refractive index distribution resulting, for instance, from lanthanide doping. In such case, either the FEM or FDM would be the preferred approach. Furthermore, unlike the FEM and FDM all the three related analytical methods (MPM, SMT, and BEM) require a solution of a nonlinear eigenvalue problem, which can be quite tricky and requires a very good understanding of the technique and problem solved [205]. Most likely, this latter problem has to a large extent limited the use of MPM, SMT, and BEM and made a number of users prefer FDM and FEM solvers.

Another group of analytical methods is based on the Fourier expansion whereby the unknown fields are expanded using harmonic functions in either Cartesian [206] or cylindrical coordinate system [207]. One weak point of this approach is related to the Gibbs phenomenon, that is, the limited ability of a Fourier series to represent function discontinuities using a few terms. This problem has, however, been overcome fairly recently by applying fast Fourier factorisation. The novel Fourier series based technique proposed by Boyer et al. [208,209] compares favourably with the standard MPM.

REFERENCES

1. Sujecki, S., et al., Novel vectorial analysis of optical waveguides. *Journal of Lightwave Technology*, 1998. 16(7): p. 1329–1335.
2. Tamir, T., *Integrated Optics*. 1975, Berlin: Springer Verlag.
3. Ebeling, K.J., *Integrated Optoelectronics: Waveguide Optics, Photonics, Semiconductors*. 1993, Berlin: Springer Verlag.
4. Debnath, L. and Miskusinski, P., *Introduction to Hilbert Spaces with Applications*. 2005, London: Elsevier.
5. Kreyszig, E., *Introductory Functional Analysis*. 1978, London: Wiley.
6. Majewski, A. and S. Sujecki, Modes in rectangular fibres. *Optoelectronics Review*, 1996. 4(1/2): p. 45–50.
7. Marcatili, E.A.J., Dielectric rectangular waveguide an directional coupler for integrated optics. *Bell System Technical Journal*, 1969. 48(7): p. 2071–2102.
8. Goell, J.E., A circular harmonic computer analysis of rectangular dielectric waveguides. *Bell System Technical Journal*, 1969. 48(7): p. 2133–2160.
9. Majewski, A. and S. Sujecki, Analiza wlasciwosci propagacyjnych swiatlowodów prostokatnych (in Polish). *Prace Naukowe Politechniki Warszawskiej*, 1996. 110: p. 5–35.
10. Goell, J.E., Rib waveguide for integrated optical circuits. *Applied Optics*, 1973. 12(12): p. 2797–2798.
11. Soref, R.A., J. Schmidtchen, and K. Petermann, Large single mode rib waveguides in GeSi-Si and Si-on-SiO2. *IEEE Journal of Quantum Electronics*, 1991. 27(8): p. 1971–1974.
12. Marcatili, E.A.J., Slab coupled waveguides. *Bell System Technical Journal*, 1974. 53(4): p. 645–674.
13. Dagli, N. and C.G. Fonstad, Universal design curves for rib waveguides. *Journal of Lightwave Technology*, 1988. 6(6): p. 1136–1145.
14. Pogossian, S.P., L. Vescan, and A. Vonsovici, The single mode condition for semiconductor rib waveguides with large cross section. *Journal of Lightwave Technology*, 1998. 16(10): p. 1851–1853.
15. Powell, O., Single mode condition for silicon rib waveguides. *Journal of Lightwave Technology*, 2002. 20(10): p. 1851–1855.
16. Lousteau, J., et al., The single mode condition for silicon on insulator optical rib waveguides with large cross section. *Journal of Lightwave Technology*, 2004. 22(8): p. 1923–1929.
17. Majewski, A. and S. Sujecki, Rib lightguides and couplers. *Bull. Polish Academy of Sciences: Technical Sciences*, 1997. 45: p. 145–150.
18. Majewski, A. and S. Sujecki, Polarisation charcteristics of optical rib waveguides. *Optica Applicata*, 1998. 28(4): p. 76–82.
19. Chiang, K.S. and W.P. Wong, Rib waveguides with degenerate polarised modes. *Electronics Letters*, 1996. 32(12): p. 1098–1099.
20. Petermann, K., Properties of optical rib-guides with large cross-section. *Archiv fuer Elektrische Uebertragung*, 1976. 30(3): p. 139–140.
21. Duguay, M.A., et al., Antiresonant reflecting optical waveguides in SiO2-Si multilayer structure. *Applied Physics Letters*, 1986. 49(7): p. 13–15.
22. Gerces, I., J. Subias, and R. Alonso, Analysis of the modal solutions of rib antiresonant reflecting optical waveguides. *Journal of Lightwave Technology*, 1999. 17(9): p. 1566–1574.
23. Zmudzinski, C., et al., Three-core ARROW type diode laser: Novel high power, novel high power, single mode device, and effective master oscillator for flared antiguided MOPAs. *IEEE Journal of Selected Topics in Quantum Electronics*, 1995. 2(2): p. 129–137.

24. MacFayden, D.N., C.R. Stanley, and C.D.W. Wilkinson, Waisted-rib optical waveguides in GaAs. *Electronics Letters*, 1980. 16(11): p. 440.

25. Petermann, K., *Laser Diode Modulation and Noise*. 1991, London: Kluwer Academic Publishers.

26. Qiu, B.C., et al., Monolitic fabrication of 2x2 crosspoint switches in InGaAs-InAlGaAs multiple quantum wells. *IEEE Photonics Technology Letters*, 2001. 13(12): p. 1292–1294.

27. Polman, A., et al., European Patent 98830592.6, US Patent 09/415,022.

28. Liu, A., et al., A high-speed silicon optical modulator based on a metal-oxide-semiconductor capacitor. *Nature*, 2004. 427(6975): p. 615–618.

29. Rong, H., et al., An all-silicon Raman laser. *Nature*, 2005. 433(7023): p. 292–294.

30. Wooten, L., et al., A review of lithium niobate modulators for fiber optic communications systems. *IEEE Journal of Selected Topics in Quantum Electronics*, 2000. 6(1): p. 69–81.

31. Brinkman, R., et al., Erbium-doped single- and double-pass Ti:LiNbO$_3$ waveguide amplifiers. *IEEE Journal of Quantum Electronics*, 1994. 30(10): p. 2357–2360.

32. Suzuki, S., et al., High density integrated planar lightwave circuits using SiO2-GeO2 waveguides with high refractive index difference. *Journal of Lightwave Technology*, 1994. 12(5): p. 790–796.

33. Wang, W., et al., 40-GHz polymer electrooptic phase modulators. *IEEE Photonics Technology Letters*, 1995. 7(6): p. 638–640.

34. Sloo, L.H., et al., Optical properties of lissamine functionalized Nd^{3+} complexes in polymer waveguides and solution. *Optical Materials*, 2000. 14(2): p. 101–107.

35. Hunsperger, R.G., *Integrated Optics*. 2002, Berlin: Springer Verlag.

36. Maerz, R., *Integrated Optics: Design and Modelling*. 1995, London: Artech House.

37. Petermann, K., et al., Integrated Optical Waveguides in Silicon. *Archiv fuer Elektrische Uebertragung (AEU)*, 1991. 45(5): p. 273–278.

38. Fishbeck, G., et al., Singlemode optical waveguides using a high temperature stable polymer with low losses in the 1.55 mm range. *Electronics Letters*, 1997. 33(518–519).

39. Yan, Y.C., et al., Erbium-doped phosphate glass waveguide on silicon with 4.1 dB/cm gain at 1.535 mm. *Applied Physics Letters*, 1997. 71(20): p. 2922–2924.

40. Marcuse, D., *Light Transmission Optics*. 1972, New York: van Nostrand Reinhold.

41. Marcuse, D., *Theory of Dielectric Optical Waveguides*. 1974, San Diego: Academic Press.

42. Rozzi, T. and M. Mongiardo, *Open Electromagnetic Waveguides*. 1997, London: IEE.

43. Anemogiannis, E. and E.N. Glytsis, Multilayer wave-guides - efficient numerical-analysis of general structures. *Journal of Lightwave Technology*, 1992. 10(10): p. 1344–1351.

44. Majewski, A., Numerical analysis of the multistep-index fiber having the uniaxial anisotropy. *Bulletin of the Polish Academy of Sciences, Technical Sciences*, 1984. 32(11–12): p. 709–714.

45. Ding, H. and K.T. Chan, Solving planar dielectric waveguide equations by counting the number of guided modes. *IEEE Photonics Technology Letters*, 1997. 9(2): p. 215–217.

46. Anemogiannis, E., E.N. Glytsis, and T.K. Gaylord, Efficient solution of eigenvalue equations of optical wave-guiding structures. *Journal of Lightwave Technology*, 1994. 12(12): p. 2080–2084.

47. Anemogiannis, E., E.N. Glytsis, and T.K. Gaylord, Determination of guided and leaky modes in lossless and lossy planar multilayer optical waveguides: Reflection pole method and wavevector density method. *Journal of Lightwave Technology*, 1999. 17(5): p. 929–941.

48. Sujecki, S., Arbitrary truncation order three-point finite difference method for optical waveguides with stepwise refractive index discontinuities. *Optics Letters*, 2010. 35(24): p. 4115–4117.

49. Smith, G.D., *Numerical Solution of Partial Differential Equations: Finite Difference Method*. 1988, Oxford: Oxford University Press.

50. Press, W.H., Teukolsky, S.A., Vetterling, W.T., and Flannery, B.P., *Numerical Recipes in C++: The Art of Scientific Computing*. 2002, Cambridge: Cambridge University Press.

51. Vassallo, C., Improvement of finite difference methods for step-index optical wave-guides. *IEE Proceeding J*, 1992. 139(2): p. 137–142.

52. Stern, M.S., Semivectorial polarised H field solutions for dielectric waveguides with arbitrary index profiles. *IEE Proceeding J*, 1988. 135(5): p. 333–338.

53. Chiou, Y.-P. and C.-H. Du, Arbitrary-order interface conditions for slab structures and their applications in waveguide analysis. *Optics Express*, 2010. 18(5): p. 4088–4102.

54. Sujecki, S., Accuracy of three-point finite difference approximations for optical wave-guides with step-wise refractive index discontinuities. *Opto-Electronics Review*, 2011. 19(2): p. 145–150.

55. Snyder, A.W., Love, J.D., *Optical Waveguide Theory*. 1983, London: Chapman and Hall.

56. Lee, S.L., et al., On leaky mode-approximations for modal expansion in multilayer open wave-guides. *IEEE Journal of Quantum Electronics*, 1995. 31(10): p. 1790–1802.

57. Hu, J. and C.R. Menyuk, Understanding leaky modes: Slab waveguide revisited. *Advances in Optics and Photonics*, 2009. 1(1): p. 58–106.

58. Jablonski, T.F., Complex-modes in open lossless dielectric wave-guides. *Journal of the Optical Society of America a-Optics Image Science and Vision*, 1994. 11(4): p. 1272–1282.

59. Yevick, D., et al., Stability issues in vector electric-field propagation. *IEEE Photonics Technology Letters*, 1995. 7(6): p. 658–660.

60. Xie, H., W.T. Lu, and Y.Y. Lu, Complex modes and instability of full-vectorial beam propagation methods. *Optics Letters*, 2011. 36(13): p. 2474–2476.

61. Deng, H. and D. Yevick, The nonunitarity of finite-element beam propagation algorithms. *IEEE Photonics Technology Letters*, 2005. 17(7): p. 1429–1431.

62. Ramaswamy, W., Strip-loaded film waveguides. *Bell System Technical Journal*, 1974. 53(4): p. 697–704.

63. Benson, T.M., R.J. Bozeat, and P.C. Kendall, Rigorous effective index method for semiconductor rib wave-guides. *IEE Proceedings-J Optoelectronics*, 1992. 139(1): p. 67–70.

64. Chiang, K.S., Effective-index method for the analysis of optical wave-guide couplers and arrays - an asymptotic theory. *Journal of Lightwave Technology*, 1991. 9(1): p. 62–72.

65. Chiang, K.S., Analysis of rectangular dielectric wave-guides - effective-index method with built-in perturbation. *Electronics Letters*, 1992. 28(4): p. 388–390.

66. Chiang, K.S., K.M. Lo, and K.S. Kwok, Effective-index method with built-in perturbation correction for integrated optical waveguides. *Journal of Lightwave Technology*, 1996. 14(2): p. 223–228.

67. Robertson, M.J., et al., The weighted index method - a new technique for analyzing planar optical-waveguides. *Journal of Lightwave Technology*, 1989. 7(12): p. 2105–2111.

68. Vassallo, C. and Y.H. Wang, A new semirigorous analysis of rib waveguides. *Journal of Lightwave Technology*, 1990. 8(1): p. 56–65.

69. McIlroy, P.W.A., M.S. Stern, and P.C. Kendall, Spectral index method for polarized modes in semiconductor rib waveguides. *Journal of Lightwave Technology*, 1990. 8(1): p. 113–117.

70. Robson, P.N. and P.C. Kendall, *Rib Waveguide Theory by the Spectral Index Method*. 1990, London: Research Studies Press.

71. Sudbo, A.S., Numerically stable formulation of the transverse resonance method for vector mode-field calculations in dielectric waveguides. *IEEE Photonics Technology Letters*, 1993. 5(3): p. 342–344.

72. Peng, S.T. and A.A. Oliner, Guidance and leakage properties of a class of open dielectric waveguides: Part I—mathematical formulation. *IEEE Transactions on Microwave Theory and Techniques*, 1981. 29(9): p. 843–855.

73. Sudbo, A.S., Film mode matching: a versatile numerical method for vector mode field calculations in dielectric waveguides. *Pure and Applied Optics*, 1993. 2(3): p. 211–233.

74. Rozzi, T., et al., Variational analysis of the dielectric rib waveguide using the concept of 'transition function' and including edge singularities. *IEEE Transactions on Microwave Theory and Techniques*, 1991. 39(2): p. 247–257.

75. Dagli, N. and C.G. Fonstad, Analysis of rib dielectric waveguides. *IEEE Journal of Quantum Electronics*, 1985. 21(4): p. 315–321.

76. Koshiba, M. and M. Suzuki, Vectorial wave analysis of dielectric waveguides for optical-integrated circuits using equivalent network approach. *Journal of Lightwave Technology*, 1986. 4(6): p. 656–664.

77. Payne, F.P., A new theory of rectangular optical waveguides. *Optical and Quantum Electronics*, 1982. 14(6): p. 525–537.

78. Berry, G.M., et al., Analysis of multilayered dielectric waveguides: Variational treatment. *Electronics Letters*, 1994. 30(24): p. 2029–2030.

79. M. Reed, et al., Free space radiation mode analysis of rectangular dielectric waveguides. *Optical and Quantum Electronics*, 1996. 28(9): p. 1175–1179.

80. McIlroy, P.W.A., M.S. Stern, and P.C. Kendall, Spectral index method for polarised modes in semiconductor rib waveguides. *Journal of Lightwave Technology*, 1990. 8(1): p. 113–117.

81. Smartt, C.J., T.M. Benson, and P.C. Kendall, Exact transcendental equation for scalar modes of rectangular dielectric waveguides. *Optical and Quantum Electronics*, 1994. 26(6): p. 641–644.

82. Kendal, P.C., M.S. Stern, and P.N. Robson, Analysis of discrete spectral index method for rib waveguides. *Optical and Quantum Electronics*, 1990. 22(6): p. 555–560.

83. Vassallo, C. and Y.H. Wang, A new semirigorous analysis of rib waveguides. *Journal of Lightwave Technology*, 1990. 9(1): p. 56–65.

84. Cullen, A.L., O. Ozkan, and L.A. Jackson, Point-matching technique for rectangular-cross-section dielectric rod. *Electronics Letters*, 1971. 7(17): p. 497–499.

85. Bienstman, P., Two-stage mode finder for waveguides with a 2D cross-section. *Optical and Quantum Electronics*, 2004. 36(1–3): p. 5–14.

86. Henry, C.H. and B.H. Verbeek, Solution of the scalar wave equation for arbitrarily shaped dielectric waveguides by two-dimensional Fourier analysis. *Journal of Lightwave Technology*, 1989. 7(2): p. 308–313.

87. Hewlett, S.J. and F. Ladouceur, Fourier decomposition method applied to mapped infinite domains: scalar analysis of dielectric waveguides down to modal cutoff. *Journal of Lightwave Technology*, 1995. 13(3): p. 375–383.

88. Smartt, C.J., T.M. Benson, and P.C. Kendall. Exact operator method for the analysis of dielectric waveguides with application to integrated optics devices and laser facets. In *IEE Second International Conference on Computation in Electromagnetics*. 1995.

89. Berry, G.M., et al., Exact and variational Fourier-transform methods for analysis of multilayered planar wave-guides. *Iee Proceedings-Optoelectronics*, 1995. 142(1): p. 66–75.

90. Smartt, C.J., et al., Exact polarised rib waveguide analysis. *Electronics Letters*, 1994. 30(7): p. 1127–1128.

91. Hugonin, J.P., et al., Fourier modal methods for modeling optical dielectric waveguides. *Optical and Quantum Electronics*, 2005. 37(1–3): p. 107–119.

92. Kim, C.M. and R.V. Ramaswamy, Modelling of graded-index channel waveguides using nonuniform finite difference method. *IEEE Journal of Lightwave Technology*, 1989. 7(19): p. 1581–1589.

93. Schultz, N., et al., Finite difference method without spurious solutions for the hybrid-mode analysis of diffused channel waveguides. *IEEE Transactions on Microwave Theory and Techniques*, 1990. 38(6): p. 722–729.

94. Seki, S., T. Yamanaka, and K. Yokoyama, Two-dimensional analysis of optical waveguides with non-uniform finite difference method. *IEE Proceeding J*, 1991. 138(2): p. 123–127.

95. Nehrbass, J.W. and R. Lee, Optimal finite-difference sub-gridding techniques applied to the Helmholtz equation. *IEEE Transactions on Microwave Theory and Techniques*, 2000. 48(6): p. 976–984.

96. Chiou, Y.P., Y.C. Chiang, and H.C. Chang, Improved three-point formulas considering the interface conditions in the finite-difference analysis of step-index optical devices. *Journal of Lightwave Technology*, 2000. 18(2): p. 243–251.

97. Liu, P.-L. and B.J. Li, Full-vectorial mode analysis of rib waveguides by iterative Lanczos reduction. *IEEE Journal of Quantum Electronics*, 1993. 29(12): p. 2859–2863.

98. Wijnands, F., et al., Efficient semivectorial mode solvers. *Ieee Journal of Quantum Electronics*, 1997. 33(3): p. 367–374.

99. Ramm, K., P. Lusse, and H.G. Unger, Multigrid eigenvalue solver for mode calculation of planar optical waveguides. *Ieee Photonics Technology Letters*, 1997. 9(7): p. 967–969.

100. Matin, M.A., et al., New technique for finite difference analysis of optical waveguide problems. *International Journal of Numerical Modelling*, 1994. 7(1): p. 25–33.

101. Benson, T.M., et al., Highly accurate vector correction for optical waveguide propagation constants. *IEE Proceedings-J*, 1993. 140(2): p. 93–97.

102. Xu, C.L., et al., Full-vectorial mode calculations by finite difference method. *IEE Proceeding J*, 1994. 141(5): p. 281–286.

103. Stern, M.S., Semivectorial polarised finite difference method for optical dielectric waveguides with arbitrary index profiles. *IEE Proceeding J*, 1988. 135(1): p. 56–63.

104. Gallick, A.T., T. Kerkhoven, and U. Ravaioli, Iterative solution of the eigenvalue problem for a dielectirc waveguide. *IEEE Transactions on Microwave Theory and Techniques*, 1992. 40(2): p. 699–705.

105. Hadley, G.R., Low truncation error finite difference equations for photonics simulation I: beam propagation. *Journal of Lightwave Technology*, 1998. 16(1): p. 134–141.

106. Hadley, G.R., Low truncation error finite difference equations for photonics simulation II: vertical cavity surface emitting lasers. *Journal of Lightwave Technology*, 1998. 16(1): p. 142–151.

107. Hadley, G.R., High accuracy finite difference equations for dielectric waveguide analysis I: uniform regions and dielectric interfaces. *Journal of Lightwave Technology*, 2002. 20(7): p. 1210–1218.

108. Hadley, G.R., High accuracy finite difference equations for dielectric waveguide analysis II: dielectric corners. *Journal of Lightwave Technology*, 2002. 20(7): p. 1219–1231.

109. Luesse, P., et al., Analysis of vectorial mode fields in optical waveguides by a new finite difference method. *Journal of Lightwave Technology*, 1994. 12(3): p. 487–494.

110. Decoctignie, J.D., O. Parriaux, and F. Gardiol, Birefringence properties of twin-core fibres by finite differences. *Journal of Optical Communications and Networking*, 1982. 3(1): p. 8–12.

111. Chiang, Y.C., Y.P. Chiou, and H.C. Chang, Improved full-vectorial finite-difference mode solver for optical waveguides with step-index profiles. *Journal of Lightwave Technology*, 2002. 20(8): p. 1609–1618.

112. Chiou, Y.-P. and C.-H. Du, Arbitrary-Order Full-Vectorial Interface Conditions and Higher Order Finite-Difference Analysis of Optical Waveguides. *Journal of Lightwave Technology*, 2011. 29(22): p. 3445–3452.

113. Xia, J. and J. Yu, New finite-difference scheme for simulations of step-index waveguides with tilt interfaces. *IEEE Photonics Technology Letters*, 2003. 15(9): p. 1237–1239.

114. Ando, T., et al., Eigenmode analysis of optical waveguides by a Yee-mesh-based imaginary-distance propagation method for an arbitrary dielectric interface. *Journal of Lightwave Technology*, 2002. 20(8): p. 1627–1634.

115. Chiou, Y.-P., et al., Finite-Difference Modeling of Dielectric Waveguides With Corners and Slanted Facets. *Journal of Lightwave Technology*, 2009. 27(12): p. 2077–2086.

116. Benson, T.M., R.J. Bozeat, and P.C. Kendall, Complex finite difference method applied to the analysis of semiconductor lasers. *IEE Proceeding J*, 1994. 141(2): p. 97–101.

117. Schweig, E. and W.B. Bridges, Computer-analysis of dielectric waveguides - a finite-difference method. *IEEE Transactions on Microwave Theory and Techniques*, 1984. 32(5): p. 531–541.

118. Patrick, S.S. and K.J. Webb, A variational finite difference analysis for dielectric waveguides. *IEEE Transactions on Microwave Theory and Techniques*, 1992. 40(4): p. 692–698.

119. Lagu, R.K. and R.V. Ramaswamy, A variational finite-difference method for analysing channel waveguides with arbitrary index profiles. *IEEE Journal of Quantum Electronics*, 1986. 22(6): p. 968–976.

120. Schultz, U. and R. Pregla, A new technique for the analysis of the dispersion characteristics of planar waveguides. *Archiv fuer Elektrische Uebertragung*, 1980. 34(4): p. 169–173.

121. Diestel, H., A method for calculating the guided modes of strip-loaded optical waveguides with arbitrary index profile. *IEEE Journal of Quantum Electronics*, 1984. 20(11): p. 1288–1293.

122. Hoekstra, H.J.W.M., An economic method for the solution of the scalar wave equation for arbitrary shaped optical waveguides. *Journal of Lightwave Technology*, 1990. 8(5): p. 789–793.

123. Pregla, R., *Analysis of Electromagnetic Fields and Waves: The Method of Lines*. 2008, Chichester: Wiley.

124. Obaya, S.S.A., B.M.A. Rahman, and H.A. El-Mikati, New full vectorial numerically efficient propagation algorithm based on the finite element method. *Journal of Lightwave Technology*, 2000. 18(3): p. 409–415.

125. Mabay, N., P.E. Lagasse, and P. Vandenblucke, Finite element analysis of optical waveguides. *IEEE Transactions on Microwave Theory and Techniques*, 1981. 29(6): p. 600–605.

126. Lee, J.F., D.K. Sun, and Z.J. Cendes, Full wave analysis of dielectric waveguides using tangential vector finite elements. *IEEE Transactions on Microwave Theory and Techniques*, 1991. 39(8): p. 1262–1271.

127. Koshiba, M. and K. Inoue, Simple and efficient finite element analysis of microwave and optical waveguides. *IEEE Transactions on Microwave Theory and Techniques*, 1992. 40(2): p. 371–377.

128. Hayata, K., et al., Vectorial finite element method without any spurious solutions for dielectric waveguideing problems using transverse magnetic field components. *IEEE Transactions on Microwave Theory and Techniques*, 1986. 34(11): p. 1120–1123.

129. Koshiba, M., K. Hayata, and M. Suzuki, Improved finite element formulation in terms of the magnetic field vector for dielectric waveguides. *IEEE Transactions on Microwave Theory and Techniques*, 1985. 33(3): p. 227–233.

130. Silveira, M. and A. Gophinat, Analysis of dielectric guides by transverse magnetic finite element penalty method. *IEEE Journal of Lightwave Technology*, 1995. 13(3): p. 442–446.

131. Fernandez, F.A., et al., Sparse matrix eigenvalue solver for finite element solution of dielectric waveguides. *Electronics Letters*, 1991. 27(20): p. 1824–1826.

132. Saad, Y., Numerical methods for large eigenvalue problems. 2011, SIAM: http://www-users.cs.umn.edu/~saad/eig_book_2ndEd.pdf.

133. Li, D.U. and H.C. Chung, An efficient full vectorial finite element modal analysis of dielectric waveguides incorporating inhomogenous elements across dielectric discontinuities. *IEEE Journal of Quantum Electronics*, 2000. 36(11): p. 1251–1261.

134. Themistos, C., B.M.A. Rahman, and K.T.V. Grattan, Finite element analysis for lossy optical waveguides using perturbation techniques. *IEEE Photonics Technology Letters*, 1994. 6(4): p. 537–539.
135. Koshiba, M., K. Hayata, and M. Suzuki, Approximate scalar finite element analysis of anisotropic optical waveguides with off-diagonal elements in permittivity tensor. *IEEE Transactions on Microwave Theory and Techniques*, 1984. 32(6): p. 587–593.
136. Wang, X.H., *Finite Element Methods for Nonlinear Optical Waveguides*. 1995, Amsterdam: Gordon and Breach Science Publishers.
137. Fernandez, F.A. and Y. Lu, *Microwave and Optical Waveguide Analysis by the Finite Element Method*. 1996, New York: John Wiley & Sons Inc.
138. Koshiba, M., *Optical Waveguide Analysis*. 1990, New York: McGraw-Hill, Inc.
139. Rahman, B.M.A. and A. Agrawal, *Finite Element Modeling Methods for Photonics*. 2013, Norwood: Artech House.
140. Lu, T. and D. Yevick, A vectorial boundary element method analysis of integrated optical waveguides. *Journal of Lightwave Technology*, 2003. 21(8): p. 1793–1807.
141. Lu, T. and D. Yevick, Comparative evaluation of a novel series approximation for electromagnetic fields at dielectric corners with boundary element method applications. *Journal of Lightwave Technology*, 2004. 22(5): p. 1426–1432.
142. Lu, W. and Y.Y. Lu, Waveguide mode solver based on Neumann-to-Dirichlet operators and boundary integral equations. *Journal of Computational Physics*, 2012. 231(4): p. 1360–1371.
143. Yu, B., J. Wang, and X. Sun, A Bi-boundary FEM-BEM approach for open structure optical waveguide problems. *Journal of Lightwave Technology*, 2009. 27(14): p. 2765–2770.
144. Boriskina, S.V., et al., Highly efficient full-vectorial integral equation solution for the bound, leaky, and complex modes of dielectric waveguides. *IEEE Journal of Selected Topics in Quantum Electronics*, 2002. 8(6): p. 1225–1232.
145. Sudbo, A.S., Why are accurate computations of mode fields in rectangular dielectric wave-guides difficult. *Journal of Lightwave Technology*, 1992. 10(4): p. 418–419.
146. Hadley, G.R., High-accuracy finite-difference equations for dielectric waveguide analysis II: Dielectric corners. *Journal of Lightwave Technology*, 2002. 20(7): p. 1219–1231.
147. Thomas, N., P. Sewell, and T.M. Benson, A new full-vectorial higher order finite-difference scheme for the modal-analysis of rectangular dielectric waveguides. *Journal of Lightwave Technology*, 2007. 25(9): p. 2563–2570.
148. Lui, W.W., et al., Full vectorial mode analysis with considerations of field singularities at corners of optical waveguides. *Journal of Lightwave Technology*, 1999. 17(8): p. 1509–1513.
149. Vassallo, C., Mode solvers 1993–1995 optical mode solvers. *Optical and Quantum Electronics*, 1997. 29(2): p. 95–114.
150. Hondros, D. and P. Debye, Electromagnetic waves in dielectrical wires. *Annalen Der Physik*, 1910. 32(8): p. 465–476.
151. Okoshi, T., *Optical Fibres*. 1982, London: Academic Press.
152. Majewski, A., *Theory and Design of optical fibres (in Polish)*. 1991, Warsaw: Wydawnictwo Naukowo-Techniczne.
153. Okamoto, K., *Fundamentals of Optical Waveguides*. 2006, London: Elsevier.
154. Black, R.J. and Gagnon, L., *Optical Waveguide Modes: Polarization, Coupling and Symmetry*. 2010, London: McGraw-Hill.
155. Yeh, C., Elliptical dielectric waveguides. *Journal of Applied Physics*, 1962. 33(11): p. 3235.
156. Dyott, R.B., *Elliptical Fiber Waveguides*. 1996, Boston: Artech House.
157. Majewski, A. and H.G. Unger, Single-mode w-fibers with profile-threshold for low dispersion over a wide spectral range. *Aeu-Archiv Fur Elektronik Und Ubertragungstechnik-International Journal of Electronics and Communications*, 1983. 37(9–10): p. 336–338.

158. Temelkuran, B., et al., Wavelength-scalable hollow optical fibres with large photonic bandgaps for $CO(2)$ laser transmission. *Nature*, 2002. 420(6916): p. 650–653.
159. Ibanescu, M., et al., An all-dielectric coaxial waveguide. *Science*, 2000. 289(5478): p. 415–419.
160. Noda, J., K. Okamoto, and Y. Sasaki, Polarization-maintaining fibers and their applications. *Journal of Lightwave Technology*, 1986. 4(8): p. 1071–1089.
161. Birks, T.A., J.C. Knight, and P.S. Russell, Endlessly single-mode photonic crystal fiber. *Optics Letters*, 1997. 22(13): p. 961–963.
162. Bjarklev, A., Bjarklev, A., and Broeng, J., *Photonic Crystal Fibres*. 2003, London: Kluwer Academic Publishers.
163. Poli, F., Cucinotta, A., and Selleri, S., *Photonic Crystal Fibres*. 2007: Springer.
164. Zolla, F., Reneversez, G., Nicolet, A., Kuhlmev, B., Guenneau, S., Felbacq, D., *Foundations of Photonic Crystal Fibres*. 2005, London: Imperial College Press.
165. Lin, A., et al., Solid-core tellurite glass fiber for infrared and nonlinear applications. *Optics Express*, 2009. 17(19): p. 16716–16721.
166. L. Sojka, Z.T., N.C. Neate, D. Furniss, S. Sujecki, T.M. Benson, and A.B. Seddon, Broadband, mid-ifrared emission from r3+ doped GeAsGaSe chalcogenide fiber, optically clad. *Optics Material*, 2014. 36(6): p. 1076–1082.
167. Shalem, S., et al., Silver halide single-mode fibers for the middle infrared. *Applied Physics Letters*, 2005. 87(9).
168. Ishigure, T., et al., Formation of the refractive index profile in the graded index polymer optical fiber for gigabit data transmission. *Journal of Lightwave Technology*, 1997. 15(11): p. 2095–2100.
169. Richardson, K., Krol, D., and Hirao, K., Glasses for photonic applications. *International Journal of Applied Glass Science*, 2010. 1(1): p. 74–86.
170. Harrington, J.A., *Infrared Fibres and Their Applications*. 2004, Bellingham: SPIE.
171. El-Amraoui, M., et al., Strong infrared spectral broadening in low-loss As-S chalcogenide suspended core microstructured optical fibers. *Optics Express*, 2010. 18(5): p. 4547–4556.
172. Lian, Z.G., et al., Solid Microstructured Chalcogenide Glass Optical Fibres for the Near- and Mid-Infrared Spectral Regions. *IEEE Photonics Technology Letters*, 2009. 21(24): p. 1804–1806.
173. Limpert, J., et al., Extended single-mode photonic crystal fiber lasers. *Optics Express*, 2006. 14(7): p. 2715–2720.
174. Du, C.-H. and Y.-P. Chiou, Higher-Order Full-Vectorial Finite-Difference Analysis of Waveguiding Structures With Circular Symmetry. *IEEE Photonics Technology Letters*, 2012. 24(11): p. 894–896.
175. Lu, Y.-C., et al., Improved full-vector finite-difference complex mode solver for optical waveguides of circular symmetry. *Journal of Lightwave Technology*, 2008. 26(13–16): p. 1868–1876.
176. Yariv, A., Optical electronics in modern communications. 1997, New York: Oxford University Press.
177. Abramowitz, A. and I.A. Stegun, *Handbook of Mathematical Functions: with Formulas, Graphs, and Mathematical Tables*. 1970, New York: Dover Publications.
178. Gloge, D., Weakly guiding fibers. *Applied Optics*, 1971. 10(10): p. 2252–2258.
179. Riishede, J., N.A. Mortensen, and J. Laegsgaard, A 'poor man's approach' to modelling micro-structured optical fibres. *Journal of Optics a-Pure and Applied Optics*, 2003. 5(5): p. 534–538.
180. Pozirikidis, C., *Numerical Computation in Science and Engineering*. 1998, Oxford: Oxford University Press.
181. Torres, P., V.H. Aristizabal, and M.V. Andres, Modeling of photonic crystal fibers from the scalar wave equation with a purely transverse linearly polarized vector potential. *Journal of the Optical Society of America B-Optical Physics*, 2011. 28(4): p. 787–791.

182. Obayya, S.S.A., B.M.A. Rahman, and K.T.V. Grattan, Accurate finite element modal solution of photonic crystal fibres. *Iee Proceedings-Optoelectronics*, 2005. 152(5): p. 241–246.

183. Koshiba, M., Full-vector analysis of photonic crystal fibers using the finite element method. *Ieice Transactions on Electronics*, 2002. E85C(4): p. 881–888.

184. Saitoh, K. and M. Koshiba, Numerical modeling of photonic crystal fibers. *Journal of Lightwave Technology*, 2005. 23(11): p. 3580–3590.

185. De Francisco, C.A., B.V. Borges, and M.A. Romero, A semivectorial method for the modeling of photonic crystal fibers. *Microwave and Optical Technology Letters*, 2003. 38(5): p. 418–421.

186. Fogli, F., et al., Full vectorial BPM modeling of index-guiding photonic crystal fibers and couplers. *Optics Express*, 2002. 10(1): p. 54–59.

187. Zhu, Z.M. and T.G. Brown, Full-vectorial finite-difference analysis of microstructured optical fibers. *Optics Express*, 2002. 10(17): p. 853–864.

188. Chiou, Y.P. and C.H. Du, Arbitrary-order full-vectorial interface conditions and higher order finite-difference analysis of optical waveguides. *Journal of Lightwave Technology*, 2011. 29(22): p. 3445–3452.

189. Saitoh, K. and M. Koshiba, Full-vectorial imaginary-distance beam propagation method based on a finite element scheme: Application to photonic crystal fibers. *IEEE Journal of Quantum Electronics*, 2002. 38(7): p. 927–933.

190. Dangui, V., M.J.F. Digonnet, and G.S. Kino, A fast and accurate numerical tool to model the modal properties of photonic-bandgap fibers. *Optics Express*, 2006. 14(7): p. 2979–2993.

191. Mortensen, N.A., et al., Modal cutoff and the V parameter in photonic crystal fibres. *Optics Letters*, 2003. 28(20): p. 1879–1881.

192. Koshiba, M. and K. Saitoh, Applicability of classical optical fiber theories to holey fibers. *Optics Letters*, 2004. 29(15): p. 1739–1741.

193. Li, Y.F., C.Y. Wang, and M.L. Hu, A fully vectorial effective index method for photonic crystal fibers: Application to dispersion calculation. *Optics Communications*, 2004. 238(1–3): p. 29–33.

194. Rastogi, V. and K.S. Chiang, Holey optical fiber with circularly distributed holes analyzed by radial effective-index method. *Optics Letters*, 2003. 28(24): p. 2449–2451.

195. Mogilevtsev, D., T.A. Birks, and P.S. Russell, Localized function method for modeling defect modes in 2-D photonic crystals. *Journal of Lightwave Technology*, 1999. 17(11): p. 2078–2081.

196. Monro, T.M., et al., Holey optical fibres: an efficient modal model. *Journal of Lightwave Technology*, 1999. 17(6): p. 1093–1102.

197. Monro, T.M., et al., Modelling large air fraction holey optical fibres. *Journal of Lightwave Technology*, 2000. 18(1): p. 50–56.

198. Knudsen, E. and A. Bjarklev, Modelling photonic crystal fibres with Hermite-Gaussian functions. *Optics Communications*, 2003. 222(1–6): p. 155–160.

199. Zhi, W., G.B. Ren, and S.Q. Lou, A novel supercell overlapping method for different photonic crystal fibers. *Journal of Lightwave Technology*, 2004. 22(3): p. 903–916.

200. White, T.P., et al., Multipole method for microstructured optical fibers. I. Formulation. *Journal of the Optical Society of America B-Optical Physics*, 2002. 19(10): p. 2322–2330.

201. Hochman, A. and Y. Leviatan, Analysis of strictly bound modes in photonic crystal fibers by use of a source-model technique. *Journal of the Optical Society of America a-Optics Image Science and Vision*, 2004. 21(6): p. 1073–1081.

202. Guan, N., et al., Boundary element method for analysis of holey optical fibers. *Journal of Lightwave Technology*, 2003. 21(8): p. 1787–1792.

203. Sotsky, A.B. and L.I. Sotskaya, Method of integral equations in the theory of microstructured optical fibers. *Technical Physics*, 2004. 49(2): p. 174–182.

204. Cheng, H., et al., Fast, accurate integral equation methods for the analysis of photonic crystal fibers - I: Theory. *Optics Express*, 2004. 12(16): p. 3791–3805.
205. Hochman, A. and Y. Leviatan, Efficient and spurious-free integral-equation-based optical waveguide mode solver. *Optics Express*, 2007. 15(22): p. 14431–14453.
206. Szpulak, M., et al., Comparison of different methods for rigorous modeling of photonic crystal fibers. *Optics Express*, 2006. 14(12): p. 5699–5714.
207. Issa, N.A. and L. Poladian, Vector wave expansion method for leaky modes of microstructured optical fibers. *Journal of Lightwave Technology*, 2003. 21(4): p. 1005–1012.
208. Boyer, P., et al., A new differential method applied to the study of arbitrary cross section microstructured optical fibers. *Optical and Quantum Electronics*, 2006. 38(1–3): p. 217–230.
209. Boyer, P., et al., Improved differential method for microstructured optical fibres. *Journal of Optics a-Pure and Applied Optics*, 2007. 9(7): p. 728–740.

4 Beam Propagation Method

Photonic devices operate on the phenomenon of light guiding along optical wave-guides. Optical waveguides are used in laser diodes, directional couplers, fibre lasers, amplifiers, and so on [1–3]. For the study of light propagation in the waveguid-ing structures that are used in photonic devices, the numerical techniques that were discussed in Chapter 3 are often insufficient. This is because the waveguiding structures used in photonic devices are neither perfectly straight nor infinitely long. For instance, the typical length of an edge emitting laser diode is around several hundreds of micrometres while in the case of a fibre laser the length of the optical waveguide can equal several metres. Furthermore, an optical beam whilst propa-gating along an optical waveguide can encounter abrupt air–dielectric interfaces. For example, cleaved semiconductor mirrors are commonly used in laser diodes to provide a feedback mechanism for the generation of the optical radiation.

Generally, in dielectric media, Maxwell equations can be solved numerically, using, for instance, either finite differences or finite elements [4]. These techniques calculate the field distribution in the considered domain subject to a distribution of sources and imposed boundary conditions. However, when applied to the opti-cal waveguiding structures that are used in photonic devices, they suffer from low computational efficiency because the dimensions of optical waveguiding structures are often very large when compared with the wavelength of the guided light. Hence, the matrices storing all field sampling points within the computational domain have to be very large to allow for sufficient accuracy. For instance, 20 samples per wave-length for an optical waveguide that is 10,000 wavelengths long (a typical length of a high power edge emitting laser diode) would result in 200,000 samples just for a one-dimensional analysis. Furthermore, optical waveguides, when compared with the operating wavelength, can have large transverse dimensions whilst incor-porating nanometre-scale features in the transverse cross-section. For instance, the optical waveguide of a broad area laser operating around 1 μm wavelength can be 200 μm wide, while its quantum well can be 10 nm thick. Therefore, many numerical techniques specially tailored for studying the light propagation in the waveguiding structures of photonic devices have been developed during the last three decades. In this context, the beam propagation method (BPM) proved to be particularly useful. Unlike the direct Maxwell equations solution methods, BPM allows calculating the properties of the output beam subject to the input beam shape (Figure 4.1), without the necessity of storing the intermediate field samples.

The central assumption underpinning the BPM is that within the device a pre-ferred direction of optical field propagation can be identified. The preferred direc-tion is defined as the direction along which the majority of the optical power is transported, while the power flow in the direction that is perpendicular to the preferred one is negligibly small (Figure 4.2). Once the preferred direction is selected, the field can be decomposed into backward and forward travelling waves with respect to the preferred direction. Such an approach has been found applicable to a large class of photonic devices and hence the BPM plays an important role in photonics as a design tool.

The theory of the BPM will be presented in a systematic fashion in this chapter. The hierarchy of approximations applicable to various waveguiding structures is derived and presented here along with the techniques used to handle the square root and exponential operators. In the first section, an introduction to the theory of BPM is given. The second section discusses the unidirectional BPM algorithms. In the third section, the bidirectional BPM is presented. The fourth section gives the discussion of various aspects of the BPM numerical implementation, while the fifth one discusses the application of the BPM to the study of optical propagation in selected waveguiding structures. For the sake of completeness, in the last section, we discuss briefly the time domain BPM and the travelling wave approximation.

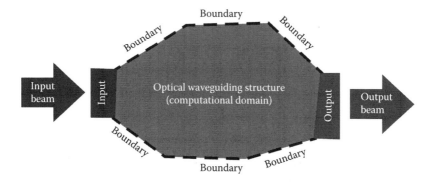

FIGURE 4.1   Schematic representation of the planar optical waveguiding structure.

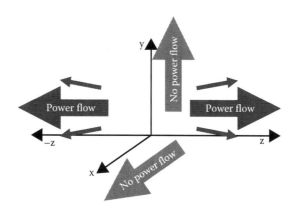

FIGURE 4.2   The power flow in a device modelled by directional decomposition method.

INTRODUCTION

In this section, we show how the optical field can be formally decomposed into forward and backward propagating waves, and we will derive an equation that forms the basis for the derivation of BPM algorithms.

In the following derivations, we assume that the z-axis (Figure 4.3) can be aligned with the preferred direction along which most of the energy is transported, while the energy flow in the transverse direction can be neglected. In the rectangular coordinate system (Figure 4.3), one obtains from 3.5a the following set of three equations:

$$\frac{\partial^2 E_x}{\partial y^2} + \frac{\partial}{\partial x}\left(\frac{1}{\varepsilon}\frac{\partial(\varepsilon E_x)}{\partial x}\right) + \omega^2 \mu_0 \varepsilon E_x + \frac{\partial^2 E_x}{\partial z^2} + \frac{\partial}{\partial x}\left(\frac{1}{\varepsilon}\frac{\partial(\varepsilon E_y)}{\partial y}\right) - \frac{\partial^2 E_y}{\partial x \partial y}$$

$$+ \frac{\partial}{\partial x}\left(\frac{1}{\varepsilon}E_z\frac{\partial \varepsilon}{\partial z}\right) = 0 \tag{4.1a}$$

$$\frac{\partial^2 E_y}{\partial x^2} + \frac{\partial}{\partial y}\left(\frac{1}{\varepsilon}\frac{\partial(\varepsilon E_y)}{\partial y}\right) + \omega^2 \mu_0 \varepsilon E_y + \frac{\partial^2 E_y}{\partial z^2} + \frac{\partial}{\partial y}\left(\frac{1}{\varepsilon}\frac{\partial(\varepsilon E_x)}{\partial x}\right) - \frac{\partial^2 E_x}{\partial y \partial x}$$

$$+ \frac{\partial}{\partial y}\left(\frac{1}{\varepsilon}E_z\frac{\partial \varepsilon}{\partial z}\right) = 0 \tag{4.1b}$$

$$\frac{\partial^2 E_z}{\partial x^2} + \frac{\partial^2 E_z}{\partial y^2} + \omega^2 \mu_0 \varepsilon E_z + \frac{\partial}{\partial z}\left(\frac{1}{\varepsilon}\frac{\partial(\varepsilon E_z)}{\partial z}\right) + \frac{\partial}{\partial z}\left(\frac{1}{\varepsilon}E_x\frac{\partial \varepsilon}{\partial z}\right) + \frac{\partial}{\partial z}\left(\frac{1}{\varepsilon}E_y\frac{\partial \varepsilon}{\partial z}\right) = 0 \tag{4.1c}$$

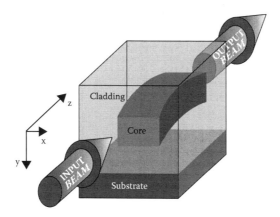

FIGURE 4.3 A longitudinally slowly varying structure in a rectangular coordinate system.

Similarly from 3.5b one obtains:

$$\frac{\partial^2 H_x}{\partial x^2} - \varepsilon \frac{\partial}{\partial y}\left(\frac{1}{\varepsilon}\frac{\partial H_x}{\partial y}\right) + \omega^2 \mu_0 \varepsilon H_x + \varepsilon \frac{\partial}{\partial z}\left(\frac{1}{\varepsilon}\frac{\partial H_x}{\partial z}\right) + \frac{1}{\varepsilon}\left(\frac{\partial \varepsilon}{\partial y}\frac{\partial H_y}{\partial x} + \frac{\partial \varepsilon}{\partial z}\frac{\partial H_z}{\partial x}\right) = 0$$

(4.2a)

$$\varepsilon \frac{\partial}{\partial x}\left(\frac{1}{\varepsilon}\frac{\partial H_y}{\partial x}\right) + \frac{\partial^2 H_y}{\partial y^2} + \omega^2 \mu_0 \varepsilon H_y + \varepsilon \frac{\partial}{\partial z}\left(\frac{1}{\varepsilon}\frac{\partial H_y}{\partial z}\right) + \frac{1}{\varepsilon}\left(\frac{\partial \varepsilon}{\partial x}\frac{\partial H_x}{\partial y} + \frac{\partial \varepsilon}{\partial z}\frac{\partial H_z}{\partial y}\right) = 0$$

(4.2b)

$$\varepsilon \frac{\partial}{\partial x}\left(\frac{1}{\varepsilon}\frac{\partial H_z}{\partial x}\right) + \varepsilon \frac{\partial}{\partial y}\left(\frac{1}{\varepsilon}\frac{\partial H_z}{\partial y}\right) + \omega^2 \mu_0 \varepsilon H_z + \frac{\partial^2 H_z}{\partial z^2} + \frac{1}{\varepsilon}\left(\frac{\partial \varepsilon}{\partial x}\frac{\partial H_x}{\partial z} + \frac{\partial \varepsilon}{\partial y}\frac{\partial H_y}{\partial z}\right) = 0$$

(4.2c)

Optical waveguiding structures typically vary slowly along the direction of the wave propagation. Consequently, it is assumed that along z the dielectric constant distribution ε varies slowly, that is, $(\partial \varepsilon/\partial z) \approx 0$ [5]. As a consequence of this assumption, the transverse components of the electromagnetic field decouple from the longitudinal ones and hence for the transverse field components the set of Equation 4.1 can be approximated by the following equation:

$$\frac{\partial \vec{E}_t}{\partial z^2} = -\left(A + \omega^2 \mu_0 \varepsilon\right) \vec{E}_t$$

(4.3a)

while Equation 4.2 by

$$\frac{\partial \vec{H}_t}{\partial z^2} = -\left(B + \omega^2 \mu_0 \varepsilon\right) \vec{H}_t$$

(4.3b)

where $\vec{E}_t = \vec{i}_x E_x + \vec{i}_y E_y$ and $\vec{H}_t = \vec{i}_x H_x + \vec{i}_y H_y$. The definitions of the matrices A and B are given by 3.13.

Equation 4.3 is used in deriving most of the BPM algorithms. There are only few exceptions when this is not the case and the longitudinal components are used together with the transverse ones [6–8]. Similarly as in Chapter 3, Equation 4.3 can be further simplified for waveguiding structures, which support polarised modes. Similarly as for the case of optical waveguide modal analysis, for the x-polarised modes 4.3 reduces to:

$$\frac{\partial^2 E_x}{\partial z^2} = -\left(a_{11} + \omega^2 \mu_0 \varepsilon\right) E_x$$

(4.4a)

and

$$\frac{\partial^2 H_y}{\partial z^2} = -\left(b_{22} + \omega^2 \mu_0 \varepsilon\right) H_y \tag{4.4b}$$

while for y-polarised modes they reduce to:

$$\frac{\partial^2 E_y}{\partial z^2} = -\left(a_{22} + \omega^2 \mu_0 \varepsilon\right) E_y \tag{4.4c}$$

and

$$\frac{\partial^2 H_x}{\partial z^2} = -\left(b_{11} + \omega^2 \mu_0 \varepsilon\right) H_x \tag{4.4d}$$

The operators a_{ji} and b_{ji} are given in Tables 3.1 and 3.2.

Similarly as in the case of an optical waveguide, if ε varies slowly with the transverse coordinates, one can use the scalar approximation:

$$\frac{\partial^2 \Phi}{\partial z^2} = -\left(s + \omega^2 \mu_0 \varepsilon\right) \Phi \tag{4.5}$$

where Φ is the scalar potential and s is given by 3.16.

Furthermore, in the case of a slab waveguide structure (Figure 3.2), which extends infinitely along the y direction, the following equations are obtained, by neglecting the derivatives with respect to y in 4.4:

$$\frac{\partial^2 E_x}{\partial z^2} = -\left(a_{TE} + \omega^2 \mu_0 \varepsilon\right) E_x \tag{4.6a}$$

$$\frac{\partial^2 H_y}{\partial z^2} = -\left(b_{TE} + \omega^2 \mu_0 \varepsilon\right) H_y \tag{4.6b}$$

for TE modes and for TM modes:

$$\frac{\partial^2 E_y}{\partial z^2} = -\left(a_{TM} + +\omega^2 \mu_0 \varepsilon\right) E_y \tag{4.6c}$$

$$\frac{\partial^2 H_x}{\partial z^2} = -\left(b_{TM} + \omega^2 \mu_0 \varepsilon\right) H_x \tag{4.6d}$$

The operators a_{TE}, a_{TM}, b_{TE}, and b_{TM} are given in Table 3.5.

If ε varies slowly with the transverse coordinates, one can again use the scalar approximation:

$$\frac{\partial^2 \Phi}{\partial z^2} = -\left(s_{1D} + \omega^2 \mu_0 \varepsilon\right) \Phi \tag{4.7}$$

where Φ is the scalar potential and

$$s_{1D} = \frac{\partial^2}{\partial y^2}.$$

The effective index method described in the section "Effective Index Method," when applicable, can be used to obtain a two-dimensional (2D) approximation to a three-dimensional (3D) structure.

It can be observed that Equations 4.3 through 4.7 have the same generic form, namely:

$$\frac{\partial^2 F}{\partial z^2} = -\left(L + \omega^2 \mu_0 \varepsilon\right) F \tag{4.8}$$

In principle, Equation 4.8 can be solved directly using either finite elements or finite differences. In many cases of practical importance, however, such solution results in inefficient algorithms. One way of improving efficiency consists in decomposing F into forward and backward propagating waves. This step is usually accomplished by formally factoring 4.8:

$$\left(\frac{\partial}{\partial z} - j\sqrt{L + \omega^2 \mu_0 \varepsilon}\right)\left(\frac{\partial}{\partial z} + j\sqrt{L + \omega^2 \mu_0 \varepsilon}\right) F = 0$$

If only one of the product components is considered, the so called one-way wave equation is obtained:

$$\frac{\partial F}{\partial z} = -j\sqrt{L + \omega^2 \mu_0 \varepsilon} \, F \tag{4.9}$$

The name "one-way wave equation" comes from the fact that the operator acting on the function F on the right-hand side of 4.9 annihilates all waves propagating along one selected z direction. This property of 4.9 is used, for example, to construct annihilating operator boundary conditions to absorb scattered light in numerical simulations of optical and RF devices [9].

After the introduction of the slowly varying (with z) envelope function Φ, defined by $F = \Phi(x,y,z)*\exp(-j\beta_r z)$, where β_r is a suitably selected reference propagation constant, 4.9 transforms into:

$$\frac{\partial \Phi}{\partial z} = j\beta_r\left(-\sqrt{1 + \frac{L + \omega^2 \mu_0 \varepsilon - \beta_r^2}{\beta_r^2}} + 1\right) \Phi \tag{4.10}$$

The solution of Equation 4.10 can be formally expressed in the following way:

$$\Phi\left(z_0 + \Delta z\right) = \exp\int\limits_{z=z_0}^{z=z_0+\Delta z}\left[j\beta_r\left(-\sqrt{1+\frac{L+\omega^2\mu_0\varepsilon-\beta_r^2}{\beta_r^2}}+1\right)\right]\Phi\left(z_0\right)dz \quad (4.11)$$

If a medium is divided into a concatenation of sections with a z-independent refractive index distribution, and within each section rectangle numerical integration rule is applied, then Equation 4.11 reduces to:

$$\Phi\left(z_0 + \Delta z\right) = \exp\left[j\beta_r\left(-\sqrt{1+\frac{L+\omega^2\mu_0\varepsilon-\beta_r^2}{\beta_r^2}}+1\right)\right]\Delta z\ \Phi\left(z_0\right) \quad (4.12)$$

Equation 4.12 implies an algorithm that calculates the value of the field distribution at the position $z_0 + \Delta z$ subject to the known field distribution at the previous position z_0. The main problem in the application of Equation 4.12 is finding suitable approximations for the operator L, and exponential and square root operators, which allow for an efficient calculation of $\Phi(z_0+\Delta z)$ when $\Phi(z_0)$ is known.

Figure 4.4 summarises the main approximation techniques used in the construction of efficient BPM algorithms. There are three main techniques of handling the exponential operator: the split-step method, the eigenvalue expansion method, and the matrix expansion method. With respect to the square root operator, the BPM algorithms are divided into two categories: paraxial algorithms and wide angle algorithms. Finally, there are four main ways of approximating the operator L: using the finite difference method (FDM), finite element method (FEM), mode matching

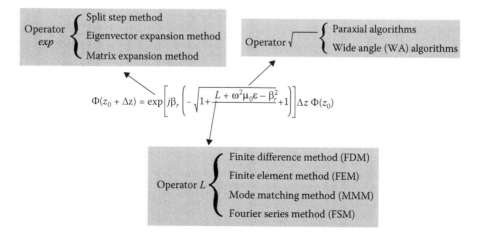

FIGURE 4.4 Diagram listing main approximations used in BPM algorithms.

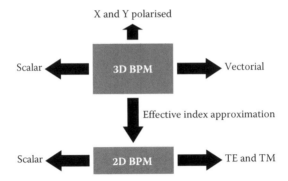

FIGURE 4.5   Diagram showing classification of BPM algorithms.

method (MMM), or Fourier series method (FSM). We discuss the application of these techniques in more detail in the next section.

It is noted that Equation 4.12 implies also a hierarchy of BPM algorithms depending on which approximation is used to represent the operator L, that is, vectorial, polarised, scalar, and so on. In order to avoid confusion, Figure 4.5 summarises the corresponding approximations and the resulting BPM algorithms.

BPM ALGORITHMS

In this section, we provide a review of selected BPM algorithms that have been presented in the literature. We class the algorithms according to the way in which the exponential operator in 4.12 is approximated. Following these classification criteria, BPM techniques are divided into three groups: eigenmode expansion, matrix expansion, and split operator techniques. The split operator techniques result in very efficient algorithms but are mainly limited to the analysis of low refractive index contrast structures under the paraxial approximation. Eigenmode expansion and matrix expansion techniques, on the other hand, are straightforward to extend to 3D polarised and vectorial propagation analysis and allow for a straightforward implementation of wide-angle (WA) propagation algorithms. When compared with matrix expansion techniques, the eigenmode expansion techniques are more efficient if the analysed structure consists of concatenations of longitudinally homogenous waveguide elements. The matrix expansion techniques are more appropriate for the analysis of the light propagation in continuously longitudinally varying structures. Typically, eigenmode expansion techniques are used with fast mode matching or Fourier method–based mode solvers. The matrix expansion techniques, on the other hand, are mainly used with the FDM and FEM. It is also possible to combine the advantages of various approaches by using different methods to treat slowly varying and abrupt sections. Hayashi et al. combined a finite element eigenmode expansion method (for treatment of discontinuities) and a finite element matrix expansion BPM (for propagation in media slowly varying along the direction of propagation) [10].

Split Operator BPM

Historically, the first beam propagation algorithm was a split operator method based on the Fourier split-step technique [11–14]. This BPM algorithm consists in splitting the exponential operator in such a way that the diagonalisation of the resulting operators may be performed easily. A standard splitting consists in separating the operator L, which contains spatial derivatives, from the operator $\omega^2 \mu \varepsilon$. This splitting applied to 4.12 under the paraxial approximation yields [11,12]:

$$\Phi(z_0 + \Delta z) \approx \exp\left[\mp j\beta_r \frac{1}{2} \frac{L}{\beta_r^2}\right] \exp\left[\mp j\beta_r \frac{1}{2} \frac{\omega^2 \mu_0 \varepsilon - \beta_r^2}{\beta_r^2}\right] \Phi(z_0)$$

In order to decrease the error resulting from the operator splitting (because the operators involved do not commute), the splitting is recast into a symmetrised form:

$$\Phi(z_0 + \Delta z) \approx \exp\left[\mp j\beta_r \frac{1}{4} \frac{L}{\beta_r^2}\right] \exp\left[\mp j\beta_r \frac{1}{2} \frac{\omega^2 \mu_0 \varepsilon - \beta_r^2}{\beta_r^2}\right] \exp\left[\mp j\beta_r \frac{1}{4} \frac{L}{\beta_r^2}\right] \Phi(z_0)$$

$$(4.13)$$

In the 2D scalar case, applying the Fourier transform to handle the operator L in 4.13 results in a very efficient BPM algorithm known as the split-step FFT BPM. It takes an advantage of the fact that in the Fourier space the first operator on the left (and the last on the right) in 4.13 are diagonal. So, for the calculation of the field distribution for the next step, one needs to carry out subsequent Fourier transforms, vector multiplications, and inverse Fourier transforms [11]. The split-step FFT method has also been extended to cylindrical coordinate systems [15]. However, the extension of this algorithm to wide-angle analysis of waveguides with a large refractive index contrast is not straightforward [16,17].

A paraxial, fully vectorial 3D split-step BPM algorithm can be obtained by combining the FFT method with the FD method [18]. Also by applying the alternating directions implicit (ADI) method, a split operator technique that can be used for a full vectorial (mainly paraxial) 3D analysis, can be obtained. These algorithms allow replacing a solution of a 3D problem with a sequential solution of 2D problems. Consequently, instead of inverting a sparse matrix with a complicated pattern, only an inversion of three diagonal matrices is required. This can be carried out very efficiently using, for instance, the Thomas algorithm. The ADI split-operator techniques have successfully been implemented to polarised [19] and vectorial cases [20–22], and extended also to lateral wide-angle propagation [23]. An efficient vectorial and fully wide-angle algorithm can be obtained if the ADI method is used only as a preconditioner for a subspace iteration method [24]. It must be noted, however, that ADI split-operator techniques suffer from instabilities [20–22].

Sharma and Agrawal [25] suggested an approach to a split-operator BPM method that is applicable to 3D wide-angle analysis. In order to obtain a wide-angle algorithm as suggested by them, 4.8 is first recast into:

$$\frac{\partial^2 G}{\partial z^2} = H\ G \tag{4.14}$$

where

$$G = \begin{bmatrix} F \\ \partial F / \partial z \end{bmatrix}$$

$$H = \begin{bmatrix} 0 & 1 \\ -\left(L + \omega^2 \mu_0 \varepsilon\right) & 0 \end{bmatrix}$$

Then the operator H is split into two parts. There are several ways in which this can be done [25–27]. For instance the operator H can be split into two parts:

$$H = H_1 + H_2 = \begin{bmatrix} 0 & 1 \\ -\left(L + \beta_r^2\right) & 0 \end{bmatrix} + \begin{bmatrix} 0 & 0 \\ \left(\beta_r^2 - \omega^2 \mu_0 \varepsilon\right) & 0 \end{bmatrix}$$

and then, similarly to the split-step FFT BPM, after splitting the exponential operators, the solution of 4.14 can be formally expressed using the symmetrised form:

$$G\left(z + \Delta z\right) = PQPG\left(z + \Delta z\right) + O\left(\left(\Delta z\right)^3\right) \tag{4.15}$$

where $P = e^{1/2 H_1 \Delta z}$ and $Q = e^{H_2 \Delta z}$. In 4.15 the operator P governs the propagation in a homogenous space, which can be easily evaluated whilst the evaluation of Q is straightforward since the operator H_2 is idempotent. Recently, a modification of the algorithm originating from Sharma and Agrawal [25] was suggested that uses the Magnus expansion and allows for a significant improvement of the accuracy [28].

EIGENMODE EXPANSION BPM

The eigenmode expansion techniques make it straightforward to implement wide angle propagation which also extend to polarised and full vectorial analysis. The eigenmode expansion techniques are based on the fact that the operator $L + \omega^2 \mu\ \varepsilon$ can be diagonalised using its eigenvector set. This can be formally expressed by recasting Equation 4.8 into the following form:

$$\frac{\partial^2 F_t}{\partial z^2} = -\Gamma\ F_t \tag{4.16}$$

where the $F_t = V^{-1} F$, $\Gamma = V^{-1}(L + \omega^2 \mu \, \varepsilon)V$ and gives the eigenvalues of $L + \omega^2 \mu_0 \, \varepsilon$, while the columns of V contain the eigenvectors of $L + \omega^2 \mu_0 \, \varepsilon$. Equation 4.16 can be formally solved:

$$F_t(z) = \exp\left(\int (-\sqrt{\Gamma}z)dz\right) F_t^F(z) + \exp\left(\int (\sqrt{\Gamma}z)dz\right) F_t^B(z) \qquad (4.17)$$

where the superscripts F and B correspond to forward and backward propagating waves, respectively. For the one-way propagation over a waveguide section, Equation 4.17 reduces to:

$$F_t(z_0 + \Delta z) = \exp\left(-\sqrt{\Gamma}\Delta z\right) F_t^F(z_0) \qquad (4.18)$$

In the form 4.18, the numerical evaluation of the square root and exponential operators' action on F_t is straightforward, because the Γ operator is diagonal. So, the whole difficulty lies in the efficient calculation of the operators Γ, V, and V^{-1} (for Hermitian matrices representing $L + \omega^2 \mu_0 \, \varepsilon$, $V^{-1} = V^H$, otherwise $V^{-1} = V^A$, where V^A are the eigenvectors of the adjoint operator). As Γ and V contain the eigenvalues and eigenvectors of $L + \omega^2\mu_0\varepsilon$, the whole problem reduces to an efficient evaluation of the field distributions and propagation constants of the waveguide modes at a particular cross-section. Consequently, the most efficient eigenmode expansion BPM methods are based on analytical and semianalytical methods that have been discussed in Chapter 3, for example, the MMM [29–31] or method of lines [32]. These techniques hence inherit the deficiencies of MMM and MOL; that is, they become inefficient for waveguides with continuously changing refractive index distribution. In such cases the eigenmode expansion techniques, which use the FDM [33] or FEM [34], are more efficient. The eigenmode expansion BPM, which uses FDM, is often referred to as the method of lines BPM [35,36]. Similarly, the eigenmode expansion technique, which uses the FEM, is often referred to as the FEM of lines BPM [37]. The eigenmode expansion BPM needs very large computer memory for storing all the eigenvectors. To address this issue several algorithms for the reduction of the number of the stored eigenvectors were suggested. For this purpose, Gerdes proposed the selection of a number of most significant eigenvalues and eigenvectors [38]. This technique is practically equivalent to selecting a WA order in a matrix expansion BPM, whereby the accuracy of the calculations depends on the number of modes retained in the approximation.

A method that is akin to mode expansion techniques and relies on using only a part of the eigenmode spectrum to study numerically an optical beam propagation is the BPM based on the Lanczos reduction [39,40]. The main shortcoming of Lanczos reduction BPM is an inability to model accurately both the propagating and the evanescent parts of the modal spectrum [41,42]. In order to remedy this deficiency, the Arnoldi method can be used instead of the Lanczos method for the generation of the orthogonal basis [43].

When applying the modal expansion BPM, spurious reflection from the boundary of the analysis window can introduce a large error. To avoid this problem, appropriate boundary conditions must be used when calculating the waveguide propagation constants and field distributions of the modes. For instance, the complex coordinate stretching with PML has been developed specially to reduce the spurious reflection in modal expansion BPM [44,45]. An issue that is linked with the spurious reflection phenomenon is finding an accurate approximation of the continuous spectrum of radiation and evanescent modes. Usually this is achieved by truncation of the analysis window and imposing the boundary conditions, which results in a discrete representation of the continuous part of the L operator spectrum in Equation 4.14. However, for low refractive index contrast structures, the continuous radiation mode spectrum can be approximated with the radiation mode spectrum of free space with an appropriately chosen effective index. The radiation mode spectrum can be then integrated analytically, and its effect upon the propagating field distribution can then be accurately incorporated into the algorithm [31].

The set of eigenmodes can also be calculated very efficiently using the Fourier method [46,47] or a mapped barycentric Chebyshev differentiation matrix method [48]. An additional advantage of applying the Fourier method to the calculation of the modal spectrum is that it yields a reciprocal BPM algorithm if the modelled structure consists of a set of concatenated waveguide sections each having a different refractive index distribution [49]. Originally, the reciprocal eigenmode expansion BPM method was restricted to 2D analysis. However, this technique was recently extended to 3D analysis [50].

The eigenmode expansion method has been also applied to the analysis of optical tapers [51,52] and bends [53,54] in a cylindrical coordinate system and for the study of optical propagation in nonlinear optical waveguides [33]. Eigenmode expansion BPM can also be used for the media with gain or loss but at the expense of reduced efficiency [55].

MATRIX EXPANSION BPM

With few exceptions [56], the matrix expansion BPM relies mainly either on the FDM or on the FEM to provide algebraic representations of the operator L in 4.10. Therefore, we focus the following discussion on these two techniques only.

Matrix expansion finite difference BPM (FD-BPM) algorithms are derived directly from Equations 4.10 and 4.12. In order to solve 4.12 using finite differences, the operator L, square root, and exponential operators have to be approximated. In general, the same finite difference formulations can be used to represent the operator L in 4.12 as are used for FDM eigenmode calculations. Since, these FD methods have been discussed in Chapter 3 we refrain from repeating the discussion here. However, we note that due to reciprocity and stability (cf. the section "Numerical implementation of BPM") of the resulting algorithms symmetrical matrix representations of the L operator in 4.12 are preferred. In the scalar case, the derivatives can be represented by the standard three-point (five-point in 3D case) FD stencil, which yields a symmetrical matrix representation of L. However, in the polarised and vectorial cases, and also in the 2D TM case, there are field discontinuities at the interfaces between dielectrics with different values of the refractive index. The application of a FDM in such circumstances results in a lack of symmetry in the FD coefficient matrix [57–59]. This problem also appears in the 2D TE case when higher order FD approximations are used. However, in the TE case a five-point formula can be derived, which is fourth order accurate and results in a symmetrical matrix [60]. In the 2D TM case an application of the Sigmoid function was suggested to approximate a step wise refractive index distribution by a smooth function [61], and hence to eliminate the discontinuities. However, there are no FD formulations that result in a symmetrical coefficient matrix for the polarised and vectorial 3D BPM analyses.

In the FD-BPM the exponential operator is approximated either by a Padé series [62], in which case the Padé (1,1) approximation corresponds to the very popular Crank-Nicolson (CN) scheme [63], or a Taylor expansion that corresponds to Padé (n,0) [64]. The square root operator can be approximated by Padé approximants using a recurrent formula [65,66] or more convenient direct expressions [67], Thiele [68] approximants (that is, Padé using continued fraction representation), and Taylor expansion, which if retaining only the first order term gives the very popular paraxial scheme. Alternatively, 4.12 can be recast into [69]:

$$\Phi(z_0 + \Delta z) = \frac{P\left(\sqrt{L + \omega^2 \mu_0 \varepsilon - \beta_r^2}\right)}{P\left(\sqrt{L + \omega^2 \mu_0 \varepsilon - \beta_r^2}\right)*} \Delta z \; \Phi(z_0)$$

where

$$P\left(\sqrt{L + \omega^2 \mu_0 \varepsilon - \beta_r^2}\right) = \cos\left(\frac{1}{2}\sqrt{L + \omega^2 \mu_0 \varepsilon - \beta_r^2}\right) - j\frac{1}{2}\sqrt{L + \omega^2 \mu_0 \varepsilon - \beta_r^2}$$

$$\times \frac{\sin\left(\frac{1}{2}\sqrt{L + \omega^2 \mu_0 \varepsilon - \beta_r^2}\right)}{\frac{1}{2}\sqrt{L + \omega^2 \mu_0 \varepsilon - \beta_r^2}} \tag{4.19}$$

Equation 4.19 can be expanded using Taylor series resulting in an unconditionally stable algorithm without using rational approximations [69].

A very popular choice, when applying a FD-BPM, is the use of the CN scheme and the paraxial approximation. This algorithm was implemented in polarised [70] and vectorial BPM simulations, using either three-component [6] or more efficient two-component [71] formulations.

The efficiency of paraxial FD-BPM based on the CN scheme can be improved by the introduction of the Douglas scheme [72]. The difficulty with the Douglas scheme, though, is that the algorithm developed for the two-dimensional TE case cannot be used in the TM case, as it does not result in the cancellation of truncation errors. So, the Douglas scheme algorithms have to be derived separately for each particular case. This, however, is not a major problem, since many Douglas scheme–based

BPM algorithms are available in the literature, for example, Douglas schemes for nonuniform meshes [73], 3D polarised, and 3D vectorial analysis [74,75].

The computational efficiency of the CN and Douglas scheme BPM depends critically on how the resulting sets of linear algebraic equations are solved. In the 2D case Thomas algorithm is typically applied, while in the 3D case modern iterative methods offer an optimum performance [76].

Under the paraxial approximation, very efficient algorithms can be obtained using explicit methods. Additional advantage comes from the fact that the computational efficiency of explicit methods can be further improved by using parallel processing. Explicit FD techniques are claimed to outperform the FFT based methods in the 2D case [77,78]. Particularly, efficient explicit BPMs are based on the DuFort-Frankel algorithm [79]. The main shortcoming of the DuFort-Frankel algorithm is that it suffers from spurious modes [80] and instabilities when applied to structures with loss or gain [81]. Another, very efficient, explicit algorithm relies on the iterated CN scheme. Unlike other algorithms, this technique can be extended to include WA analysis using the Taylor expansion of the square root operator [82].

The efficiency of FD-BPM can also be improved by the introduction of nonuniform meshes [83], nonorthogonal coordinate systems [84–88], and non-Cartesian orthogonal coordinate systems [89–91]. We discuss the nonorthogonal coordinate systems in more detail in the section "Selected examples of BPM application."

Finite element matrix expansion BPM (FE-BPM) formulations rely on Equation 3.4b. So far, three approaches have been suggested for vectorial 3D finite element BPM. All of them are based on CN scheme. In Montanari et al. [92], the authors used a three-component magnetic field formulation and complemented it with a zero divergence condition to suppress spurious modes. In another three-component formulation [93], the spurious modes were suppressed by use of edge elements. The two component formulation was suggested [94]. It formulates the equations for x and y magnetic field components using Equation 4.1b and eliminating z component of the magnetic field using the zero divergence condition, which results in simultaneously suppressing the spurious modes. All FE-BPMs developed so far use the recurrent algorithm for generating Padé approximations to the square root operator. One of the advantages of FE-BPM is that it is fairly straightforward to apply in anisotropic media [95–98].

The advantage of FE-BPM over the FD-BPM is that it allows for the use of adaptive meshing [99]. Some authors also claim that vectorial FE-BPM, unlike the vectorial FD-BPM, conserves the total power carried by the optical beam. However, a rigorous analysis performed in Deng and Yevick [100] contradicts this claim. The FD-BPM, on the other hand, allows implementing efficient (nonrecurrent) wide angle algorithms and higher order approximations for the exponential operator.

BIDIRECTIONAL BPM

In planar photonic devices, the guided optical beam encounters an abrupt dielectric discontinuity when a planar optical waveguide is coupled with another planar optical waveguide, a planar optical waveguide is coupled with an optical fibre, or when a planar optical waveguide is terminated (that is, a dielectric facet or coated facet problem). In such cases, the BPM cannot be applied directly. However, the BPM can be extended to include the reflection and transmission of optical waves at abrupt discontinuities.

HANDLING ABRUPT DISCONTINUITIES

To derive expressions relating impinging, transmitted, and reflected waves at a dielectric discontinuity, we consider the problem that is depicted in Figure 4.6. A wave incident upon a discontinuity is partially reflected and transmitted. To calculate the reflected and transmitted wave distribution, the relationship between the transverse components of electric and magnetic fields of the incident and reflected field has to be derived. For this purpose, from Equation 3.1, the following relationship between the transverse components of the electric and magnetic fields is obtained:

$$\frac{\partial}{\partial z}\begin{bmatrix} H_y \\ -H_x \end{bmatrix} = \begin{bmatrix} r_{e11} & r_{e12} \\ r_{e21} & r_{e22} \end{bmatrix}\begin{bmatrix} E_x \\ E_y \end{bmatrix} \tag{4.20a}$$

$$\frac{\partial}{\partial z}\begin{bmatrix} E_x \\ E_y \end{bmatrix} = \begin{bmatrix} r_{h11} & r_{h12} \\ r_{h21} & r_{h22} \end{bmatrix}\begin{bmatrix} H_y \\ -H_x \end{bmatrix} \tag{4.20b}$$

Equation 4.20 can be recast into a more compact form:

$$\frac{\partial \vec{H}_t}{\partial z} = j R_{\mathrm{E}}\, \vec{E}_t \tag{4.21a}$$

$$\frac{\partial \vec{E}_t}{\partial z} = j R_{\mathrm{H}}\, \vec{H}_t \tag{4.21b}$$

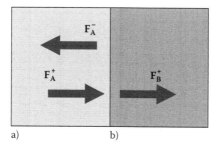

a)   b)

FIGURE 4.6 Reflection of an incident wave at a dielectric discontinuity between two media a) and b).

where $\vec{E}_t = \vec{i}_x E_x + \vec{i}_y E_y$ and $\vec{H}_t = \vec{i}_x H_x + \vec{i}_y H_y$. The elements of matrices R_E and R_H are given in Tables 4.1 through 4.3.

For the 3D polarised TE case and 2D TE case, $\vec{E}_t$ and $\vec{H}_t$ reduce to E_x and H_y, respectively. Similarly, for the 3D polarised TM case and 2D TM case, $\vec{E}_t$ and $\vec{H}_t$ reduce to E_y and H_x.

From Equations 4.21a and 4.3b, one obtains a relationship between the transverse components of the electric and magnetic fields:

TABLE 4.1
Elements of R_E

r_{eji}	$i = 1$	$i = 2$
$j = 1$	$\omega\varepsilon + \dfrac{1}{\omega\varepsilon}\dfrac{\partial^2}{\partial y^2}$	$-\dfrac{1}{\omega\mu_0}\dfrac{\partial^2}{\partial y\partial x}$
$j = 2$	$-\dfrac{1}{\omega\mu_0}\dfrac{\partial^2}{\partial x\partial y}$	$\omega\varepsilon + \dfrac{1}{\omega\varepsilon}\dfrac{\partial^2}{\partial x^2}$

TABLE 4.2
Elements of R_H

r_{hji}	$i = 1$	$i = 2$
$j = 1$	$\omega\mu_0 + \dfrac{1}{\omega}\dfrac{\partial}{\partial x}\left(\dfrac{1}{\varepsilon}\dfrac{\partial}{\partial x}\right)$	$\dfrac{1}{\omega}\dfrac{\partial}{\partial x}\left(\dfrac{1}{\varepsilon}\dfrac{\partial}{\partial y}\right)$
$j = 2$	$\dfrac{1}{\omega}\dfrac{\partial}{\partial y}\left(\dfrac{1}{\varepsilon}\dfrac{\partial}{\partial x}\right)$	$\omega\mu_0 + \dfrac{1}{\omega}\dfrac{\partial}{\partial y}\left(\dfrac{1}{\varepsilon}\dfrac{\partial}{\partial y}\right)$

TABLE 4.3
Elements of Matrices (Scalars) R_E and R_H for 3D Polarised and 2D Approximations

	3D Polarised TE	3D Polarised TM	2D TE	2D TM
R_E	$\omega\varepsilon + \dfrac{1}{\omega\varepsilon}\dfrac{\partial^2}{\partial y^2}$	$\omega\varepsilon + \dfrac{1}{\omega\varepsilon}\dfrac{\partial^2}{\partial x^2}$	$\omega\varepsilon + \dfrac{1}{\omega\varepsilon}\dfrac{\partial^2}{\partial y^2}$	$\omega\varepsilon$
R_H	$\omega\mu_0 + \dfrac{1}{\omega}\dfrac{\partial}{\partial x}\left(\dfrac{1}{\varepsilon}\dfrac{\partial}{\partial x}\right)$	$\omega\mu_0 + \dfrac{1}{\omega}\dfrac{\partial}{\partial y}\left(\dfrac{1}{\varepsilon}\dfrac{\partial}{\partial y}\right)$	$\omega\varepsilon$	$\omega\mu_0 + \dfrac{1}{\omega}\dfrac{\partial}{\partial y}\left(\dfrac{1}{\varepsilon}\dfrac{\partial}{\partial y}\right)$

$$\vec{E}_t = Z_E \vec{H}_t = \frac{1}{Y_E} \vec{H}_t = \frac{\sqrt{B + \omega^2 \mu_0 \varepsilon}}{R_E} \vec{H}_t$$

From Equations 4.21b and 4.3a, one obtains an alternative relationship between the transverse components of the electric and magnetic fields:

$$\vec{E}_t = Z_E \vec{H}_t = \frac{1}{Y_E} \vec{H}_t = \frac{R_H}{\sqrt{A + \omega^2 \mu_0 \varepsilon}} \vec{H}_t$$

Applying the continuity of the electric and magnetic fields at the interface between the regions A and B (Figure 4.6), the reflection and transmission operators are obtained:

$$F_A^- = \frac{Y_A - Y_B}{Y_A + Y_B} F_A^+ = \frac{Z_B - Z_A}{Z_A + Z_B} F_A^+ = r_{AB} F_A^+ \tag{4.22a}$$

$$F_B^+ = \frac{2Y_A}{Y_A + Y_B} F_A^+ = \frac{2Z_B}{Z_A + Z_B} F_A^+ = t_{AB} F_A^+ \tag{4.22b}$$

In the scalar case, one requires the continuity of the scalar potential and its derivative, which leads to the following relationships between the incident, reflected, and transmitted waves:

$$F_A^- = \frac{\sqrt{[L_A]} - \sqrt{[L_B]}}{\sqrt{[L_A]} + \sqrt{[L_B]}} F_A^+ = r_{AB} F_A^+ \tag{4.23a}$$

$$F_B^+ = \frac{2\sqrt{[L_A]}}{\sqrt{[L_A]} + \sqrt{[L_B]}} F_A^+ = t_{AB} F_A^+ \tag{4.23b}$$

The expressions 4.22 and 4.23 can be evaluated using techniques compatible with either the FFT split-step algorithm or mode expansion BPM and matrix expansion BPM.

The technique compatible with the FFT BPM has been suggested by Kaczmarski and Lagasse [101] and then extended to the analysis of angled facets by Jin et al. [102]. In this method the incident wave is first Fourier transformed and the reflected wave contributions are calculated for each plane wave separately using a reflection coefficient for an interface between suitably selected uniform reference media. Then the inverse Fourier transform is performed, on the reflected wave thus calculated, and multiplied by a space dependent coefficient [101,102]. So far, this technique is limited to 2D scalar TE case only.

The mode expansion techniques, on the other hand, can be used to evaluate 4.22 and 4.23 in the scalar, polarised, and full vectorial 3D cases. The most straightforward way of calculating the reflectivity using the modal expansion is by expressing the field distributions on both sides of the discontinuity using a linear combination of local eigenmodes. Then, by applying the orthogonality relations, matrix equations are established that relate the linear combination coefficients on both sides of the discontinuity [103,104]. In the case of media with loss or gain, the modal sets are not orthogonal and "adjoint" waveguide modes need to be used [105], similarly as for mode expansion BPM. The continuous spectrum of radiation modes is usually discretised using the techniques developed for mode calculation. A particularly efficient technique, known as the free space radiation method (FSRM), can be obtained if the media on the both sides of the discontinuity either are homogenous or have low refractive index contrast. In such cases, the radiation mode spectrum can be approximated with the radiation mode spectrum of free space with an appropriately chosen effective index [106]. More recently, the application of FSRM was extended to the analysis of air clad structures, for example, air clad rib waveguides [107], and also to arbitrary waveguide facet analysis [108] by combining matrix expansion BPM with the FSRM method.

In the matrix expansion technique, the square root operators are estimated using Padé rational expansions. The resulting equations can be solved using, for instance, a von Neumann expansion [109]. However, the von Neumann method does not always converge. Problems with convergence are typically encountered when the refractive index discontinuity is large. In such cases an effective remedy may consist in either adding a thin intermediate dielectric layer [109] or by using the BiCGSTAB method [110].

Initially, algorithms for calculating the wave reflection from a discontinuity were available for the 2D TE case only. The extension to the 2D TM case was given in Chiou and Chang [111], while the formalism for the extension of matrix expansion method to the calculation of facet reflectivities in the 3D vectorial case was derived in Helfert and Pregla [112]. The results of the matrix expansion method depend significantly on the choice of the reference refractive index on both sides of the discontinuity. The choice of the real reference refractive index usually results in spurious field oscillations at the edges of the computational domain. Therefore, when calculating the reflected wave distribution, using complex Padé approximants [110,113] is necessary (cf. the section "Dispersion Characteristics").

Both the mode expansion and matrix expansion methods require an application of the staircasing approximation when a curved interface between two media is encountered by the incident wave (cf. the section "Staircasing Approximation"). This may result in a very slow convergence of the results, with an increase in either the number of the sampling points or the number of the eigenmodes. This problem can be alleviated by applying the local normal approach (LNA) [114]. The LNA consists in matching at the interface the field and its derivative calculated in the direction that is normal to the curved surface. The results presented in Petruskevicius [114] show that the application of the LNA results in a significant improvement of the convergence rate.

Another approach that can be used when the interfaces are not parallel to the coordinate system axes is the application of the direct wave equation solvers. The dielectric discontinuities, unlike long planar slowly varying planar optical waveguiding structures, have a fairly compact structure. Hence, they can also be relatively efficiently modelled by direct wave equation solvers that use standard numerical techniques, including the FEM and FDM [115,116].

HANDLING MULTIPLY REFLECTED WAVES

The BPM algorithms calculating the light transmission and reflection at a discontinuity can be extended to provide an efficient technique for the analysis of slowly varying planar optical waveguiding structures with a number of local discontinuities. This technique, commonly referred to as bidirectional BPM, is schematically illustrated in Figure 4.7.

In fact, bidirectional BPM can be derived by combining Equation 4.11 with the expressions 4.22 or 4.23, which results in:

$$\begin{pmatrix} F_{\mathrm{OUT}}^+ \\ F_{\mathrm{OUT}}^- \end{pmatrix} = M \begin{pmatrix} F_{\mathrm{IN}}^+ \\ F_{\mathrm{IN}}^- \end{pmatrix} \qquad\qquad M = P_n T_{n-1,n} \cdots P_2 T_{12} P_1 \qquad\qquad (4.24)$$

where P_i is given by:

$$P_i = \begin{bmatrix} \exp\left(-\int\limits_{z=z_i}^{z=z_{i+1}} L_i dz \right) & 0 \\ 0 & \exp\left(\int\limits_{z=z_i}^{z=z_{i+1}} L_i dz \right) \end{bmatrix}$$

while

$$T_{i,j} = \frac{1}{2} \begin{bmatrix} 1 + L_j^{-1} L_i & 1 - L_j^{-1} L_i \\ 1 - L_j^{-1} L_i & 1 + L_j^{-1} L_i \end{bmatrix}$$

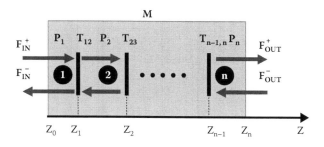

FIGURE 4.7 Schematic diagram of a multiple interface structure; P_i propagation matrix in region i. T_{ij} is the interface matrix for the interface between regions i and j. M is the overall transfer matrix.

where

$$L_i = \mathrm{j}\beta_r \left(-\sqrt{1 + \frac{L + \omega^2 \mu_0 \varepsilon - \beta_r^2}{\beta_r^2}} + 1 \right)$$

The transmission matrix $T_{i,j}$ can be derived from scattering matrix S for a dielectric interface using the relationship between the transmission matrix T and the scattering matrix S [3]:

$$S = \begin{bmatrix} r_{12} & t_{21} \\ t_{12} & r_{21} \end{bmatrix}$$

$$T = \frac{1}{t_{12}} \begin{bmatrix} 1 & -r_{21} \\ r_{12} & t_{12}t_{21} - r_{21}r_{21} \end{bmatrix}$$

where r_{ij} and t_{ij} are given by 4.22 through 4.23:

Equation 4.24 allows for the direct calculation of F_{OUT} when F_{IN} is given. However, when the wave is incident from the left side in Figure 4.7, 4.24 needs to be solved for given F^+_{IN} while $F^-_{OUT} = 0$. Obviously, one can equally well consider the problem of a wave incident from the right side in Figure 4.7. Hence, recasting 4.24 into a scattering matrix problem makes calculating the solution straightforward. However, the matrix M *is* not always known explicitly, which turns the finding of an algorithm that would calculate the scattering matrix into a nontrivial task. Therefore, two main types of techniques exist. One of them uses an iterative procedure to calculate the unknown scattered fields without trying to calculate the scattering matrix. The other approach consists in calculating the scattering matrix explicitly first and then obtaining the scattered fields subject to the known incident fields.

Without compromising the generality, the solution of Equation 4.24 can be discussed for the case when F^+_{IN} is not equal to zero. The methods of solving 4.24 practically consist in combining BPM with algorithms solving Equations 4.24 or 4.23. Hence, similarly to BPM, bidirectional BPM algorithms solving 4.24 use split operator methods or mode expansion or matrix expansion methods. The matrix expansion BPM is usually combined with iterative techniques while with the eigenmode expansion methods either matrix M in 4.24 or the scattering matrix is calculated explicitly.

When using mode expansion techniques the matrix M in 4.24 can be calculated explicitly. The reflected wave F^-_{IN} can be then obtained from the equation: $m_{21}F^+_{IN} = m_{22}F^-_{IN}$, where m_{22} and m_{21} are the elements of the matrix M, that is,

$$M = \begin{bmatrix} m_{11} & m_{12} \\ m_{21} & m_{22} \end{bmatrix}.$$

Finally, the transmitted wave is calculated by multiplying M by F_{IN} [38,104,117]. An alternative approach has been suggested by Liu and Chew [118]. In this algorithm the matrix representations of the total reflection and transmission operators are calculated recursively. Once these matrices are known, the distributions of the reflected and transmitted waves can be obtained by a simple matrix vector multiplication.

In the case of matrix expansion techniques, the explicit forming of matrix M destroys the matrix sparsity pattern (as these techniques, similarly to matrix expansion BPM, use finite differences or finite elements and hence the resulting matrices are sparse), which results in a very inefficient algorithm. A practical algorithm can be formed, however, by starting from an initial incident field distribution and a guess of the reflected wave distribution. Then the distributions of the scattered waves are updated iteratively until convergence is reached [117]. Unfortunately, this simple algorithm can be unstable for structures with large refractive index contrast. In such a case, more robust algorithms can be applied, which uses modern subspace projection based iterative matrix solvers, for example, the BiCGSTAB method, which can iteratively calculate the scattered fields, without destroying the sparsity pattern [117,119]. There are also iterative bidirectional BPM algorithms, which do not rely on the transfer matrix formulation 4.24. One example of such a technique is the Bremmer series method (BSM). The BSM relies on an integral equation formulation of the problem rather than the transfer matrix approach 4.24. The solution is obtained iteratively applying the von Neumann series method. This corresponds to an algorithm, in which in the first pass the reflections are determined at each discontinuity. The backward BPM is then used to add up these reflections and to form the first approximation to the reflected wave, and to calculate the new set of reflections in the forward direction (cf. the section "Transmission Through Thin Optical Elements"). This procedure is applied iteratively until convergence is reached [68].

In the 2D case, it is also possible to use with a matrix expansion BPM an algorithm that calculates (similar to Liu and Chew [118]) recursively the matrix representations of the total reflection and transmission operators [120]. There are also bidirectional BPM algorithms that are tailored for a particular class of problems. For instance, when only within a small subdomain, the directional decomposition method is not applicable a combination of BPM with direct wave equation solvers can result in a very efficient algorithm. Such an algorithm was suggested by Yoneta et al. and proved successful in the analysis of integrated optical T-shaped and cross-shaped beam splitters [116]. Another representative example is a periodic structure. In this case, the determination of the Floquet modes allows for significant improvement of an algorithm's efficiency and stability [47,121–123].

NUMERICAL IMPLEMENTATION OF BPM

In this section, we consider various practical aspects of a BPM algorithm implementation. In order to bring this discussion into the context of an actual BPM algorithm, we discuss first as, an example, a scalar, paraxial FD-BPM algorithm.

In order to derive the paraxial wave equation, we identify first in the solution the slowly varying complex amplitude $\tilde{A}(x,z)$:

$$\tilde{U}(x,z) = \tilde{A}(x,z) * e^{-jn_{ref}kz} \tag{4.25}$$

where n_{ref} is a reference refractive index and k is the wave number. Substituting 4.25 into the scalar wave Equation 2.3 yields:

$$\frac{\partial \tilde{A}(x,z)}{\partial z} = -\frac{j}{2n_{\text{eff}}k}\left[\frac{\partial^2}{\partial x^2} + \left(n^2(x,z)k^2 - n_{\text{eff}}^2 k^2\right)\right]\tilde{A}(x,z) \qquad (4.26)$$

whereby in 4.26 we neglected the second derivative of $\tilde{A}(x,z)$ due to the slowly varying envelope assumption. We note that here we derived 4.26 directly from the scalar wave equation. However, Equation 4.26 could be also obtained from 4.10 by applying the paraxial approximation.

Since, the refractive index distribution is a function of x we cannot use the FFT algorithm from Chapter 2 to solve 4.26. Although, the numerical solution of 4.2 using FFT can be accomplished using a split-step FFT BPM, we will not discuss split-step FFT BPM here since it is very akin to the split-step FFT method discussed in Chapter 9, and, in detail, by Yamauchi [17] and Poon and Kim [124]. Instead we present a FD-BPM approach, cf. [125].

If in 4.26 we use the standard central finite difference approximation for the second order partial derivative with respect to x, select the size of the computational window along the x-axis, and within this window choose a set of equidistant sampling points x_i spaced by the distance Δx, Equation 4.26 yields a set of ordinary differential equations at the selected sampling points x_i, $i = 1,\dots N$:

$$\frac{\mathrm{d}\tilde{A}(x_i,z)}{\mathrm{d}z} = -\frac{j\Delta z}{2n_{\text{eff}}k\,\Delta x^2}\Big[\tilde{A}(x_i - \Delta x, z)$$
$$+ \tilde{A}(x_i + \Delta x, z) + \left(n^2(x_i,z)k^2 - n_{\text{eff}}^2 k^2 - 2\right)\tilde{A}(x_i,z) \qquad (4.27)$$

At the edge of the computational window we need samples of $\tilde{A}(x,z)$ that are not within the selected set. We assume that these samples are equal to zero, that is, we impose the Dirichlet boundary condition. Finally, we can represent 4.27 in a compact form using matrix algebra:

$$\frac{\mathrm{d}\bar{A}(z+\Delta z)}{\mathrm{d}z} = \bar{M}\,\bar{A}(z) \qquad (4.28)$$

where the vector $\bar{A}(z)$ and matrix $\bar{M}$ are defined as:

$$\bar{A}(z) = \begin{bmatrix} \tilde{A}(x_1,z) \\ \tilde{A}(x_2,z) \\ \tilde{A}(x_3,z) \\ \vdots \end{bmatrix}$$

$$\bar{M} = \frac{j}{2n_{\text{eff}} \, k\Delta x^2} \begin{bmatrix} \left(n_1^2 - n_{\text{eff}}^2\right)k^2 - 2 & 1 & 0 & \cdots \\ 1 & \left(n_2^2 - n_{\text{eff}}^2\right)k^2 - 2 & 1 & \cdots \\ 0 & 1 & \left(n_3^2 - n_{\text{eff}}^2\right)k^2 - 2 & 1 \\ \vdots & \vdots & 1 & \ddots \end{bmatrix}$$

and $n_i^2 = n^2\left(x_i, z\right)$

In principle, all standard numerical algorithms for solving sets of ordinary differential equations can be applied to 4.28. A tempting way of solving numerically 4.28 is the application of the Euler method, which uses a simple approximation for the first derivative with respect to z:

$$\frac{d\tilde{A}\left(x_i, z\right)}{dz} = \frac{\tilde{A}\left(x, z + \Delta z\right) - \tilde{A}\left(x, z\right)}{\Delta z}$$

This results in a so called explicit algorithm, which is a well-known way of solving the diffusion equation [126], in which case it is conditionally stable [126]. However, Equation 4.26 only resembles the diffusion equation and unfortunately, for 4.26, the explicit scheme is inherently unstable, [78] although the effect of the instability is very subtle when Δz is small.

Here we use the CN scheme, which is another standard technique used for the solution of the diffusion equation [126]. Unlike the explicit scheme, the CN scheme is unconditionally stable for 4.26. The application of the CN scheme to 4.28 yields:

$$\left(1 - \frac{1}{2}\Delta z\bar{M}\right)\bar{A}\left(z + \Delta z\right) = \left(1 + \frac{1}{2}\Delta z\bar{M}\right)\bar{A}\left(z\right) \tag{4.29}$$

The numerical solution of 4.29 involves a multiplication of the matrix

$$1 + \frac{1}{2}\Delta z\bar{M}$$

by the vector $\bar{A}\left(z\right)$ followed by the inversion of the matrix

$$1 - \frac{1}{2}\Delta z\bar{M}.$$

The entire algorithm for a CN FD-BPM is conveniently summarised next (algorithm 4.1).

The only issue that still needs attention is the selection of the reference refractive index. The accuracy of BPM calculations strongly depends on the selection of n_{ref} value. There are some general guidelines on how this parameter should be selected. The rule of thumb is that one should try to select n_{ref} so that the product $n_{\text{ref}} \, k$ matches as closely as possible the propagation constant "seen" by the propagating wave. If the refractive index contrast in the waveguiding structure is small, selecting the

ALGORITHM 4.1   CN FD-BPM

1. Start
2. Set the free space wavelength λ and define the structure
3. Set the propagation distance L
4. Select N equidistant FD sampling points
5. Initialise $\tilde{A}(x,z)$
6. Obtain the refractive index distribution at z, select n_{ref} and calculate $\tilde{A}(x,z+\Delta z)$ from 4.29
7. $z = z + \Delta z$
8. If $z < L$, go to 6
9. Stop

core refractive index might be sufficiently accurate. Otherwise, either the effective index of a waveguide mode or Rayleigh quotient:

$$k^2 n_{ref}^2 = \sum_{i=1}^{N} \tilde{A}^*\left(x_i,z\right)\left[\tilde{A}\left(x_i - \Delta x,z\right) + \tilde{A}\left(x_i + \Delta x,z\right) + \left(n^2\left(x_i,z\right)k^2\right)\tilde{A}\left(x_i,z\right)\right]/$$
$$\tilde{A}^*\left(x_i,z\right)\tilde{A}\left(x_i,z\right)$$

is used. We discuss this issue in more detail in the section "Dispersion Characteristics."

Using the Rayleigh quotient and in particular the modal effective index, involves however, an additional computational overhead. The code given below shows a MATLAB implementation of the Algorithm 4.1.

```
% CN FDBPM propagation in two core slab waveguide
%
% *********************************************************
% calculation modes for 2 core slab waveguide using FDs;
clear
format long e
pi = 3.141592653589793e+000;
i = sqrt(-1);

lam = 1.55;% wavelength in [micrometres]
k0 = 2*pi/lam;% wave number k0
nf = 3.3704;% refractive index in core 1&2
nf12 = 3.2874;% refractive index between core 1&2
ns = 3.2874;% refractive index in cladding
nc = 3.2874;% refractive index in substrate

dist_c = 4;
% distance to edge of the window in cladding [micrometres]
dist_s = 4;
% distance to edge of the window in substrate [micrometres]
t1 = 2;% core 1 thickness in [micrometres]
t2 = 2;% core 2 thickness in [micrometres]
dist_t12 = 2;% core spacing in [micrometres]
```

```
No_t1 = 10;% number of mesh points in waveguide core 1
dx = t1/No_t1;% dx in [micrometres]
No_c = dist_c/dx;% number of mesh points in cladding
No_s = dist_s/dx;% number of mesh points in substrate
No_t2 = t2/dx;% number of mesh points in waveguide core 2
No_t12 = dist_t12/dx;% number of mesh points between core 1 & 2
N_tot = No_c+No_t1+No_t12+No_t2+No_s;% total number of mesh
points

x = (1:N_tot)*dx;%vector with positions of FD nodes

c = 1/(dx*dx);
A = zeros(N_tot);% initialisation of matrix A
% setting off diagonals
for j = 1:N_tot
a1(j) = c;
a3(j) = c;
end
% setting up vector storing the values of refractive index
n = ns*ones(N_tot,1);
for j = 1:No_c
n(j) = nc;
end
for j = No_c+1:No_c+No_t1
n(j) = nf;
end
for j = No_c+No_t1+1:No_c+No_t1+No_t12
n(j) = nf12;
end
for j = No_c+No_t1+No_t12+1:No_c+No_t1+No_t12++No_t1
n(j) = nf;
end

% setting main diagonal of the matrix
shift = nf*nf*k0*k0;
a2 = ((n.*n*k0*k0-2*c)-shift);

A = spdiags([a1' a2 a3'], -1:1, N_tot, N_tot);

SIGMA = 'SM';
[X,D] = eigs(A,2,SIGMA);

'propagation constants of supermodes'
beta = sqrt(diag(D)+shift)
'effective refractive indices of supermodes'
neff = beta/k0

% ****************************************************************
% CN FDBPM

nref = nf; % reference refractive index
dz = 0.01;% propagation step [micrometres]
% setting main diagonal of the matrix M
shift = nref*nref*k0*k0;
a2 = ((n.*n*k0*k0-2*c)-shift);

M = spdiags([a1' a2 a3'], -1:1, N_tot, N_tot);
```

```
M = (-i/(2*nref*k0))*M;

M_prop1 = sparse(diag(ones(N_tot,1)))+0.5*dz*M;
M_prop2 = sparse(diag(ones(N_tot,1)))-0.5*dz*M;
N_steps = 10000;
E_init = X(:,1)/max(abs(X(:,1)));%initial field distribution
E = E_init;
for j = 1:N_steps
E = M_prop1*E;
E = M_prop2\E;
power(j) = sum(abs(E).^2);
j
end
```

The MATLAB code allows for studying the beam propagation in a test structure, which consists of a 10 μm long section of a two core slab waveguide with the
thickness of the core equal to 2 μm and the distance between cores equal to 2 μm.
The operating wavelength is 1.55 μm while the refractive index in the core and the
cladding is 3.3704 and 3.2874, respectively. It uses the Dirichlet condition at the
boundary of the computational domain. The initial field distribution corresponds to
the fundamental mode of the two core slab waveguide.

Figure 4.8 shows the dependence of the sum of the absolute values squared of
$\tilde{A}(x,z)$ samples:

$$\left|\overline{A}\right|^2 = \sum_{i=1}^{N} \tilde{A}^*\left(x_i,z\right)\tilde{A}\left(x_i,z\right)$$

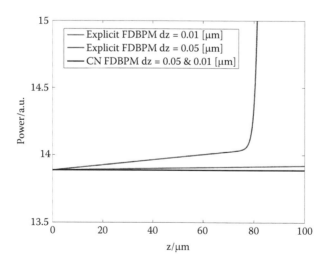

FIGURE 4.8 The dependence of the optical power on the propagation distance. The two
core slab waveguide is 10 μm long. The thickness of each core is equal to 2 μm while the
distance between the cores is 2 μm. The operating wavelength is 1.55 μm whereas the refractive index in the core and the cladding is 3.3704 and 3.2874, respectively. The initial field
distribution corresponds to the fundamental mode.

on the propagation distance. Within the scalar approximation, this sum represents the total optical power that is guided by the waveguide. One would therefore expect that in a lossless waveguide this figure of merit does not change with the distance, when the Dirichlet boundary condition is applied. Figure 4.1 shows the dependence of total optical power on the distance for CN-FDBPM. For comparison, the dependence of power on distance for the explicit scheme was added. Two values of the propagating step were considered $\Delta z = 0.05$ and $\Delta z = 0.01$ μm. The value of $\left|\overline{A}\right|^2$ is constant only for the C-N scheme. For the explicit scheme, the value of $\left|\overline{A}\right|^2$ is approximately constant when Δz equals 0.01 μm. However, for $\Delta z = 0.05$ μm after propagating approximately 80 μm, the total optical power starts to grow quickly, which is a manifestation of the lack of stability of the explicit algorithm. It should be noted that if the propagation continued sufficiently long with $\Delta z = 0.01$ μm, the optical power would eventually start to grow quickly. The fact that the CN scheme preserves strictly the value of $\left|\overline{A}\right|^2$ is very important and is one of the reasons why this particular scheme is used in many BPM codes.

In subsequent sections we discuss the essential issues that relate to the implementation of BPM algorithms on a computer. These include handling of the artifacts generated by the limited dimensions of the computational domain (see Figure 2.10), accuracy analysis of a BPM algorithm, and the application of nonorthogonal coordinate systems in order to eliminate the staircasing approximation.

Boundary Condition

The boundary conditions allow the truncation of the calculation window (Figure 4.9). This is necessary when applying numerical methods to obtain a solution of the electromagnetic field boundary value problems. The application of Dirichlet or Neumann boundary condition for the truncation of the calculation window in BPM simulations results in spurious reflections (see Chapter 2). Hence, it was necessary to develop formulations applicable to the finite difference, FEM, mode matching or method of lines, which will allow waves incident on the boundary of the calculation window

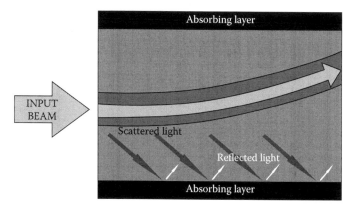

FIGURE 4.9 Schematic illustration of absorbing scattered light by an absorbing layer situated at the boundary of the calculation domain.

to leave the computational domain without producing spurious reflected waves. In general, there are three types of boundary conditions used in BPM: transparent boundary condition (TBC), radiation (or absorbing) boundary condition (RBC), and absorbing layers.

The TBC uses the information about the field distribution at the current longitudinal position to predict the field behaviour at the next step. The TBC was first proposed by Hadley [127] for paraxial FD-BPM calculations. It relies on the assumption that the field has a simple nearly plane wave exponential behaviour near the boundary of the computational domain. For a number of years, it was the most popular boundary condition used in BPM codes. It was very simple to apply and did not require additional nodes to be added to the FD mesh. However, this technique was not very effective for more complicated wave patterns like, for instance, a combination of two divergent Gaussian beams. A generalisation of the TBC is described in Song [128]. Instead of using the field distribution at the current longitudinal position, it uses field samples from all previous longitudinal positions to predict the field behaviour at the next step. However, this approach has so far been limited to the 2D case only.

A much more robust boundary condition can be imposed by using radiation boundary conditions. The RBCs use specially designed annihilation operators. The annihilation operators are constructed to annihilate the following: back reflected waves for a set of selected incidence angles [129]; reflections for waves, whose incidence angle is around a selected reference angle [9]; or several terms in the asymptotic far field expansion [9]. The application of RBCs to FD-BPM changes the sparsity pattern of the matrix L in 4.10 as additional nodes are included in the finite difference equations at a given node. As in the 2D case, this takes place only in the first and the last row of the matrix, the efficient direct LU decomposition based solver can still be efficiently applied [129]. The radiation boundary conditions have been so far applied to 2D and 3D scalar FD BPM analysis [130]. Recently, a novel RBC using complementary operators has been developed for FD-BPM. This technique relies on propagating the field twice, using two complementary operators as boundary conditions each time. Then the two calculated fields are averaged. The complementary operators are designed so that the spurious reflections cancel out. The disadvantage of this technique is that the calculation time is twice as long as it takes the simulation to be repeated, with the complementary operator as the boundary condition.

Alternatively the scattered light can be suppressed by applying specially designed absorbing layers. The absorbing layers are added outside of the computational domain to absorb spurious reflections caused by back-reflected outgoing waves (Figure 4.9). The absorbing layers have to be designed so that the attenuation of the layer is large enough to sufficiently suppress the outgoing waves. Simultaneously, the absorbing layer has to be matched to the medium so that it itself does not produce spurious reflections. The first absorbing layers were simply layers of complex refractive index. The real part of the refractive index matched the refractive index of the adjacent medium while the imaginary part was varied to suppress spurious reflections [131]. Much more efficient suppression of the reflected waves can be achieved by the application of Berenger's perfectly matched layers (PMLs) [132]. A PML introduces an anisotropic area with loss outside of the computational window [133]. This technique proved to be very robust and very effective at attenuating the waves impinging upon the boundary at large incidence angles. The waves leaving the computational domain experience no (minor) reflection and undergo attenuation whilst propagating in the PML. In fact, a PML should rather be understood as a mapping of real transverse coordinates onto the complex plane [129] or expressed alternatively as a complex coordinate stretching [45]. Although a PML requires additional nodal points to be included during the calculations, which results in an increased demand on the computer memory needed to store the coefficient matrices, it is the standard boundary condition used in today's BPM software, due to its robustness and simplicity of implementation. A PML is equally efficient when used for BPM propagation, discontinuities analysis, or modal calculations [44]. The PML is also straightforward to extend to 3D polarised and vectorial analysis.

Analytical expressions that are helpful in designing an optimal boundary condition for BPM simulations can be found in Vassallo and Collino [134] and Yevick et al. [135].

DISPERSION CHARACTERISTICS

When applying BPM to study optical propagation in planar waveguiding structures, the actual dispersion characteristics, implicit when using a particular BPM algorithm, are only approximating the real dispersion characteristics for the medium. This discrepancy stems from the fact that BPM algorithms rely on approximate representations of the exponential, square root, and L operators in 4.12.

In this section, a discussion of the numerical dispersion characteristics of BPM algorithms is given. In a homogenous medium, the dependence of the longitudinal component of the propagation vector on the transverse one is given by the square root function (cf. 2.9 in Chapter 2). However, if only a (n,n) Padé approximant in a series form is used to approximate the square root, then the following approximate relationship is obtained (cf. the section "Paraxial and Wide Angle Approximations"):

$$k_z = k\sqrt{1 - \frac{k_x^2}{k^2}} \approx k\left(1 + \sum_{i=1}^{n} \frac{-a_{\text{in}}\frac{k_x^2}{k^2}}{1 - b_{\text{in}}\frac{k_x^2}{k^2}}\right) \tag{4.29}$$

where the coefficients a_{in} and b_{in} are given by:

$$a_{\text{in}} = \frac{2}{2n+1}\sin^2\left(\frac{i\pi}{2n+1}\right), \quad b_{\text{in}} = \cos^2\left(\frac{i\pi}{2n+1}\right)$$

The quality of (n,n) Padé approximation depends on whether we consider the propagating or evanescent modes. Alternative methods include Padé expansion in the product form [136] and Thiele approximants [68] or formulations that avoid the square root operator [25,137]. We perform a fairly detailed analysis of the dispersion characteristics for the propagating modes first, and then discuss the evanescent part of the modal spectrum.

We use the dispersion analysis, which is based on the plane wave decomposition [138]. The starting point is Equation 4.12, which relates the field in the next and current longitudinal positions, that is:

$$\phi^{n+1} = \underline{Q}\ \phi^n \tag{4.30}$$

where Q is the operator on the right-hand side of 4.12 acting on $\Phi(z_0)$ while $\varphi^{n+1} = \Phi(z_0 + \Delta z)$ and $\varphi^n = \Phi(z_0)$. Substituting $\Phi(z) = \exp(k_z z + jk_x x)$ into 4.30 yields an eigenvalue problem:

$$\nu\ \phi^n = Q\ \phi^n \tag{4.31}$$

The eigenvalues ν of Equation 4.31 can be calculated analytically and related to the the longitudinal component of the propagation vector k_z:

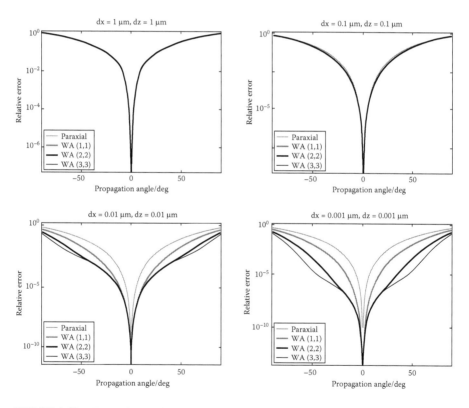

FIGURE 4.10 Dependence of the relative error in the z component of the wave vector on the propagation angle for selected values of longitudinal and transverse mesh size and paraxial and wide angle (WA) orders approximations of the square root operator. The refractive index of the medium is 3.262, while the reference refractive index matches that of the medium. The operating wavelength is 0.732 μm. A (1,1) Padé approximation to the exponential operator and a three-point finite difference scheme for the L operator were used.

$$k_z = -1 / \Delta z \ \mathrm{Arg}(\nu) \qquad (4.32)$$

Figure 4.10 shows the dependence of the relative error in the z-component of the wave vector on the propagation angle calculated from 4.32 for a 2D FD-BPM using CN scheme (i.e., [1,1] Padé approximation of the exponential operator for the representation of the exponential operator) and paraxial and (n,n) Padé approximations for the square root operator. The relative error has been calculated as $(k_z - k_{\mathrm{exact}})/k$ where k_z and k_{exact} are the calculated and exact values of the z-component of the wave vector and k stands for the medium wave number. The second derivative has been approximated using the standard central three point finite difference scheme. The analysis has been carried out for a medium with the refractive index of 3.262, which is a typical value of the effective index for a semiconductor laser epitaxy [139]. The operating wavelength is 732 nm. The wave propagation angle was calculated from the ratio of the transverse components of the wave vector. Figure 4.8 shows that higher order WA approximations to the square root operator achieve accurate results for wider range of propagation angles. However, if the FD mesh resolution is low, there is no noticeable advantage in using high order WA schemes. The difference between the WA BPM and the paraxial scheme becomes pronounced only if a sufficiently fine FD mesh is applied. Consequently, higher order WA schemes have to be applied with a sufficiently fine FD mesh in order to achieve the maximum accuracy that is possible within the given WA approximation order.

The accuracy at a given transverse mesh size can be improved by applying higher order finite difference approximations to the differential operator L in Equation 4.12. In order to show the advantages of the higher order FD approximations, we recalculated the results from Figure 4.10 using the central difference 5 point FD approximation of the second derivative in the operator L. In order to calculate the central difference 5 point FD approximation, we used the following series expansion [140]:

$$D^2 = \frac{1}{\Delta x^2}\left(\delta^2 - \frac{1}{12}\delta^4 + \frac{1}{90}\delta^6 - \frac{1}{560}\delta^8 + \cdots \right)$$

where D is the derivative operator, $-d/dx$, while δ is the central difference operator. We present the dispersion characteristics obtained in Figure 4.11. These results show that the higher order FD approximations improve the accuracy especially when the FD mesh is relatively low. However, if the FD mesh is sufficiently fine no further improvement in accuracy is achieved either through applying higher order FD schemes or by refining the FD mesh. This observation leads to a conclusion that there is a value of FD mesh size for which the maximum possible accuracy is attained within the selected FD approximation order. The results from Figure 4.11 show that the numerical value of this FD mesh size depends also on the WA approximation order.

The accuracy of a BPM algorithm for a given longitudinal step can be improved by using higher order Padé approximations for the exponential operator. Figure 4.12 compares the dispersion characteristics of a 2D FD-BPM using a (2,2) Padé approximation

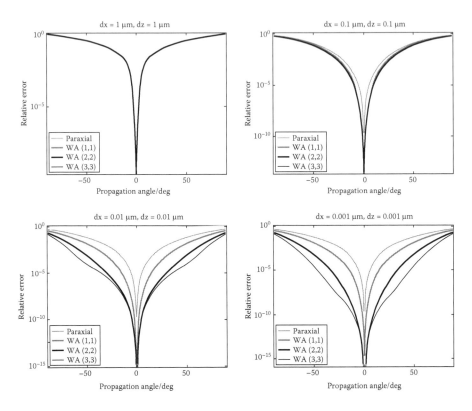

FIGURE 4.11 Dependence of the relative error in the z component of the wave vector on the propagation angle for selected values of longitudinal and transverse mesh size and paraxial and wide angle (WA) orders approximations of the square root operator. The refractive index of the medium is 3.262, while the reference refractive index matches that of the medium. The operating wavelength is 0.732 μm. A (1,1) Padé approximation to the exponential operator and a five-point finite difference scheme for the L operator were used.

for the representation of the exponential operator with the standard (1,1) one. All other simulation parameters are the same as the results presented in Figure 4.10. Again improvement in accuracy is achieved for a large mesh size. However, with a fine FD mesh, no advantage is offered by the higher order Padé approximation.

In the next example, we consider the sensitivity of the error to the reference refractive index. This problem is important because in many practical situations, it is difficult to calculate the reference refractive index in such a way that all longitudinal phase variations are eliminated. For instance, in 3D BPM simulations, the calculation of the local mode effective index is very costly in terms of computation time and is hence (best) avoided. Also when studying multimode propagation, for example, when studying the optical properties of MMI couplers, it is impossible to match the longitudinal phase variations for all modes using a single value of the reference refractive index. Figure 4.13 shows the relative error in the z component of the wave vector of a 2D FD-BPM when the reference refractive index does not match the refractive index

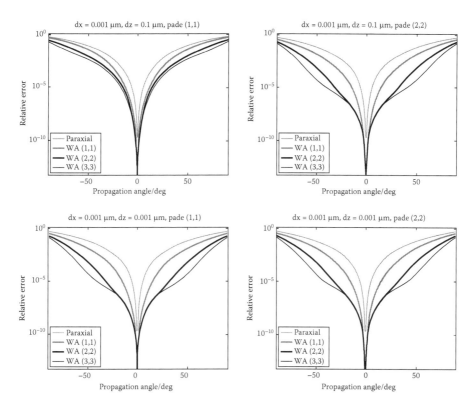

FIGURE 4.12 Comparison of dependence of the relative error in the z component of the wave vector on the propagation angle for (1,1) and (2,2) Padé approximation of the exponential operator. A three-point finite difference scheme was used. The refractive index of the medium is 3.262, while the reference refractive index matches that of the medium. The operating wavelength is 0.732 μm.

of the medium and equals 1. All other simulation parameters are the same as in the example from Figure 4.10. In this example, the difference between the refractive index of the medium and the reference refractive index has been exaggerated on purpose in order to show clearly the consequences of the reference refractive index mismatch. The results from Figure 4.13 confirm that if the reference refractive index does not match the refractive index of the medium the error might be large, particularly for waves propagating along the z-axis. Interestingly, for the paraxial scheme this error can even increase when the mesh is refined. The only way of reducing the error resulting from the reference refractive index mismatch is the application of high order wide angle (WA) Padé approximations to the square root operator together with a fine FD mesh.

The results from Figures 4.10 through 4.12 show that the reduction of the error for beams propagating at a large angle with respect to z-axis can only be achieved with large WA approximation orders and a fine FD mesh. Applying high WA orders and fine FD mesh, however, is not numerically efficient. On the other hand, the results from Figure 4.13 show that it is possible to obtain a small value of error at large

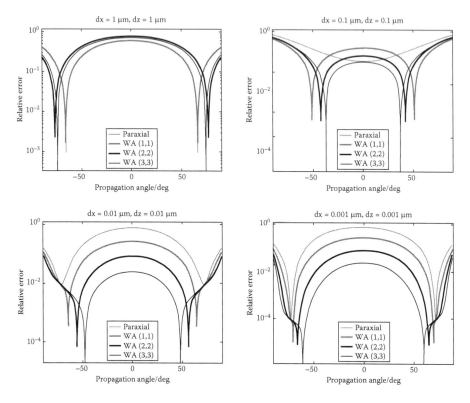

FIGURE 4.13 Dependence of the relative error in the z component of the wave vector on the propagation angle for selected values of longitudinal and transverse mesh size and paraxial and wide angle (WA) orders approximations of the square root operator. The refractive index of the medium is 3.262, while the reference refractive index equals 1. The operating wavelength is 0.732 μm. A (1,1) Padé approximation, to the exponential operator and a 3 point finite difference scheme for the L operator were used.

values of the propagation angle, even using the paraxial approximation. This property of the BPM can be exploited conveniently by applying the Generalised–BPM (G-BPM) algorithm [141]. In the G-BPM, the envelope function is extracted in such a way that the preferred angle of propagation θ can be selected in an arbitrary way with respect to the z-axis [141]:

$$F = \Phi \, \exp\left(-j\beta\cos\theta \; z - j\beta\sin\theta \; x\right).$$

Substituting the so-defined envelope function into 4.9 yields the G-BPM algorithm [141]. Figure 4.14 shows the dependence of the relative error in the z-component of the wave vector on the propagation angle of the G-BPM for the envelope angle $\theta = 40° - a$) and $\theta = 80° - b$). The longitudinal and transverse mesh size is equal to

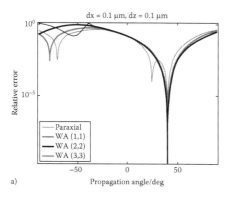
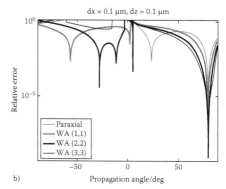

FIGURE 4.14 Dependence of the relative error in the z component of the wave vector on the propagation angle of the generalised rectangular BPM for the envelope angle a) $\theta = 40°$ and b) $\theta = 80°$. The longitudinal and transverse mesh size is equal to 0.1 μm. The refractive index of the medium is 3.262, while the reference refractive index matches that of the medium.

0.1 μm. The refractive index of the medium is 3.262, while the reference refractive index matches that of the medium. Figure 4.14 shows that the error is minimised around the propagation angles that are equal to θ, even with low FD mesh resolution and at low WA approximation orders.

So far, we have discussed the errors that result from applying BPM to study the propagating mode evolution. However, in many cases, in order to provide predictive results, BPM needs to model accurately both propagating and evanescent modes. This task cannot be accomplished using standard WA Padé approximations for the square root operator with real expansion coefficients, cf. the section "Paraxial and Wide Angle Approximations." This results from the fact that standard Padé approximants to the square root operator (and paraxial approximation) do not model correctly the evanescent part of the modal spectrum. Consequently, the evanescent modes propagate as if they were propagation modes, cf. the section "Paraxial and Wide Angle Approximations." This results in particularly large errors when handling wave reflections [113,142]. In order to remedy this deficiency, a number of techniques were suggested, mostly for studying the propagation of acoustic waves [143–148]. In the field of photonics, the approach that found most wide spread application consists in using Padé approximants with real expansion coefficients whilst using either a complex value of the reference refractive index (for instance, by rotation in the complex plane) or rotating the square root branch cut. Both techniques have been shown to be equivalent in Rao et al. [142]. The main advantage of such an approach is that BPM algorithms with arbitrary high approximation order can be easily implemented. A technique similar in nature that consists of shifting in the complex plane the zeros and poles of the standard Padé series was presented in Petruskevicius [114].

More recently techniques have been proposed that use Padé approximants directly for the exponential and square root operators [147]. One of such techniques relies on the use of $[(p - 1)/p]$ Padé approximants [149], which are found more accurate than the standard $[p/p]$ approximation for the evanescent part of the modal spectrum. However, $[(p - 1)/p]$ Padé approximants reduce the accuracy for the propagating part of the spectrum. Alternatively, a CN scheme with the θ parameter near but not equal to ½ can be used [150]. Similarly to the study done by Lu and Ho [149], this method is capable of reproducing the evanescent wave behaviour, but was found to suppress propagating modes [150].

A separate issue that is linked with the use of Padé approximants is the stability of the BPM algorithm. There are two separate problems that can be encountered in this context. One of them is the mapping of the matrix eigenvalues by a particular BPM algorithm that uses Padé approximants. As such algorithms typically result in a solution of systems of algebraic equations, it is in principle possible that shifts to the eigenvalues applied by Padé expansion coefficients bring one of the coefficient matrix eigenvalues near to zero. This renders the matrix badly conditioned, which in turn can result in significant numerical errors being incurred during the matrix inversion. Such errors can build up and lead to numerical instabilities [151]. The other problem is linked with the propagation of the complex modes, and the mapping of their propagation constants, by the Padé approximants. The properties of the complex modes in lossless optical waveguides have been studied by Jablonski [152]. In fact, the complex eigenvalues can be brought into BPM simulations through a nonsymmetric matrix representation of the operator L in 4.10. Such a situation is even possible in the 2D TE case when applying high order accurate FD approximations [153,154]. The presence of such eigenvalues can render a BPM algorithm unstable [155]. This issue is particularly difficult to handle in the case of full vectorial FD-BPM algorithms whereby symmetrical FD representations are not available [155,156]. Several methods for overcoming this problem have been suggested in [155].

Lastly, the problem of BPM algorithm instability should not be confused with the problem of power conservation by BPM algorithms. The power conservation in vectorial BPM simulations can be formally imposed by propagating both the electric and magnetic fields [157]. However, even when both fields are propagated the BPM algorithm may still be unstable due to the presence of the complex eigenvalues [158]. The issue of power conservation in BPM simulations is discussed in Yamauchi [17] and Vassallo [159].

STAIRCASING APPROXIMATION

When the boundaries between the regions with different values of the refractive index are not parallel to the coordinate system axes, it is necessary to use the staircasing approximation. Such situation occurs, for example, when trying to use BPM in the rectangular coordinate system to analyse optical beam propagation along an obliquely positioned optical waveguide, bent optical waveguide, or an optical taper. Figure 4.15 illustrates the idea of the staircasing approximation whereby a curved interface between two regions with different values of the refractive index is approximated using a series of stairs. When the staircasing approximation is used in BPM, spurious scattering can be observed in the calculated filed distribution [87]. Several techniques have been developed to suppress this effect. One approach consists of the

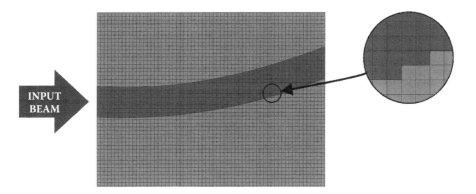

FIGURE 4.15   Schematic illustration of staircasing in the case of a bent optical waveguide.

application of advanced FD-BPM schemes [160]. Alternatively, specially designed coordinate systems can be introduced that allow the changes of the refractive index distribution to be followed exactly during the beam propagation. Both orthogonal and nonorthogonal coordinate systems were used for this purpose. The orthogonal coordinate systems, are relatively easy to implement, but cannot be directly interfaced with the standard rectangular coordinate system [51]. This deficiency can be removed by applying a conformal mapping. However, conformal mapping does not eliminate staircasing completely and is limited to the 2D case only [161]. More recently an energy conserving BPM algorithm was developed that approximates curved boundaries with slanted walls [162]. However, so far it can be applied in the 2D TE case only. The application of BPM in nonorthogonal coordinate systems on the other hand eliminates the need for staircasing while allowing for direct interfacing with standard BPM in rectangular coordinate system. Further, nonorthogonal coordinates can be easily implemented in 3D BPM algorithms and allow for adapting the size and position of the computational window during the propagation [87], which is difficult to implement in rectangular coordinate system (so far only shifting of the computational window was demonstrated [163]).

There are two nonorthogonal coordinate systems which are most widely used in the analysis of the planar waveguiding structures, that is, the tapered coordinate system and the oblique coordinate system. In this section, BPM algorithms in both these coordinate systems will be derived and analysed. Not to obscure the main idea, we limit the discussion to the 2D TE case only. The discussion of the extension of this technique to 3D BPM can be found in Sujecki et al. [87].

First, we consider BPM in the tapered coordinate system. The concept of the tapered coordinate system BPM stems from Sewell et al. [87]. The tapered coordinates t and z are related to the rectangular ones x and z by: $x = tz$ and $z = z$ (Figure 4.16). After changing the variables in 4.9 the one-way wave equation in the tapered coordinate system is obtained [164]:

$$\left(\frac{\partial}{\partial z} - \frac{t}{z} \frac{\partial}{\partial t} \right) \psi(t,z) = \pm j \sqrt{L_t + \mu_0 \varepsilon \omega^2} \; \psi(t,z) \qquad (4.33)$$

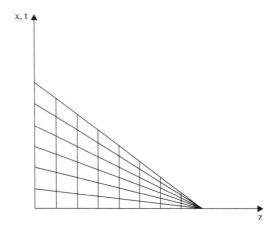

FIGURE 4.16   Tapered coordinate system.

where L_t is given by:

$$L_t = \frac{1}{z^2}\frac{\partial^2}{\partial t^2}$$

After introducing an envelope function Φ, which is related to ψ through: $\psi = \Phi$ exp $(-j\,\beta\,z)$, Equation 4.33 takes the form:

$$\frac{\partial}{\partial z}\varphi(t,z) = \left(\frac{t}{z}\frac{\partial}{\partial t} - j\beta\left[\sqrt{1+\frac{L_t+\mu_0\varepsilon\omega^2}{\beta^2}}-1\right]\right)\varphi(t,z) \qquad (4.34)$$

Equation 4.34 can be used to derive BPM algorithms using the standard methods that were elaborated for the BPM in the rectangular coordinate system [87,165]. It is noted that other envelope functions can also be used in the tapered coordinate system to better fit the wave front phase variations [87].

Similarly, in the oblique coordinate system (Figure 4.17), variables t and z are related to the rectangular ones x and z by: $t = x - z \tan \theta$ and $z = z$ [166]. After changing the variables in 4.9, the one-way wave equation in the oblique coordinate system is obtained [166]:

$$\left[\sec^2\theta\frac{\partial^2}{\partial t^2} - 2\tan\theta\frac{\partial^2}{\partial z\partial t} - 2j\beta\cos\theta\frac{\partial}{\partial z} + \frac{\partial^2}{\partial z^2} + k^2 - \beta^2\right]\psi(t,z) = 0 \quad (4.35)$$

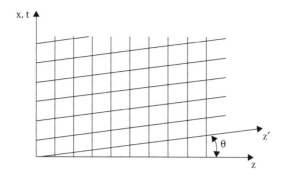

FIGURE 4.17 Oblique coordinate system.

We note that an alternative formulation of BPM in the oblique coordinate system can be obtained applying the following variable change: $t = x - z \sin \theta$ and $w = z \cos \theta$ [86]. The Equation 4.35 can be recast into a more convenient form [165]:

$$\left(\frac{\partial}{\partial z} - \tan(\theta) \frac{\partial}{\partial t} \right) \psi(t,z) = \pm j \sqrt{L' + \mu_0 \varepsilon \omega^2} \ \psi(t,z) \tag{4.36}$$

where the operator L' is given by:

$$L' = \frac{\partial^2}{\partial t^2}$$

After introducing an envelope function Φ, which is related to ψ by

$$\psi = \phi \exp \left[-j \left(\frac{\beta}{\cos(\theta)} \right) + \beta \sin(\theta) t \right]$$

we obtain from 4.36 the equation for the envelope function Φ:

$$\frac{\partial}{\partial z} \varphi(t,z) = \left(M_o - j \left[\sqrt{-M_o^2 + L_o} \right] \right) \varphi(t,z) \tag{4.37}$$

where the operators L_o and M_o are:

$$M_o = j\beta \cos(\theta) + \tan(\theta) \frac{\partial}{\partial t}$$

$$L_o = \left(1 + \tan(\theta) \right) \frac{\partial^2}{\partial t^2} + \mu_0 \varepsilon \omega^2 - \beta^2$$

Again, Equation 4.37 can be used to derive BPM algorithms using the standard techniques developed for BPM in the rectangular coordinate system [167].

It should be also noted that nonorthogonal coordinate systems can be designed to follow an arbitrary shape of the boundary between two dielectrics [85,87,88,168]. As such, BPM in the nonorthogonal system can be considered as a generalisation of the standard rectangular BPM. A number of application examples for BPM in the nonorthogonal coordinate system are given in the section "Selected Examples of BPM Application."

Similarly, as in the rectangular coordinate system, the dispersion analysis based on the plane wave decomposition [138] can be carried out for BPM algorithms in the oblique coordinate system. Again, setting the solution in the form:

$$\varphi^n(u,z) = e^{-jsu} = e^{-js(x-z\tan\theta)}, \quad s = k(\sin\varphi - \sin\theta)$$

into the equation relating the field distribution at the current and next position yields an eigenvalue problem:

$$\nu \underline{P} \ \varphi^n = Q \ \varphi^n \qquad\qquad (4.38)$$

The eigenvalues of Equation 4.38 relate to the longitudinal component of the wave vector via:

$$\beta = -1/dz \ \mathrm{Arg}(\nu).$$

Hence solving 4.38 yields the dispersion characteristics.

In order to illustrate the advantages of using BPM in the oblique coordinate system we calculated the dependence of the longitudinal (z) component of the wave vector on the propagation angle for a) $\theta = 20°$, b) $\theta = 50°$ and c) $\theta = 90°$, using an oblique FD-BPM algorithm [167] (Figure 4.18). The longitudinal and transverse mesh size is equal to 0.01 μm. The refractive index of the medium is 3.262, while the reference refractive index matches that of the medium. The results from Figure 4.18 show that, by an appropriate selection of the oblique coordinate angle θ and of the wide angle approximation order (similarly to G-BPM), the BPM in the oblique coordinate system can be optimised to cover any angular range using low-order wide angle approximations and moderate finite difference mesh size. The discussion of the other properties of the BPM algorithms in the nonorthogonal coordinate system is given in the section "Selected Examples of BPM Application."

SELECTED EXAMPLES OF BPM APPLICATION

Each photonic device can be usually decomposed into a concatenation of several simpler elements. Each element can then be then designed separately by specially tailored modelling techniques. Such an approach allows significant reduction in the computational resources needed. It is possible to identify several basic elements that

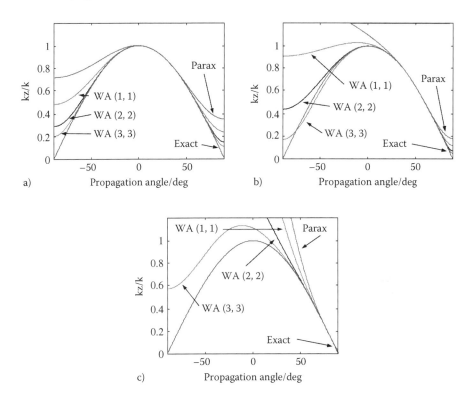

FIGURE 4.18 Dependence of z component of the wave vector on the propagation angle for a) $\theta = 20°$, b) $\theta = 50°$, and c) $\theta = 90°$ in oblique coordinate. The longitudinal and transverse mesh size is equal to 0.01 μm. The refractive index of the medium is 3.262, while the reference refractive index matches that of the medium.

are often used in photonic devices: straight and oblique waveguides, bent waveguides, tapered waveguides, and Y junctions (Figure 4.19).

An extensive literature is available on modelling straight and oblique waveguides, bent waveguides, tapered waveguides, and Y junctions. In this section, the application of BPM to the analysis of these selected elements of photonic devices is discussed. The discussion is underpinned with several illustrating examples of BPM application.

OPTICAL TAPER

Tapered optical waveguides find application as spatial mode transformers and also in high power tapered lasers and amplifiers to reduce the photon density at a given output power level. An accurate analysis of tapered structures is complicated by the nonseparability of the wave equation in standard coordinate systems. One of the ways in which optical tapers can be analysed is based on the concept of intrinsic taper modes [169]. The concept of intrinsic taper modes aids physical understanding, but the resulting analysis method is limited so far to the 2D case only. Alternatively, for

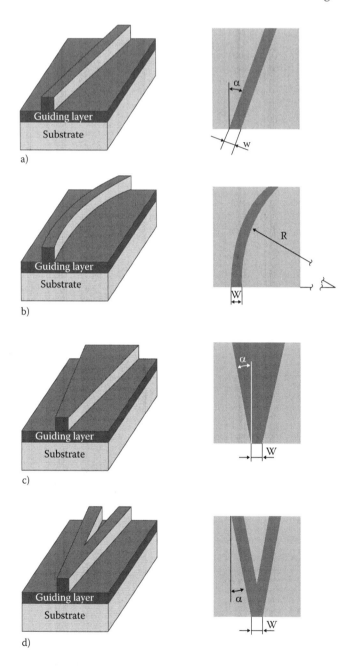

FIGURE 4.19 Examples of basic elements of photonic devices realised in planar technology: a) oblique rib waveguide, b) bent rib waveguide, c) tapered rib waveguide, and d) Y junction.

some taper structures, a conformal mapping approach can be employed, but again this approach is limited to the 2D case only [83]. The classical approach to the analysis of tapered structures consists in approximating it with a concatenation of straight wave-guides of varying width. Such an approximate structure can be then analysed using the normal mode approach; in which, the field is decomposed into the sum of guided modes of a given cross-section while the transmission through interfaces between the waveguides of different width is estimated by calculating overlap integrals [170]. The normal mode approach is efficient only when the taper transitions are nearly adia-batic. Similarly, the mode expansion BPM is not efficient when applied to a general optical taper structure because the full modal spectrum must be calculated at each step, which is particularly demanding in terms of computing resources, especially in the 3D case. The matrix expansion BPM, on the other hand, is straightforward to apply and efficient for both uniformly and nonuniformly tapered structures.

The efficiency of matrix expansion BPM may be further improved by the applica-tion of the nonorthogonal tapered coordinate system. By choosing to formulate the problem, and subsequently implementing the algorithm, in a nonorthogonal coordi-nate system, which follows exactly the boundaries between dielectrics of different refractive index, it is possible to describe sufficiently accurately the physical geometry of the problem with less sampling points [84]. This allows a substantially coarser transverse mesh and a larger propagation step to be used, which results in a significant reduction in the computational overhead required to solve the problem. Moreover the numerical noise present in the results of BPM simulations in the rectangular coor-dinate system is significantly suppressed when a tapered coordinate system is used. Alternatively, an orthogonal cylindrical coordinate system can be applied to analyse optical tapers. However, it is not straightforward to concatenate the BPM in the cylin-drical coordinate system with the standard BPM in the rectangular coordinate system.

Figure 4.20 compares the field intensity distributions calculated by rectangular and tapered BPMs. The air clad taper is decreased from a width of 0.8 to 0.4 µm over a distance of 22.9 µm (Figure 4.21). The refractive indices in the core and sub-strate are 3.3 and 3.17, respectively. The operating wavelength is 1.55 µm. The results presented are normalised with respect to the maximum field intensity of the input field distribution. In the contour plots, the lines of the constant field intensity are given starting at 0.1 of the maximum value with an increment of 0.2. The unphysi-cal numerical noise, which is present in the results obtained by the rectangular WA-BPM, can only be suppressed by applying a very fine transverse mesh size. However, in the results obtained by the tapered WA-BPM, the numerical noise is significantly reduced even when the FD mesh is coarse [87,164,165].

Oblique and Bent Waveguides

Optical bends are used to change locally the direction of the optical beam propagation. An optical beam experiences optical loss, off axis shift, and a polarisation rotation while propagating through a bent waveguide [171,172]. A number of approximate techniques have been developed to calculate the bending loss of uniform waveguides. A useful recent review can be found in Kim et al. [173]. The main shortcoming of approximate methods is that they can be applied to uniform, circular bends only.

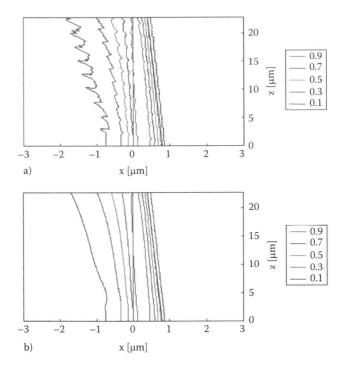

FIGURE 4.20   Contour plot of field intensity for air clad tapered structure from Figure 4.20 obtained by rectangular (1,1) a) WA-BPM and tapered (1,1) b) WA-BPM. $H = 0.8$ μm, $h = 0.4$μm, $L = 22.9$ μm, $n_1 = 1$, $n_2 = 3.3$, $n_3 = 3.17$, and $\lambda = 1.55$ μm, while the longitudinal and transverse mesh sizes are equal to 0.229 and 0.04 μm, respectively. (From Sujecki, S., New beam propagation algorithm for optical tapers. *International Journal of Electronics and Communications (AEU)*, 2001. 55(3): p. 185–190.)

For the circular bends, an eigenvalue matrix equation can be formulated and solved for the propagation constants of the vectorial eigenmodes [174]. This technique is relatively efficient but limited to regular, circular bends only. In the case of low refractive index contrast structures, the free space radiation mode method results in a very efficient algorithm for circular bends [175]. The BPM method, on the other hand, can be easily applied to the analysis of arbitrarily shaped bends using curvilinear coordinate systems [176]. Circular bends can also be efficiently analysed using BPM, in either the local cylindrical or the sectorial coordinate system [168,177]. In fact, for a large radius and low loss bends, a straight waveguide approximation is typically used. This approximation reduces the analysis of optical beam propagation in a bent waveguide to the optical beam propagation in a straight waveguide with an appropriately modified transverse refractive index distribution [178].

When a low refractive index contrast waveguide is turned only by a small angle, the most efficient technique is the standard BPM in the rectangular coordinate system. Such bends are used for instance in bent waveguide semiconductor laser cavities, and have been shown to improve the beam quality and increase the maximum

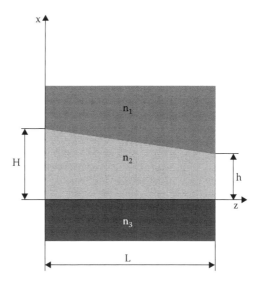

FIGURE 4.21 Asymmetric 2D taper structure.

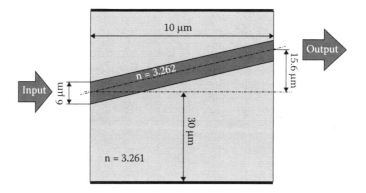

FIGURE 4.22 Example oblique waveguide structure. (From Sujecki, S., Generalised rectangular finite difference beam propagation method. *Applied Optics*, 2008. 47(23): p. 4280–4286.)

bias current of kink-free operation [179]. Instead of staircasing, a bent waveguide can be approximated by a concatenation of oblique waveguides (Figure 4.22) and using a BPM algorithm in a nonorthogonal coordinate system.

Oblique optical waveguides are used in photonic devices, for example, directional couplers. Modelling of an optical beam propagation in an oblique waveguide is not straightforward due to the existence of the numerical dispersion [180]. When the tilt angle is large, high order Padé approximations for the square root operator must be used. Moreover, spurious numerical noise may be present in the field distributions if staircasing approximation is used. The suppression of such noise is difficult, especially, in the high refractive index contrast waveguides. However, when applying an oblique coordinate system, modelling of an optical beam propagation in an oblique waveguide can be accomplished using paraxial BPM algorithms even at large wave-guide tilt angles [86,166].

As an example, we consider a tilted waveguide structure shown in Figure 4.22. The fundamental mode is launched along the waveguide, that is, its propagation angle is equal to the waveguide tilt angle. It is therefore expected that for an accurate calculation the mode field distribution maximum should stay in the centre of the waveguide. The width of the window is 60 μm, the waveguide width is 6 μm, and the propagation distance measured along the z-axis is 10 μm. The waveguide offset at the output is approximately equal to 15.6 μm (waveguide tilt angle is 1 rad). The operating wavelength λ is 732 nm, while the refractive index in the core is 3.262. The refractive index in the cladding is 3.261. In all calculations, zero field was assumed at the boundary of the calculation window, while the window itself was selected wide enough so that the results were not significantly affected.

Figure 4.23 shows the dependence of the field offset at the output of the test structure (Figure 4.22) on the WA order and oblique coordinate system angle θ (Figure 4.17) for selected values of the mesh size, whereby θ = 0 (Figure 4.23a) cor-responds to the rectangular coordinate system and θ = 0.5 rad corresponds to the oblique coordinate system (Figure 4.23b). The transverse and longitudinal mesh size used in these calculations is 0.05 μm. The field offset is defined as the distance of the position of the light intensity maximum from the centre of the calculation window. (For an accurate result, the field maximum should be positioned in the centre of the waveguide.) The results shown in Figure 4.23 confirm that the error in the angle of propagation cannot be reduced by simply increasing the BPM wide angle order. In order to take advantage of higher BPM wide angle orders, to improve the accuracy, it is necessary also to refine the FD mesh. A faster convergence of the oblique BPM, both with the FD mesh size and the WA order, demonstrates the advantage of being

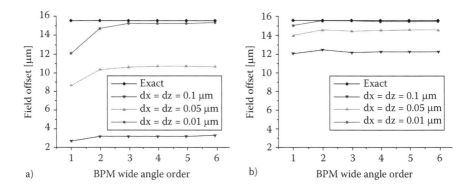

FIGURE 4.23 Dependence of field offset on the wide angle approximation order for the test waveguide structure from Figure 4.22; for a) θ = 0 rad and b) θ = 0.5 rad. (From Sujecki, S., Wide-angle, finite-difference beam propagation in oblique coordinate system. *Journal of the Optical Society of America A-Optics Image Science and Vision*, 2008. 25(1): p. 138–145.)

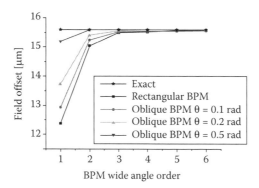

FIGURE 4.24   Dependence of field offset on the wide angle approximation order for the test waveguide structure from Figure 4.22. The longitudinal and transverse mesh size is equal to 0.005 μm. (From Sujecki, S., Wide-angle, finite-difference beam propagation in oblique coordinate system. *Journal of the Optical Society of America A-Optics Image Science and Vision*, 2008. 25(1): p. 138–145.)

able to select the preferred direction of propagation nearer to the actual power flow direction (Figure 4.23).

Figure 4.24 shows the dependence of the field offset at the output of the test structure on the WA order for the selected values of the oblique coordinate system tilt angle θ (Figure 4.17). The transverse and longitudinal mesh size used in these calculations is 0.005 μm. These results further demonstrate the advantage given by the arbitrary selection of the preferred direction of propagation. Even when the oblique coordinate system tilt angle θ is small, when compared to the waveguide tilt angle, the convergence of the results is accelerated significantly by applying the oblique coordinate system.

Y JUNCTION

Y junctions are used in power splitters and combiners, modulators, and Mach-Zehnder interferometres. A number of techniques have been applied to the numerical analysis of Y junctions. The most straightforward ones are based on the normal mode theory [181–183]. In 2D, this approach allows to derive analytical expressions for Y junctions made of step index waveguides [184]. Also volume current method was used to study Y junctions [185]. However, these techniques [181–185] are very efficient in 2D case only and, with the exception of Cascio et al. [182] and Kuznetsov [185], do not include the spectrum of radiation modes. Matrix expansion BPM, on the other hand, is a straightforward to extend the analysis of Y junctions with arbitrary refractive index distribution, and to perform their full 3D vectorial analysis. In the rectangular coordinate system, BPM cannot model exactly the oblique boundaries between dielectrics with different values of the refractive index; it has to rely on the staircasing approximation. This results in an unphysical noise present in the results calculated by the rectangular BPM. The nonorthogonal BPM algorithms on

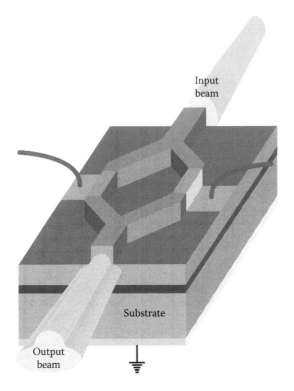

Input
beam

Substrate

Output
beam

FIGURE 4.25   Mach-Zehnder interferometre.

the other hand allow modelling exactly the oblique boundaries. As a result accurate
calculations can be performed with a lower mesh resolution [85].

In Sewell et al. [85], the bi-oblique coordinate system was applied to analyse
Y junctions. However, a more efficient algorithm can be obtained if the tapered and
bi-oblique coordinate systems are concatenated [88].

As an illustrative example, the nonorthogonal BPM has been applied to a 2D
analysis of a Mach-Zehnder interferometre (Figure 4.25). A Mach-Zehnder interferometre is usually applied as an amplitude modulator. The amplitude of the output
beam can be controlled by applying, for example, an appropriate voltage between the
arms of the interferometre. Due to an electro-optic effect, a difference in the optical
paths in both arms of the interferometre is created, which results in either a con-
structive or destructive interference at the output. A Mach-Zehnder interferometre
is a concatenation of a straight waveguide, tapered, and Y-junction sections, which
makes this device well suited for BPM analysis.

The dimensions of the Mach-Zehnder studied are, using the notation shown
in Figure 4.26, $L_1 = 4$ μm, $L_2 = L_6 = 125$ μm, $L_3 = L_5 = 1000$ μm, $L_4 = 2256$ μm,
$L_7 = 160$ μm, $W_1 = W_5 = 1.5$ μm, $W_2 = W_4 = 3$ μm, $W_3 = 3$ μm, and $O = 5.4$ μm. The
operating wavelength is 1 μm, and $n_1 = 1.01$ and $n_2 = 1.0$. FD mesh size values are
given in Figure 4.26. The device was modelled by a succession of BPM algorithms in
rectangular, tapered, bi-oblique, rectangular, bi-oblique, and rectangular coordinates.

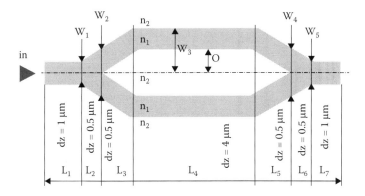

FIGURE 4.26 Dimensions and finite difference mesh size for example Mach-Zehnder interferometre.

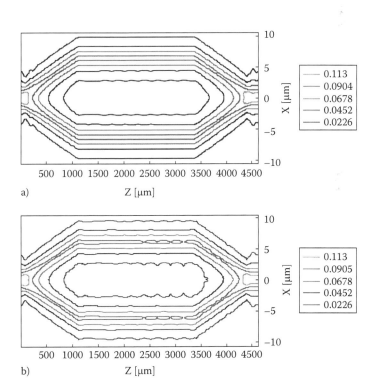

FIGURE 4.27 Contour plot of field intensity calculated for Mach-Zehnder interferometre structure from Figure 4.26: a) nonorthogonal coordinate BPM and b) rectangular coordinate BPM. (From Benson, T.M., et al., Structure related beam propagation. *Optical and Quantum Electronics*, 1999. 31(9): p. 689–703.)

Results are compared with the rectangular BPM, with the same transverse mesh at the input of the device. Figure 4.27 compares the field distribution calculated by both algorithms. It can be observed that the nonorthogonal BPM suffers much less from numerical noise at a given mesh size than the BPM in a rectangular coordinate system.

TIME DOMAIN ANALYSIS

Standard BPM algorithms that were discussed in sections "Introduction" through "Selected Examples of BPM Application" are well suited to handle large computational domains and hence have found many applications in the design and modelling of photonic devices. BPM is particularly efficient if the propagation of light is unidirectional. If many reflected waves need to be included, then the efficiency of BPM calculations quickly diminishes. In such cases either the direct solution of the wave equation in the frequency domain [186] or a time domain approach become more attractive.

The time domain methods were initially developed for calculating the electromagnetic field distributions in the radio frequency range. Several techniques have a widespread use, that is, the finite difference time domain method (FD-TDM) [4], finite element time domain method (FE-TDM) [187,188], and transmission line method (TLM) [189–191]. FE-TDM and FD-TDM were subsequently adopted for use in photonics [192–194]. The main shortcoming of FD-TDM and FE-TDM is that their application requires substantial computational resources and results in a relatively long calculation time. Therefore, a large effort has been invested in improving the computational efficiency of the time domain methods when applied in photonics. Initially, techniques were proposed that exploit the scalar and semivectorial (polarised) approximations discussed in this and Chapter 3 [195,196]. Then, BPM techniques that were discussed in this chapter were exploited to improve the efficiency of the time domain algorithms. This resulted in the development of the time domain beam propagation method (TD-BPM), which we will briefly discuss in the first subsection. Another, time domain approach that is well established in photonics, especially in the field of laser modelling, is the travelling wave approach (TWA). Hence, we discuss TWA in the second subsection.

TIME DOMAIN BPM

The time domain beam propagation method (TD-BPM) solves the wave equation for a suitably selected envelope function. Since, this is, in principle, the same idea as in the case of the standard BPM the resulting algorithms resemble those developed for the standard BPM. In order to explain the basic concepts, let's consider the vectorial wave equation for the electric field vector (chapter 9 in [17]) and substitute $\vec{E} = \widehat{E} * \exp(j\omega t)$. This yields the following equation for the envelope function $\widehat{E}$:

$$\nabla \times \nabla \times \widehat{E} = -\mu_0 \varepsilon \left(\frac{\partial^2 \widehat{E}}{\partial t^2} + 2j\omega \frac{\partial \widehat{E}}{\partial t} - \omega^2 \widehat{E} \right) \tag{4.39}$$

Applying the slowly varying envelope approximation (SVEA):

$$\left|\frac{\partial^2 \hat{E}}{\partial t^2}\right| << 2\omega\left|\frac{\partial \hat{E}}{\partial t}\right| \tag{4.40}$$

and reordering the terms yields the following equation:

$$-2\text{j}\omega\mu_0\varepsilon\frac{\partial \hat{E}}{\partial t} = \nabla\times\nabla\times\hat{E} + \omega^2\mu_0\varepsilon\hat{E} \tag{4.41}$$

which in essence is equivalent to applying the paraxial approximation to the solution of Equation 4.10. Consequently, the numerical algorithms used for the solution of 4.41 mimic the algorithms used for the solution of 4.10 in the paraxial approximation. Also, the hierarchy of approximations including the scalar, polarised, and lower dimensional ones that use the effective index method can be introduced following exactly the same steps as with deriving Equations 4.3 through 4.7. For instance, in the scalar 2D approximation, 4.41 reduces to:

$$2\text{j}\omega\mu_0\varepsilon\frac{\partial \hat{E}}{\partial t} = \frac{\partial^2 \hat{E}}{\partial x^2}\frac{\partial^2 \hat{E}}{\partial z^2} - \omega^2\mu_0\varepsilon\hat{E} \tag{4.42}$$

Formally, the solution of Equation 4.42 can be expressed as:

$$\hat{E} = \exp\left[\frac{\Delta t}{2\text{j}\omega\mu_0\varepsilon}\left(\frac{\partial^2}{\partial x^2} + \frac{\partial^2}{\partial z^2} - \omega^2\mu_0\varepsilon\right)\right]\hat{E} \tag{4.43}$$

In terms of the practical implementation of the TD-BPM, the FDM has found the widest use. Examples of the implementation of the TD-BPM using the FDM and SVEA approximation can be found in Shibayama et al. [197,198], Feng et al. [199], Masoudi et al. [200,201], Liu et al. [202], and Maes et al. [203]. The finite element version of the FE-TDM algorithm can be found in Fujisawa et al. [204], Rodriguez-Esquerre et al. [205], and Obayya [206]. It is also possible to introduce wideband approximation to TD-BPM algorithm in the way that mimics the wide angle BPM algorithms discussed in the section "BPM Algorithms" [207–209]. A comprehensive, advanced study of the finite difference TD-BPM techniques is given in Yamauchi [17].

TRAVELLING WAVE APPROACH

The main application of the travelling wave approach (TWA) is in the modelling of the lasers. In the TWA an envelope function is introduced in order to eliminate the need for a subwavelength longitudinal sampling of the electromagnetic waves. This allows for a significant reduction of the required computational resources and shortening of the calculation time. We will explain the basics of the TWA in the scalar case. We substitute $\hat{E} = \breve{E} * \exp(-j\beta z)$ into the scalar wave Equation 4.42 and obtain the following equation for $\breve{E}$:

$$2j\omega\mu_0\varepsilon\frac{\partial\breve{E}}{\partial t} = \frac{\partial^2\breve{E}}{\partial x^2} + \left(\frac{\partial^2\breve{E}}{\partial z^2} - 2j\beta\frac{\partial\breve{E}}{\partial z} - \beta^2\breve{E}\right) - \omega^2\mu_0\varepsilon\breve{E} \quad\quad (4.44)$$

After applying SVEA again:

$$\left|\frac{\partial^2\hat{E}}{\partial z^2}\right| << 2\beta\left|\frac{\partial\hat{E}}{\partial z}\right| \quad\quad (4.45)$$

we obtain the equation that forms the basis for the TWA:

$$2j\omega\mu_0\varepsilon\frac{\partial\breve{E}}{\partial t} + 2j\beta\frac{\partial\breve{E}}{\partial z} = \frac{\partial^2\breve{E}}{\partial x^2} - \left(\omega^2\mu_0\varepsilon + \beta^2\right)\breve{E} \quad\quad (4.46)$$

When compared with the TD-BPM, the solution of 4.46 no longer allows for the tracking of the reflected waves. However, when applying the TWA to studying the time evolution of the photon density within a laser cavity, this problem is circumvented by introducing a separate equation for the backward propagating wave [210,211]. Equation 4.46 can be solved efficiently either by introducing the finite difference discretisation on the characteristics of 4.46 in the z-t plane [211,212] or using a hopscotch method [210]. The application of the TWA in laser modelling will be discussed in Chapters 7 and 8.

REFERENCES

1. Ebeling, K.J., *Integrated Optoelectronics*. 1992, Berlin: Springer-Verlag.
2. Digonnet, M.J.F., *Rear-Earth-Doped Fibre Lasers and Amplifiers*. 2001, New York: Marcel Dekker.
3. Coldren, L.A. and S.W. Corzine, *Diode Lasers and Photonic Integrated Circuits*. Microwave and Optical Engineering, ed. K. Chang. 1995, New York: Wiley.
4. Taflove, A. and S.C. Hagness, *Computational Electrodynamics: The Finite-Difference Time-Domain Method*. 2005, London: Artech House.
5. Vassallo, C., Difficulty with vectorial BPM. *Electronics Letters*, 1997. 33(1): p. 61–62.
6. Clauberg, R. and P.v. Allmen, Vectorial beam propagation method for integrated optics. *Electronics Letters*, 1991. 27(8): p. 654–655.
7. Shintaku, T., Electromagnetic-field propagation based on Maxwell's propagation operator. *Optics Letters*, 1996. 21(21): p. 1727–1728.
8. Fogli, F., et al., Highly efficient full-vectorial 3-D BPM modeling of fibre to planar waveguide couplers. *Journal of Lightwave Technology*, 1999. 17(1): p. 136–143.
9. Moore, T.G., et al., Theory and application of radiation boundary operators. *IEEE Transactions on Antennas and Propagation*, 1988. 36(12): p. 1797–1812.

10. Hayashi, K., et al., Combination of beam propagation method and mode expansion propagation method for bidirectional optical beam propagation analysis. *Journal of Lightwave Technology*, 1998. 16(11): p. 2040–2045.
11. Fleck, J.A., J.R. Morris, and M.D. Feit, Time-dependent propagation of high energy laser beams through the atmosphere. *Applied Physics*, 1976. 10: p. 129–160.
12. Feit, M.D. and J.A. Fleck, Light propagation in graded-index optical fibres. *Applied Optics*, 1978. 17(24): p. 3990–3998.
13. Yevick, D. and B. Hermansson, Efficient beam propagation techniques. *IEEE Journal of Quantum Electronics*, 1990. 26(1): p. 109–112.
14. Yevick, D., C. Rolland, and B. Hermansson, Fresnel equation studies of longitudinally varying semiconductor rib waveguides: Reference wavevector dependence. *Electronics Letters*, 1989. 25(18): p. 1254–1256.
15. Feit, M.D. and J.A. Fleck, Simple spectral method for solving propagation problems in cylindrical geometry with fast Fourier transforms. *Optics Letters*, 1989. 14(13): p. 662–664.
16. Poladian, L. and F. Ladouceur, Unification of TE and TM beam propagation algorithms. *IEEE Photonics Technology Letters*, 1998. 10(1): p. 105–107.
17. Yamauchi, J., *Propagating Beam Analysis of Optical Waveguides*. 2003. Research Studies Press.
18. Kriezis, E.E. and A.G. Papagiannakis, A joint finite-difference and FFT full vectorial beam propagation scheme. *Journal of Lightwave Technology*, 1995. 13(4): p. 692–700.
19. Liu, P.L., S.L. Yang, and D.M. Yuan, The semivectorial beam propagation method. *IEEE Journal of Quantum Electronics*, 1993. 29(4): p. 1205–1211.
20. Mansour, I., A.D. Capobianco, and C. Rosa, Noniterative vectorial beam propagation method with a smoothing digital filter. *Journal of Lightwave Technology*, 1996. 14(5): p. 908–913.
21. Yamauchi, J., G. Takahashi, and H. Nakano, Full vectorial beam propagation method based on the McKee-Mitchell scheme with improved finite difference formulas. *Journal of Lightwave Technology*, 1998. 16(12): p. 2458–2464.
22. Hsueh, Y.L., M.C. Yang, and H.C. Chang, Tree dimensional noniterative full vectorial beam propagation method based on the alternating direction implicit method. *Journal of Lightwave Technology*, 1999. 17(11): p. 2389–2397.
23. Shibayama, J., et al., A three-dimensional horizontally wide-angle noniterative beam-propagation method based on the alternating-direction implicit scheme. *IEEE Photonics Technology Letters*, 2006. 18(5–8): p. 661–663.
24. Chui, S.L. and Y.Y. Lu, Wide-angle full-vector beam propagation method based on an alternating direction implicit preconditioner. *Journal of the Optical Society of America a-Optics Image Science and Vision*, 2004. 21(3): p. 420–425.
25. Sharma, A. and A. Agrawal, New method for nonparaxial beam propagation. *Journal of the Optical Society of America a-Optics Image Science and Vision*, 2004. 21(6): p. 1082–1087.
26. Bhattacharya, D. and A. Sharma, Three-dimensional finite difference split-step nonparaxial beam propagation method: New method for splitting of operators. *Applied Optics*, 2009. 48(10): p. 1878–1885.
27. Clark, C.D., III and R.J. Thomas, Wide-angle split-step spectral method for 2D or 3D beam propagation. *Optical and Quantum Electronics*, 2009. 41(11–13): p. 849–857.
28. Hokr, B.H., et al., Higher-order wide-angle split-step spectral method for non-paraxial beam propagation. *Optics Express*, 2013. 21(13): p. 15815–15825.
29. Willems, J., J. Haes, and R. Baets. Eigenmode propagation analysis of radiation losses in waveguides with discontinuities and grating assisted couplers. In *Integrated Photonics Research*. 1993. Washington DC: OSA, p. 229–232.

30. Smartt, C.J., T.M. Benson, and P.C. Kendall. Exact operator method for the analysis of dielectric waveguides with application to integrated optics devices and laser facets. In *IET Second International Conference on Computation in Electromagnetics*. 1995, 335–338.

31. Smartt, C.J., T.M. Benson, and P.C. Kendall, Free space radiation mode method for the analysis of propagation in optical waveguide devices. *IEE Proceedings J*, 1993. 140(1): p. 56–61.

32. Gerdes, J., S. Helfert, and R. Pregla, Three-dimensional vectorial eigenmode algorithm for nonparaxial propagation in reflecting optical waveguide structures. *Electronics Letters*, 1995. 31(1): p. 65–66.

33. Bertolotti, M., P. Masciulli, and C. Sibilia, MoL numerical analysis of nonlinear planar waveguide. *Journal of Lightwave Technology*, 1994. 12(5): p. 784–789.

34. Mustieles, F.J., E. Ballesteros, and F. Hernandez-Gil, Multimodal analysis method for the design of passive TE/TM converters in integrated waveguides. *IEEE Photonics Technology Letters*, 1993. 5(7): p. 809–811.

35. Lee, P.C., D. Schultz, and E. Voges, Tree-dimensional finite difference beam propagation algorithms for photonic devices. *Journal of Lightwave Technology*, 1992. 10(12): p. 1832–1838.

36. Ahlers, E. and R. Pregla, 3-D modelling of concatenations of straight and curved waveguides by MoL-BPM. *Optical and Quantum Electronics*, 1997. 29: p. 151–156.

37. Kawano, K., et al., Bidirectional finite-element method-of-line beam propagation method (FE-MOL-BPM) analysing optical waveguides with discontinuities. *IEEE Photonics Technology Letters*, 1998. 10(2): p. 244–245.

38. Gerdes, J., Bidirectional eigenmode propagation analysis of optical waveguides based on method of lines. *Electronics Letters*, 1994. 30(7): p. 550–551.

39. Ratowsky, R.P. and J.A. Fleck, Accurate numerical-solution of the Helmholtz-equation by iterative Lanczos reduction. *Optics Letters*, 1991. 16(11): p. 787–789.

40. Ratowsky, R.P., J.A. Fleck, and M.D. Feit, Accurate solution of the Helmholtz-equation by Lanczos orthogonalization for media with loss or gain. *Optics Letters*, 1992. 17(1): p. 10–12.

41. Ratowsky, R.P., J.A. Fleck, and M.D. Feit, Accurate description of ultrawide-angle beam-propagation in homogeneous media by Lanczos orthogonalization. *Optics Letters*, 1994. 19(17): p. 1284–1286.

42. Hermansson, B., et al., A comparison of Lanczos electric-field propagation methods. *Journal of Lightwave Technology*, 1992. 10(6): p. 772–776.

43. Luo, Q. and C.T. Law, Nonparaxial propagation of a cylindrical beam with Lanczos reduction. *Optics Letters*, 2000. 25(12): p. 869–871.

44. Bienstman, P. and R. Baets, Advanced boundary conditions for eigenmode expansion models. *Optical and Quantum Electronics*, 2002. 34: p. 523–540.

45. Chew, W.C., J.M. Jin, and E. Michielssen, Complex coordinate stretching as a generalised absorbing boundary condition. *Microwave and Optical Technology Letters*, 1997. 15(6): p. 363–369.

46. Ortega-Monux, A., I. Molina-Fernandez, and J.G. Wanguemert-Perez, 3D-Scalar Fourier Eigenvector Expansion Method (Fourier-EEM) for analyzing optical waveguide discontinuities. *Optical and Quantum Electronics*, 2005. 37(1–3): p. 213–228.

47. Ctyroky, J., A simple bi-directional mode expansion propagation algorithm based on modes of a parallel-plate waveguide. *Optical and Quantum Electronics*, 2006. 38(1–3): p. 45–62.

48. Mo, G.-L. and X.-C. Zhang, High accuracy modal analysis and beam propagation method for nano-waveguides. *Optical and Quantum Electronics*, 2012. 44(10–11): p. 459–470.

49. Ctyroky, J., Improved bidirectional-mode expansion propagation algorithm based on fourier series. *Journal of Lightwave Technology*, 2007. 25(9): p. 2321–2330.

50. Ctyroky, J., 3-D bidirectional propagation algorithm based on Fourier series. *Journal of Lightwave Technology*, 2012. 30(23): p. 3699–3708.

51. Helfert, S. and R. Pregla. Modelling of taper structures in cylindrical coordinates. In *Integrated Photonics Research*. 1995. Dana Point, California USA: OSA, p. 30–32.

52. Ahlers, E. and R. Pregla. New vector BPM in cylindrical coordinates based on method of lines. In *Integrated Photonics Research*. 1995. Dana Point, California, USA: OSA, 24–26.

53. Pregla, R. and E. Ahlers, Method of lines for analysis of arbitrarily curved waveguide bends. *Electronics Letters*, 1994. 30(18): p. 1478–1479.

54. Gu, J.S., P.A. Besse, and H. Melchior, Method of lines for the analysis of propagation characteristics of curved optical rib waveguides. *IEEE Journal of Quantum Electronics*, 1991. 27(3): p. 531–537.

55. Bienstman, P., Two-stage mode finder for waveguides with 2D cross-section. *Optical and Quantum Electronics*, 2004. 36: p. 5–14.

56. Lopez-Dona, J.M., J.G. Wanguemert-Perez, and I. Molina-Fernandez, Fast-Fourier-based three-dimensional full-vectorial beam propagation method. *IEEE Photonics Technology Letters*, 2005. 17(11): p. 2319–2321.

57. Vassallo, C., Improvement of finite difference methods for step-index optical wave-guides. *IEE Proceeding J*, 1992. 139(2): p. 137–142.

58. Yamauchi, J., et al., Modified finite difference formula for the analysis of semivecto-rial modes in step index optical waveguides. *IEEE Photonics Technology Letters*, 1997. 9(7): p. 961–963.

59. Y.P. Chiou, Y.C. Chiang, and H.C. Chang, Improved three point formulas considering the interface conditions in the finite difference analysis of step index optical devices. *Journal of Lightwave Technology*, 2000. 18(2): p. 243–251.

60. Stoffer, R. and H.J.W.M. Hoekstra, Efficient interface conditions based on a 5-point finite difference operator. *Optical and Quantum Electronics*, 1998. 30: p. 375–383.

61. Yamauchi, J., et al., Transverse-magnetic BPM analysis of a step-index slab waveguide expressed by a sigmoid Function. *IEEE Photonics Technology Letters*, 2009. 21(1–4): p. 149–151.

62. Chiou, Y.P. and H.C. Chang, Efficient beam propagation method based on Pade approxi-mants in the propagation direction. *Optics Letters*, 1997. 22(13): p. 949–951.

63. Chung, Y. and N. Dagli, An assessment of finite difference beam propagation method. *IEEE Journal of Quantum Electronics*, 1990. 26(8): p. 1335–1339.

64. Splett, A., M. Majd, and K. Petermann, A novel beam propagation method for large refractive index steps and large propagation distances. *IEEE Photonics Technology Letters*, 1991. 3(5): p. 466–468.

65. Hadley, G.R., Wide angle beam propagation using Pade approximant operators. *Optics Letters*, 1992. 17(20): p. 1426–1428.

66. Hadley, G.R., Multistep method for wide angle beam propagation. *Optics Letters*, 1992. 17(24): p. 1743–1745.

67. Anada, T., et al., Very-wide-angle beam propagation methods for integrated optical circuits. *IEICE Transactions on Electronics*, 1999. E82-C(7): p. 1154–1158.

68. van Stralen, M.J.N., H. Blok, and M.V. de Hoop, Design of sparse matrix representa-tions for the propagator used in the BPM and directional wave field decomposition. *Optical and Quantum Electronics*, 1997. 29(2): p. 179–197.

69. Schultz, D., C. Glingener, and E. Voges, Novel generalised finite difference beam propa-gation method. *IEEE Journal of Quantum Electronics*, 1994. 30(4): p. 1132–1140.

70. Huang, W.P. and C.L. Xu, Simulation of three dimensional optical waveguides by full vector beam propagation method. *IEEE Journal of Quantum Electronics*, 1993. 29(10): p. 2693–2649.

71. Huang, W.P., C.L. Xu, and S.K. Chaudhuri, A vector beam propagation method based on H fields. *IEEE Photonics Technology Letters*, 1991. 3(12): p. 1117–1120.

72. Yamauchi, J., et al., Improved multistep method for wide angle beam propagation. *IEEE Photonics Technology Letters*, 1996. 8(10): p. 1361–1363.

73. Shibayama, J., et al., Efficient nonuniform schemes for paraxial and wide-angle finite difference beam propagation method. *Journal of Lightwave Technology*, 1999. 17(4): p. 677–683.

74. Shibayama, J., et al., A three-dimensional multistep horizontally wide-angle beam-propagation method based on the generalized Douglas scheme. *IEEE Photonics Technology Letters*, 2006. 18(21–24): p. 2535–2537.

75. He, Y.Z. and F.G. Shi, Improved full-vectorial beam propagation method with high accuracy for arbitrary optical waveguides. *IEEE Photonics Technology Letters*, 2003. 15(10): p. 1381–1383.

76. Le, K.Q., et al., The complex Jacobi iterative method for three-dimensional wide-angle beam propagation. *Optics Express*, 2008. 16(21): p. 17021–17030.

77. Chung, Y. and N. Dagli, Analysis of z-variant semiconductor rib waveguides by explicit finite difference beam propagation method with nonuniform mesh configuration. *IEEE Journal of Quantum Electronics*, 1991. 27(10): p. 2296–2305.

78. Chung, Y. and N. Dagli, Explicit finite difference beam propagation method: Application to semiconductor rib waveguide Y-junction analysis. *Electronics Letters*, 1990. 26(11): p. 711–713.

79. Xiang, F. and G.L. Yip, An explicit and stable finite difference 2-D vector beam propagation method. *IEEE Photonics Technology Letters*, 1994. 6(10): p. 1248–1250.

80. Masoudi, H.M., Spurious modes in the DuFort-Frankel finite-difference beam propagation method. *IEEE Photonics Technology Letters*, 1997. 9(10): p. 1382–1384.

81. Sewell, P., T.M. Benson, and A. Vukovic, A stable DuFort-Frankel beam propagation method for lossy structures and those with perfectly matched layers. *Journal of Lightwave Technology*, 2005. 23(1): p. 374–381.

82. Yioultsis, T.V., G.D. Ziogos, and E.E. Kriezis, Explicit finite-difference vector beam propagation method based on the iterated Crank-Nicolson scheme. *Journal of the Optical Society of America a-Optics Image Science and Vision*, 2009. 26(10): p. 2183–2191.

83. Lee, C.T., M.L. Wu, and J.M. Hsu, Novel beam propagation algorithms for tapered optical structures. *Journal of Lightwave Technology*, 1999. 17(11): p. 2379–2388.

84. Sewell, P., et al., Tapered beam propagation. *Electronics Letters*, 1996. 32(11): p. 1025–1026.

85. Sewell, P., et al., Bi-oblique propagation analysis of symmetric and asymmetric Y-junctions. *Journal of Lightwave Technology*, 1997. 15(4): p. 688–696.

86. Yamauchi, J., J. Shibayama, and H. Nakano, Finite difference beam propagation method using the oblique coordinate system. *Electronics and Communications in Japan, Part 2*, 1995. 78(6): p. 20–27.

87. Sujecki, S., et al., Novel beam propagation algorithms for tapered optical structures. *Journal of Lightwave Technology*, 1999. 17(11): p. 2379–2388.

88. Benson, T.M., et al., Structure related beam propagation. *Optical and Quantum Electronics*, 1999. 31: p. 689–703.

89. Rivera, M., A finite difference BPM analysis of dielectric waveguides. *Journal of Lightwave Technology*, 1995. 13(2): p. 233–238.

90. Deng, H., et al., Investigations of 3-D semivectorial finite difference beam propagation method for bent waveguides. *Journal of Lightwave Technology*, 1998. 16(5): p. 915–922.

91. Yamauchi, J., et al., Finite difference beam propagation method for circularly symmetric fields. *IEICE Transactions on Electronics*, 1992. E75-C(9): p. 1093–1095.

92. Montanari, E., et al., Finite element full vectorial propagation analysis for three dimensional z varying optical waveguides. *Journal of Lightwave Technology*, 1998. 16(4): p. 703–714.

93. Schultz, D., et al., Mixed finite element beam propagation method. *Journal of Lightwave Technology*, 1998. 16(7): p. 1336–1342.

94. Obaya, S.S.A., B.M.A. Rahman, and H.A. El-Mikati, New full vectorial numerically efficient propagation algorithm based on the finite element method. *Journal of Lightwave Technology*, 2000. 18(3): p. 409–415.

95. Tsuji, Y., M. Koshiba, and N. Takimoto, Finite element beam propagation method for anisotropic optical waveguides. *Journal of Lightwave Technology*, 1999. 17(4): p. 723–728.

96. Silva, J.P.d., H.E. Hernandez-Figueora, and A.M.F. Frasson, Improved vectorial finite element BPM analysis for transverse anisotropic media. *Journal of Lightwave Technology*, 2003. 21(2): p. 567–576.

97. Saitoh, K. and M. Koshiba, Approximate scalar finite element beam propagation method with perfectly matched layers for anisotropic optical waveguides. *Journal of Lightwave Technology*, 2001. 19(5): p. 786–792.

98. Vanbrabant, P.J.M., et al., A finite element beam propagation method for simulation of liquid crystal devices. *Optics Express*, 2009. 17(13): p. 10895–10909.

99. Schmidt, F., An adaptive approach to the numerical solution of Fresnel's wave equation. *Journal of Lightwave Technology*, 1993. 11(9): p. 1425–1434.

100. Deng, H. and D. Yevick, The nonunitarity of finite-element beam propagation algorithms. *IEEE Photonics Technology Letters*, 2005. 17(7): p. 1429–1431.

101. Kaczmarski, P. and P.E. Lagasse, Bidirectional beam propagation method. *Electronics Letters*, 1988. 24(11): p. 675–676.

102. Jin, G.H., et al., Improved bidirectional beam propagation method for analysis of reflection on nonparallel interfaces. *Electronics Letters*, 1995. 31(21): p. 1867–1868.

103. Liu, Q.H. and W.C. Chew, Analysis of discontinuities in planar dielectric waveguides: An eigenmode propagation method. *IEEE Transactions on Microwave Theory and Techniques*, 1991. 39(3): p. 422–429.

104. Sztefka, G. and H.P. Nolting, Bidirectional eigenmode propagation for large refractive index steps. *IEEE Photonics Technology Letters*, 1993. 5(5): p. 554–557.

105. Bresler, A.D., G.H. Joshi, and N. Marcuvitz, Orthogonality properties for modes in passive and active uniform wave guides. *Journal of Applied Physics*, 1958. 29(5): p. 794–798.

106. Reed, M., et al., Limitations of one dimentional models of waveguide facets. *Microwave and Optical Technology Letters*, 1997. 15(4): p. 196–198.

107. Vukovic, A., et al., Advances in facet design for buried lasers and amplifiers. *IEEE Journal of Selected Topics in Quantum Electronics*, 2000. 6(1): p. 175–184.

108. Vukovic, A., et al., Novel hybrid method for efficient 3-D fibre to chip coupling analysis. *IEEE Journal of Selected Topics in Quantum Electronics*, 2002. 8(6): p. 1285–1292.

109. Yu, C. and D. Yevick, Application of bidirectional parabolic equation method to optical facets. *Journal of Optical Society of America A*, 1997. 14(7): p. 1448–1450.

110. Wei, S.H. and Y.Y. Lu, Application of BiCGSTAB to waveguide discontinuity problems. *IEEE Photonics Technology Letters*, 2002. 14(5): p. 645–647.

111. Chiou, Y.P. and H.C. Chang, Analysis of optical waveguide discontinuities using the Pade approximants. *IEEE Photonics Technology Letters*, 1997. 9(7): p. 964–966.

112. Helfert, S.F. and R. Pregla, Determining reflections for beam propagation algorithms. *Optical and Quantum Electronics*, 2001. 33: p. 343–358.

113. El-Refaei, H., I. Betty, and D. Yevick, The application of complex Pade approximants to reflection at optical waveguide facets. *IEEE Photonics Technology Letters*, 2000. 12(2): p. 158–160.

114. Petruskevicius, R., BiBPM modeling of slow wave structures. *Optical and Quantum Electronics*, 2007. 39(4–6): p. 407–418.

115. Romanova, E.A. and S.B. Gaal, Modeling of light propagation through step-like discontinuities in slab dielectric waveguides. *Microwave and Optical Technology Letters*, 2004. 41(2): p. 108–114.

116. Yoneta, S., M. Koshiba, and Y. Tsuji, Combination of beam propagation method and finite element method for optical beam propagation analysis. *Journal of Lightwave Technology*, 1999. 17(11): p. 2398–2404.

117. Rao, H., R. Scarmozzino, and R.M. Osgood, A bidirectional beam propagation method for multiple dielectric interfaces.

118. Liu, Q. and W.C. Chew, Analysis of discontinuities in planar dielectric wave-guides - an eigenmode propagation method. *IEEE Transactions on Microwave Theory and Techniques*, 1991. 39(3): p. 422–430.

119. Lu, Y.Y. and S.H. Wei, A new iterative bidirectional beam propagation method. *IEEE Photonics Technology Letters*, 2002. 14(11): p. 1533–1535.

120. Ho, P.L. and Y.Y. Lu, A stable bidirectional propagation method based on scattering operators. *IEEE Photonics Technology Letters*, 2001. 13(12): p. 1316–1318.

121. Helfert, S.F. and R. Pregla, Efficient analysis of periodic structures. *Journal of Lightwave Technology*, 1998. 16(9): p. 1694–1702.

122. Helfert, S.F., Numerical stable determination of Floquet-modes and the application to the computation of band structures. *Optical and Quantum Electronics*, 2004. 36(1–3): p. 87–107.

123. Helfert, S.F., Determination of Floquet modes in asymmetric periodic structures. *Optical and Quantum Electronics*, 2005. 37(1–3): p. 185–197.

124. Poon, T.-C. and T. Kim, *Engineering Optics with MATLAB*. 2006, London: World Scientific Publishing.

125. Wartak, M.S., *Computational Photonics: An Introduction with MATLAB*. 2013, Cambridge: Cambridge University Press.

126. Smith, G.D., *Numerical Solution of Partial Differential Equations: Finite Difference Method*. 1988, Oxford: Oxford University Press.

127. Hadley, G.R., Transparent boundary condition for the beam propagation method. *IEEE Journal of Quantum Electronics*, 1992. 28(1): p. 363–370.

128. Song, G.H., Transparent boundary conditions for beam-propagation analysis from the Green's function method. *Journal of Optical Society of America A*, 1993. 10(5): p. 896–904.

129. Vassallo, C. and F. Collino, Highly efficient absorbing boundary condition for beam propagation method. *Journal of Lightwave Technology*, 1996. 14(6): p. 1570–1577.

130. Vassallo, C. and J.M. van-der-Keur, Highly efficient transparent boundary conditions for finite difference beam propagation method at order four. *Journal of Lightwave Technology*, 1997. 15(10): p. 1958–1965.

131. Yevick, D., J. Yu, and Y. Yayon, Optimal absorbing boundary conditions. *Journal of Optical Society of America A*, 1995. 12(1): p. 107–110.

132. Berenger, J., A perfectly matched layer for the absorption of electromagnetic waves. *Journal of Computational Physics*, 1994. 114(2): p. 185–200.

133. Huang, W.P., et al., The perfectly matched layer (PML) boundary condition for the beam propagation method. *IEEE Photonics Technology Letters*, 1996. 8(5): p. 649–651.

134. Vassallo, C. and F. Collino, Highly efficient absorbing boundary conditions for the beam propagation method. *Journal of Lightwave Technology*, 1996. 14(6): p. 1570–1577.

135. Yevick, D., J. Yu, and F. Schmidt, Analytic studies of absorbing and impedance-matched boundary layers. *IEEE Photonics Technology Letters*, 1997. 9(1): p. 73–75.

136. Bamberger, A., et al., Higher-order paraxial wave-equation approximations in heterogeneous media. *SIAM Journal on Applied Mathematics*, 1988. 48(1): p. 129–154.

137. Helfert, S.F. and R. Pregla, A finite difference beam propagation algorithm based on generalized transmission line equations. *Optical and Quantum Electronics*, 2000. 32 (6–8): p. 681–690.

138. Sewell, P., et al., The dispersion characteristics of oblique coordinate beam propagation algorithms. *Journal of Lightwave Technology*, 1999. 17: p. 514–518.

139. Sujecki, S., et al., Nonlinear properties of tapered laser cavities. *IEEE Journal of Selected Topics in Quantum Electronics*, 2003. 9(3): p. 823–834.

140. Spiegel, M.R., *Calculus of Finite Differences and Difference Equations*. 1994: McGraw-Hill, Inc.

141. Sujecki, S., Generalized rectangular finite difference beam propagation method. *Applied Optics*, 2008. 47(23): p. 4280–4286.

142. Rao, H., et al., Complex propagators for evanescent waves in bidirectional beam propagation method. *Journal of Lightwave Technology*, 2000. 18(8): p. 1155–1160.

143. Collins, M.D., Higher-order pade approximations for accurate and stable elastic parabolic equations with application to interface wave-propagation. *Journal of the Acoustical Society of America*, 1991. 89(3): p. 1050–1057.

144. Collins, M.D., A 2-way parabolic equation for elastic media. *Journal of the Acoustical Society of America*, 1993. 93(4): p. 1815–1825.

145. Milinazzo, F.A., C.A. Zala, and G.H. Brooke, Rational square-root approximations for parabolic equation algorithms. *Journal of the Acoustical Society of America*, 1997. 101(2): p. 760–766.

146. Lingevitch, J.F. and M.D. Collins, Wave propagation in range-dependent poro-acoustic waveguides. *Journal of the Acoustical Society of America*, 1998. 104(2): p. 783–790.

147. Yevick, D. and D.J. Thomson, Complex Pade approximants for wide-angle acoustic propagators. *Journal of the Acoustical Society of America*, 2000. 108(6): p. 2784–2790.

148. Lu, Y.Y., A complex coefficient rational approximation of root 1+x. *Applied Numerical Mathematics*, 1998. 27(2): p. 141–154.

149. Lu, Y.Y. and P.L. Ho, Beam propagation method using a (p-1)/p Pade approximant of the propagator. *Optics Letters*, 2002. 27(9): p. 683–685.

150. Chui, S.L. and Y.Y. Lu, A propagator-theta beam propagation method. *IEEE Photonics Technology Letters*, 2004. 16(3): p. 822–824.

151. Vassallo, C., Limitations of the wide-angle beam propagation method in nonuniform systems. *Journal of the Optical Society of America a-Optics Image Science and Vision*, 1996. 13(4): p. 761–770.

152. Jablonski, T.F., Complex-modes in open lossless dielectric wave-guides. *Journal of the Optical Society of America A-Optics Image Science and Vision*, 1994. 11(4): p. 1272–1282.

153. Chiang, Y.C., Y.P. Chiou, and H.C. Chang, Improved full-vectorial finite-difference mode solver for optical waveguides with step-index profiles. *Journal of Lightwave Technology*, 2002. 20(8): p. 1609–1618.

154. Chiou, Y.P., Y.C. Chiang, and H.C. Chang, Improved three-point formulas considering the interface conditions in the finite-difference analysis of step-index optical devices. *Journal of Lightwave Technology*, 2000. 18(2): p. 243–251.

155. Yevick, D., The application of complex Pade approximants to vector field propagation. *IEEE Photonics Technology Letters*, 2000. 12(12): p. 1636–1638.

156. Xie, H., W. Lu, and Y.Y. Lu, Complex modes and instability of full-vectorial beam propagation methods. *Optics Letters*, 2011. 36(13): p. 2474–2476.

157. Nito, Y., et al., A beam-propagation method using both electric and magnetic fields. *IEEE Photonics Technology Letters*, 2011. 23(7): p. 429–431.

158. Yevick, D., W. Bardyszewski, and M. Glasner, Stability issues in vector electric field propagation. *IEEE Photonics Technology Letters*, 1995. 7(6): p. 658–660.

159. Vassallo, C., Wide-angle bpm and power conservation. *Electronics Letters*, 1995. 31(2): p. 130–131.

160. Helfert, S. and R. Pregla, Finite difference expressions for arbitrarily positioned dielectric steps in waveguide structures. *Journal of Lightwave Technology*, 1996. 14(10): p. 2414–2421.

161. Lee, C.T., M.L. Wu, and J.M. Hsu, Beam propagation analysis for tapered waveguides: Taking account of the curved phase-front effect in paraxial approximation. *Journal of Lightwave Technology*, 1997. 15(11): p. 2183–2189.

162. Hadley, G.R., Corrections to "Slanted-wall beam propagation." *Journal of Lightwave Technology*, 2009. 27(17): p. 3959–3960.

163. Ma, C. and E. Van Keuren, A three-dimensional wide-angle BPM for optical waveguide structures. *Optics Express*, 2007. 15(2): p. 402–407.

164. Sujecki, S., Optimised tapered beam propagation. *Optoelectronics Review*, 2000. 8(3): p. 269–274.

165. Sujecki, S., New beam propagation algorithm for optical tapers. *International Journal of Electronics and Communications (AEU)*, 2001. 55(3): p. 185–190.

166. Sewell, P., et al., Non standard beam propagation. *Microwave and Optical Technology Letters*, 1996. 13: p. 24–26.

167. Sujecki, S., Wide-angle, finite-difference beam propagation in oblique coordinate system. *Journal of the Optical Society of America a-Optics Image Science and Vision*, 2008. 25(1): p. 138–145.

168. Sewell, P., T.M. Benson, and P.C. Kendall, Sectorial coordinates for curved optical waveguides. *Microwave and Optical Technology Letters*, 1997. 14(4): p. 202–204.

169. Cada, M., F. Xiang, and L.B. Felsen, Intrinsic modes in tapered optical waveguides. *IEEE Journal of Quantum Electronics*, 1999. 24(5): p. 758–765.

170. Ebeling, K.J., *Integrated Optoelectronics: Waveguide Optics, Photonics, Semiconductors*. 1993, Berlin: Springer Verlag.

171. Kitoh, T., et al., Bending loss reduction in silica-based waveguides by using lateral offsets. *Journal of Lightwave Technology*, 1995. 13(4): p. 555–562.

172. Lui, W.W., et al., Polarisation rotation in semiconductor bending waveguides: A coupled-mode theory formulation. *Journal of Lightwave Technology*, 1998. 16(5): p. 929–936.

173. Kim, C.M., Y.M. Kim, and W.K. Kim, Leaky modes of circular slab waveguides: Modified Airy functions. *IEEE Journal of Selected Topics in Quantum Electronics*, 2002. 8(6): p. 1239–1245.

174. Kim, S. and A. Gopinath, Vector analysis of optical dielectric waveguide bends using finite-difference method. *Journal of Lightwave Technology*, 1996. 14(9): p. 2085–2092.

175. Sewell, P. and T.M. Benson, Efficient curvature analysis of buried waveguides. *Journal of Lightwave Technology*, 2000. 18(9): p. 1321–1329.

176. Lu, Y.Y. and P.L. Ho, Beam propagation modelling of arbitrarily bent waveguides. *IEEE Photonics Technology Letters*, 2002. 14(12): p. 1698–1700.

177. Rivera, M., A finite difference BPM analysis of bent dielectric waveguides. *Journal of Lightwave Technology*, 1995. 13(2): p. 233–238.

178. Lui, W.W., et al., Full-vectorial wave propagation in semiconductor optical bending waveguides and equivalent straight waveguide approximations. *Journal of Lightwave Technology*, 1998. 16(5): p. 910–914.

179. Scrifes, D.R., W. Streifer, and R.D. Burnham, Curved stripe GaAs: GaAlAs diode lasers and waveguides. *Applied Physics Letters*, 1978. 32(4): p. 231–234.

180. Nolting, H.P. and R. Maerz, Results of benchmark tests for different numerical BPM algorithms. *Journal of Lightwave Technology*, 1995. 13(2): p. 216–224.

181. Huang, W.P. and B.E. Little, Power exchange in tapered optical couplers. *IEEE Journal of Quantum Electronics*, 1991. 27: p. 1932–1938.

182. Cascio, L., T. Rozzi, and L. Zappelli, Radiation loss of Y-junctions in rib waveguide. *IEEE Transactions on Microwave Theory and Techniques*, 1995. 43: p. 1788–1797.

183. Yajima, H., Coupled mode analysis of dielectric planar branching waveguides. *IEEE Journal of Quantum Electronics*, 1978. 14: p. 749–755.

184. Burns, W.K. and A.F. Milton, An analytic solution in optical waveguide branches. *IEEE Journal of Quantum Electronics*, 1980. 16: p. 446–454.

185. Kuznetsov, M., Radiation loss in dielectric waveguide y branch structures. *Journal of Lightwave Technology*, 1985. 3: p. 674–677.

186. Yang, W. and A. Gopinath, A boundary integral method for propagation problems in integrated optical structures. *IEEE Photonics Technology Letters*, 1995. 7(7): p. 777–779.

187. He, Q., H. Gan, and D. Jiao, Explicit time-domain finite-element method stabilized for an arbitrarily large time step. *IEEE Transactions on Antennas and Propagation*, 2012. 60(11): p. 5240–5250.

188. Teixeira, F.L., Time-domain finite-difference and finite-element methods for Maxwell equations in complex media. *IEEE Transactions on Antennas and Propagation*, 2008. 56(8): p. 2150–2166.

189. Christopoulos, C., *The Transmission-Line Modeling Method: TLM*. 1995, New Jersey: IEEE Press.

190. Chan, Y.C., M. Premaratne, and A.J. Lowery, Semiconductor laser line width from the transmission-line laser model. *IEE Proceedings-Optoelectronics*, 1997. 144(4): p. 246–252.

191. Janyani, V., et al., TLM modelling of nonlinear optical effects in fibre Bragg gratings. *IEE Proceedings-Optoelectronics*, 2004. 151(4): p. 185–192.

192. Joseph, R.M. and A. Taflove, Spatial solution deflection mechanism indicated by FD-TD Maxwells equations modeling. *IEEE Photonics Technology Letters*, 1994. 6(10): p. 1251–1254.

193. Liang, T. and R.W. Ziolkowski, Mode conversion of ultrafast pulses by grating structures in layered dielectric waveguides. *Journal of Lightwave Technology*, 1997. 15(10): p. 1966–1973.

194. Rodriguez-Esquerre, V.F., M. Koshiba, and H.E. Hernandez-Figueroa, Finite-element time-domain analysis of 2-D photonic crystal resonant cavities. *IEEE Photonics Technology Letters*, 2004. 16(3): p. 816–818.

195. Huang, W.P., et al., A scalar finite-difference time-domain approach to guided-wave optics. *IEEE Photonics Technology Letters*, 1991. 3(6): p. 524–526.

196. Huang, W.P., S.T. Chu, and S.K. Chaudhuri, A semivectorial finite-difference time-domain method. *IEEE Photonics Technology Letters*, 1991. 3(9): p. 803–806.

197. Shibayama, J., et al., A finite-difference time-domain beam-propagation method for TE- and TM-wave analyses. *Journal of Lightwave Technology*, 2003. 21(7): p. 1709–1715.

198. Shibayama, J., et al., Efficient time-domain finite-difference beam propagation methods for the analysis of slab and circularly symmetric waveguides. *Journal of Lightwave Technology*, 2000. 18(3): p. 437–442.

199. Feng, N.N., G.R. Zhou, and W.P. Huang, An efficient split-step time-domain beam-propagation method for modeling of optical waveguide devices. *Journal of Lightwave Technology*, 2005. 23(6): p. 2186–2191.

200. Masoudi, H.M. and M.S. Akond, Efficient iterative time-domain beam propagation methods for ultra short pulse propagation: Analysis and assessment. *Journal of Lightwave Technology*, 2011. 29(16): p. 2475–2481.

201. Masoudi, H.M., M.A. Al-Sunaidi, and J.M. Arnold, Efficient time-domain beam-propagation method for modeling integrated optical devices. *Journal of Lightwave Technology*, 2001. 19(5): p. 759–771.

202. Liu, P.L., Q. Zhao, and F.S. Choa, Slow-wave finite-difference beam-propagation method. *IEEE Photonics Technology Letters*, 1995. 7(8): p. 890–892.

203. Maes, B., et al., Modeling comparison of second-harmonic generation in high-index-contrast devices. *Optical and Quantum Electronics*, 2008. 40(1): p. 13–22.

204. Fujisawa, T. and M. Koshiba, Time-domain beam propagation method for nonlinear optical propagation analysis and its application to photonic crystal circuits. *Journal of Lightwave Technology*, 2004. 22(2): p. 684–691.

205. Rodriguez-Esquerre, V.F. and H.E. Hernandez-Figueroa, Novel time-domain step-by-step scheme for integrated optical applications. *IEEE Photonics Technology Letters*, 2001. 13(4): p. 311–313.

206. Obayya, S.S.A., Efficient finite-element-based time-domain beam propagation analysis of optical integrated circuits. *IEEE Journal of Quantum Electronics*, 2004. 40(5): p. 591–595.

207. Lim, J.J., et al., Wideband finite-difference-time-domain beam propagation method. *Microwave and Optical Technology Letters*, 2002. 34(4): p. 243–247.

208. Le, K.Q., T. Benson, and P. Bienstman, Application of modified Pade approximant operators to time-domain beam propagation methods. *Journal of the Optical Society of America B-Optical Physics*, 2009. 26(12): p. 2285–2289.

209. Masoudi, H.M., A novel nonparaxial time-domain beam-propagation method for modeling ultrashort pulses in optical structures. *Journal of Lightwave Technology*, 2007. 25(10): p. 3175–3184.

210. Gehrig, E., O. Hess, and R. Wallenstein, Modeling of the performance of high-power diode amplifier systems with an optothermal microscopic spatio-temporal theory. *IEEE Journal of Quantum Electronics*, 1999. 35(3): p. 320–331.

211. Egan, A., et al., Dynamic instabilities in master oscillator power amplifier semiconductor lasers. *IEEE Journal of Quantum Electronics*, 1998. 34(1): p. 166–170.

212. Chan, R.Y. and J.M. Liu, Time-domain wave-propagation in optical structures. *IEEE Photonics Technology Letters*, 1994. 6(8): p. 1001–1003.

5 Thermal Modelling of Photonic Devices

The physical processes used for light generation in photonic devices are usually accompanied by heat generation, which in turn raises the operating temperature of the devices. An uncontrolled rise in operating temperature may lead to a reduction of the available optical power, change of output beam parameters, or permanent device damage. The thermal management in photonic devices is therefore essential for achieving their optimal performance.

The thermal modelling of photonic devices allows the temperature distribution within the device to be predicted. Once the temperature distribution within the device is known, the paths along which the heat generated by the heat sources spreads within the device can be identified. Thus thermal modelling aids the design of the device packaging so that the heat generated within the device can be effectively extracted. The thermal models therefore play an important role in the photonic device design process. In this chapter we give an introduction into the thermal modelling of photonic devices. We consider two representative photonic device examples, namely, a laser diode and a fibre laser. Because the heat transfer and dissipation is relevant to many other areas of engineering, there are many books written on this topic [1]. In this chapter we therefore focus on bringing all the theory that has been developed for other engineering applications into the context of photonic device design and modelling. Consequently, we only repeat those parts of the basic theory that are either essential for understanding of the presented material or of particular importance for photonic device design.

We start with a short description of the heat flow theory and then bring this theory into the context of the heat generation and dissipation in photonic devices. Then in the sections "Finite Difference Analysis of Heat Flow in Homogenous Media" and "Finite Difference Analysis of Heat Flow in Inhomogeneous Media," we discuss the application of the finite difference method (FDM) for the calculation of the temperature distribution in homogenous and in-homogenous media, respectively. In the last section we briefly discuss selected specific aspects relevant to the thermal modelling of photonic devices, which include the thermal boundary resistance, boundary conditions, and singularities.

Finally we note that thermal modelling of photonic devices is strongly coupled with experimental analysis. The temperature distribution within a device can be measured with low spatial resolution using thermal cameras. Better spatially resolved temperature distributions can be obtained, using more advanced techniques, for example, thermo-reflectance and micro-Raman spectroscopy. A laser diode junction temperature can be measured separately using several techniques. It is not our intention to cover these topics here. We refer an interested reader to the available literature instead [2,3].

HEAT FLOW

Let us consider a set of heat sources. If we surround all the heat sources by a fictitious sphere, an equation that relates the heat flux out of the sphere, with the internal heat generation rate by all heat sources enclosed within the sphere, and the rate of the change of the energy stored within the sphere can be formulated. From the energy conservation principle, it follows that the increase in the total energy stored within the sphere, for a time interval, has to be equal to the difference between the amount of the heat generated by all the heat sources within the sphere and the energy that is transported out of the sphere by all heat transfer processes, within the time interval (Figure 5.1). If the volume enclosed within the sphere is V and the total surface of the sphere is S, then the following equation expresses the stated energy conservation principle:

$$\iiint_V \frac{\partial E}{\partial t} dV = \iiint_V Q dV - \oiint_S \vec{q} dS \tag{5.1}$$

where E is the energy density (units [J/m³]), Q is the distribution of the heat generation rate (units [J/(s m³)]), and $\vec{q}$ is the heat flux (units [W/m²]). (Note that the heat flux vector pointing outward of the sphere surface corresponds to the outgoing heat flux, and has the same orientation as the unit outward vector that is normal to the sphere surface [Figure 5.1].)

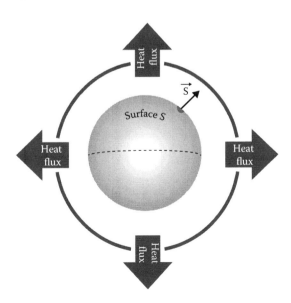

FIGURE 5.1 Schematic diagram illustrating the process of heat flow out of a sphere.

Using the divergence theorem 5.1 can be recast into:

$$\frac{\partial E}{\partial t} = Q - \nabla \vec{q}$$

(5.2)

Both the heat flux and the energy density can be related to the temperature:

$$E = cT$$

(5.3a)

$$\vec{q} = -\rho \nabla T$$

(5.3b)

where c [J/Km^3] is the volume heat capacity and ρ [W/mK] is the heat conductivity. Substituting 5.3 into 5.2 gives the final equation:

$$c\frac{\partial T}{\partial t} = Q + \nabla\left(\rho \nabla T\right)$$

(5.4)

The solution of Equation 5.4 yields the distribution of the temperature, within the considered domain, subject to the known distribution of the heat sources and the boundary conditions. If the considered domain is homogenous, 5.4 reduces to:

$$\frac{c}{\rho}\frac{\partial T}{\partial t} = \frac{Q}{\rho} + \Delta T$$

(5.5)

In the steady state, 5.5 reduces to the Poisson equation, which can further be simplified to the Laplace equation if the heat sources are absent from the considered domain. The Poisson and (especially) Laplace equations can be solved analytically for a large number of cases [4,5]. Equation 5.5 also admits a number of analytical solutions [4,5]. Furthermore, the numerical methods for 5.5 and also for the Poisson and Laplace equations are well developed and usually discussed in standard textbooks on heat transfer [1]. In some cases, one can use 5.5 instead of 5.4 when studying the heat flow in photonic devices. However, the study of the heat flow in photonic devices usually requires the solution of 5.4 rather than 5.5 since the considered medium is not homogenous. This especially applies to laser diodes, whereby the heat flows through a number of semiconductor layers with different thermal properties. With the exception of several simple cases, the calculation of the temperature distribution within a photonic device requires an application of advanced numerical methods. We discuss this topic in more detail in the next sections. However, before moving to the next section, we introduce the concept of the thermal resistance, which is very useful when analysing the heat flow in photonic devices. This derivation gives also an example of application of Equation 5.1 and helps understanding the section "Heat Flow in Photonic Devices."

We consider a rod with the transverse cross-sectional surface S and the length L (Figure 5.2).

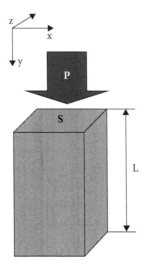

FIGURE 5.2 Heat flow through a rod.

The total power of the heat flux flowing through the rod is equal to p. The heat flux is assumed to be homogeneously distributed within the transverse cross-section. The rod is thermally insulated on the sides hence the heat flows only along the y-axis and, hence, from Equation 5.3b, we obtain an expression for the y-component of the heat flux vector:

$$q_y = -\rho \frac{dT}{dy} \tag{5.6}$$

We integrate both sides of 5.6 over the transverse surface S:

$$\iint_S q_y \, dS = -\rho \iint_S \frac{dT}{dy} dS$$

Which, keeping in mind that $\dfrac{dT}{dy}(x,z) = $ constant, yields:

$$P = -\rho \frac{dT}{dy} S$$

Integration over the entire length of the rod:

$$\int_0^L P dy = -\rho S \int_0^L \frac{dT}{dy} dy$$

yields finally:

$$PL = -\rho S \left[T\left(y = L\right) - T\left(y = 0\right) \right] \tag{5.7}$$

If we define the temperature difference $\Delta T = [T(y = L) - T(y = 0)]$ and the thermal resistance $R = L/(\rho S)$, then 5.7 can be recast into a simple equation that relates the incident heat power and the temperature difference between the ends of the considered rod:

$$P = \frac{\Delta T}{R} \tag{5.8}$$

Equation 5.8 mimics Ohm's law. Hence with help of 5.8 it is possible to analyse the heat flow in photonic devices in the same way as the current flow in electric circuits. The last observation we would like to make is that the thermal resistance is directly proportional to the length and inversely proportional to the rod's transverse cross-section surface. Hence, the longer the paths between the heat sources, present in the device, and the heat sink, the higher the expected operating temperature of the device. On the other hand, the operating temperature of the device can be reduced by spreading heat on its way to the heat sink.

HEAT FLOW IN PHOTONIC DEVICES

The knowledge of the device's operating temperature is very important since it affects the reliability, output power, and optical beam parameters of photonic devices. Usually, an elevated operating temperature has an overall negative impact on photonic device characteristics although there are also situations whereby the device temperature is increased intentionally in order to achieve a particular functionality, for example thermal switching in integrated optical devices [6].

In order to predict a photonic device operating temperature, it is necessary to identify correctly the major sources of heat and the main heat flow paths. The heat sources can be estimated from the in depth knowledge of the physical principles governing the device operation while the heat flow distribution within a photonic device depends on its internal structure and the way it is packaged. Consequently, the thermal management of photonic devices is closely linked with the packaging technology. In order to put these general statements into a practical context, we provide first a brief description of the C mount used for packaging semiconductor laser diodes.

There are three most commonly used mounts for laser diodes, that is, a C mount, a T-can mount, and a "butterfly" mount. In the case of the C mount, the internal structure can be seen without destroying the external encasing (Figure 5.3).

The laser diode is soldered on a submount that allows the mechanical stress to be reduced. The submount is placed on a heat sink, which dissipates the heat generated within the device. The heat flows from the laser diode to the heat sink through a submount. There is, therefore, a temperature difference between the laser diode and the heat sink. Furthermore, there is also an inhomogeneous temperature distribution within the laser diode semiconductor chip. This is because most of the heat in a laser

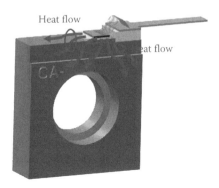

FIGURE 5.3 Schematic diagram illustrating the heat flow to the heat sink from a laser diode on a C mount.

diode is typically generated in the vicinity of the active layer [7]. Due to a very small heat conductivity of the air, the heat flows from the active region to the heat sink, and is dissipated mainly there [8]. This heat flow from the active layer to the sub-mount results in a temperature gradient within the laser diode due to a limited value of the heat conductivity of the semiconductor material. Therefore, the temperature of the active region depends not only on the parameters of the heat sink but also on the internal structure of the laser diode itself. Furthermore, the laser diode operating temperature depends on whether the p- or n-contact of the laser diode is soldered to the submount. These two cases correspond to the so called "p-side down" and "n-side down" mounting, respectively (Figure 5.4).

It is relevant for the active layer operating temperature which way the laser diode is soldered to the submount. The reason for this difference is related to the internal multilayer structure of the laser diode. In order to clarify this point, Figure 5.5 shows schematically a cross-section of an edge emitting laser diode that has been placed n-side down on a heat sink. Figure 5.5 shows typical laser dimensions whilst the red arrows indicate the major heat flow paths. There is a large difference between the distance from the quantum well (QW) to the p- and n-contact. This results from the fact that laser diodes are typically grown on n-type substrates. In one of the final stages of the fabrication process, the substrate is thinned. However, the minimum thickness of the thinned substrate is limited to about 100 μm so that the device preserves a sufficient degree of mechanical robustness. The fact that the distance between the heat source and the heat sink is larger in the case of an n-side mounted device results in a potentially larger thermal resistance between the heat source and the sink, and, consequently, in a larger temperature of the active region, when com-pared with a p-side down mounted device. More information on a laser diode pack-aging can be found on the laser diode manufacturer's web pages and in Ronnie [9].

Once the inner structure of the laser diode and the way it is mounted is known, the temperature distribution within the device can be, in principle, calculated. However, when attempting to calculate the temperature distribution within a laser diode, one is immediately confronted with a problem: the outer dimensions can be as large as 1 cm (heat sink) while the inner structure details can be on the nanometre scale (QW

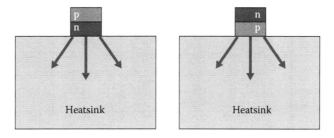

FIGURE 5.4 A schematic diagram of p- and n-side laser diode mounting.

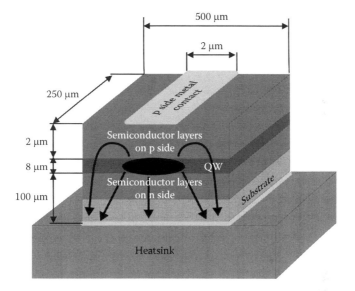

FIGURE 5.5 Schematic diagram of an edge emitting laser diode.

width of several nanometres) [7]. It is therefore convenient to identify two subdomains relevant for the thermal modelling of the laser diodes (Figure 5.6). One of them contains the laser diode itself while the other one contains the heat sink, and possibly also the submount. A separate problem can be associated with each of the subdomains. The problem associated with the subdomain 2 in Figure 5.6 consists in designing the cooling system so that it effectively extracts and dissipates the heat generated within the laser diode, using the heat exchange processes between the heat sink and the environment, that is, the thermal conduction, convection, and the heat radiation. In this case, the laser diode can be represented as a surface heat source with the total power equal to the total power of the heat that is generated within the laser diode (Figure 5.7). When designing the heat sink, a fairly accurate estimate of the total heat power generated within the laser diode can be obtained by calculating the difference between the electrical power (equal to the product of the bias voltage and the operating current) supplied and the optical power extracted from the device. All the necessary data can be usually obtained from the manufacturer's data sheet. The

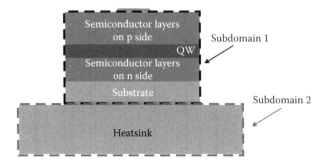

FIGURE 5.6   Two computational subdomains relevant for thermal laser diode analysis.

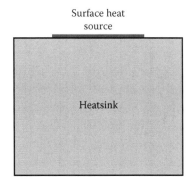

FIGURE 5.7   Problem associated with subdomain 2 from Figure 5.6.

only unknown is the spatial distribution of the surface heat source, which can be calculated once the temperature distribution is calculated self-consistently in both domains. However, in the initial analysis, a homogenous surface heat source distribution is usually assumed.

The purpose of extracting and dissipating heat by the heat sink is to maintain the temperature of the top of the heat sink within an optimal range. A high heat sink temperature will result in a high active layer temperature, and, henceforth, it may lead to a significant reduction of the optical output power, and even to an irreversible damage to the device, while low temperature may lead to water condensation on the laser diode facet.

In order to calculate the temperature distribution in the heat sink at steady state, it is necessary to solve Equation 5.4 in the absence of sources and temporal variations of the stored energy:

$$\nabla \left(\rho \nabla T \right) = 0 \tag{5.9}$$

Equation 5.9 is solved subject to appropriate boundary conditions, which can be derived if the ambient temperature and the distribution of the surface heat sources are known. This problem is usually tackled applying analytical methods.

The problem that is associated with the first subdomain consists in calculating the internal laser diode temperature distribution subject to known internal heat sources distribution and ambient temperature. The presence of the heat sink is accounted for by including a thermal resistance (Figure 5.8).

The value of the thermal resistance is usually obtained from a combination of measurements and simulations of the subdomain 2. Considering the fact that the heat conductivity within a laser diode is not homogenous, the calculation of the temperature distribution within the domain in a static case requires the solution of:

$$\nabla\left(\rho\nabla T\right) = -Q \qquad (5.10)$$

The calculation of the heat source distribution Q is necessary prior to the solution of 5.10. This is not a trivial task and requires a simultaneous calculation of the electric current distribution, electric field distribution, carrier concentrations, and photon density distribution, since the spatial distributions of all these physical quantities are in turn dependent on the temperature distribution. Further, the heat source spatial distribution also depends on the spatial distributions of all these quantities (Chapter 7).

When considering the heat flow in edge emitting diodes, the considered subdomains have a rectangular shape. This is also the case when studying the heat flow in integrated optical switches [10] and lanthanide doped slab waveguide lasers [11]. However, when studying the heat flow in vertical cavity semiconductor lasers and optical fibres, a circular symmetry is encountered (Figure 5.9) [12–14], for instance, when the heat generated in the core of a lanthanide doped fibre is conducted radially to the outer surface of the fibre. In the next sections, we will consider the study of heat flow in domains with both rectangular and circular shape.

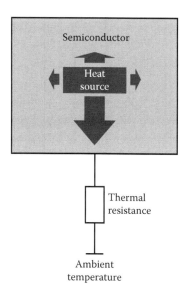

FIGURE 5.8 Problem associated with subdomain 1 from Figure 5.6.

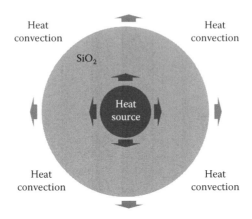

FIGURE 5.9   Schematic diagram of heat flow in a fibre laser.

The numerical methods for the analysis of the heat transfer have been studied and developed in many branches of engineering for several decades now. There are many books that are devoted to this topic [15–17]. There are two well-established numerical techniques that are used in the heat flow analysis of photonic devices, namely, the FDM and finite element method (FEM). Both techniques are robust and can, in principle, handle any problem that is related to heat flow in photonic devices. However, both techniques result in algorithms that are fairly demanding in terms of the required computer memory and calculation time. There are other methods used for the numerical solution of the heat conduction equation that result in more efficient computations. An example of such a technique is the boundary element method (BEM). In many cases, the BEM is more computationally efficient than FEM and FDM. However, BEM's efficiency drops significantly if the distribution of the heat conductivity is inhomogeneous. A comprehensive study on various numerical techniques used in the analysis of heat transfer is given in Minkowycz et al. [17]. An in-depth description of the application of the BEM to the calculation of the temperature distribution is given in Bekker [18]. An introduction to the application of the FDM and FEM in solving the heat transfer problems can be found in Croft [19], Patankar [20], and Reddy et al. [21], respectively.

In the following sections, we will outline the application of the FDM to study of the heat flow in photonic devices. In particular, we will consider the solution of Equations 5.9 and 5.10. In the next section, we study the heat flow in homogenous media. In the section "Finite Difference Analysis of Heat Flow in Inhomogeneous Media," we consider inhomogeneous media with abrupt discontinuities of the heat conductivity. A careful treatment of the heat conductivity discontinuities is necessary due to large differences between the values of the thermal conductivity of materials used for manufacturing photonic devices (see Table 5.1). The examples presented are limited to 2D analysis of the temperature distribution in the device cross-section, whereby the longitudinal heat diffusion process is neglected. Such an approach is typically followed when modelling photonic devices [7,13,22]. However, there are situations when full 3D models need to be employed, for example, modelling of the effect of bonding wires [23] and modelling of catastrophic mirror damage [24].

TABLE 5.1
Heat Conductivity of Selected Materials

Material	Application	Heat Conductivity [W/(m*K)]
Air	Ambient	0.026
Alumina (99.6% Al$_2$O$_3$)	Laser diode submount	27
GaAs	Laser diode substrate	46
Tin/lead	Solder	50
Copper	Heat sink	400
Silica glass	Fibre optics	1.38

FINITE DIFFERENCE ANALYSIS OF HEAT FLOW IN HOMOGENOUS MEDIA

In this section, we consider four simple examples that illustrate the application of the FDM to the analysis of heat flow in homogenous media. These examples illustrate the process of the derivation of the finite difference approximation for the heat conduction equation, the way, the equations are assembled together into a set of linear algebraic equations, and how these equations can be solved numerically. Particular attention is given to the process of developing an algorithm for the problem solution and converting it into a computer code that can be used to obtain the numerical values of the temperature distribution within the considered domain. The examples are selected in such a way that they are relevant to the analysis of heat flow in photonic devices.

There are two ways in which finite difference approximations for studying heat flow can be obtained. The first method attempts to solve 5.2, and is based on rigorous derivations using the Taylor series [25]. The second method attempts to solve 5.1, and is based on understanding the physical principles behind the heat flow phenomenon. This latter technique is usually referred to as the energy balance equation method [19].

For the sake of comparison, we present the application of both methods to the solution of the Poisson equation in the steady state case. We consider first the energy balance approach. We divide the entire space into a set of adjacent cubes and consider the heat flux through the sides of each cube. Figure 5.10 shows an elementary cube (marked with grey shading) and the set of nodal points. Applying Equation 5.3b, while approximating the derivative of temperature with using central difference:

$$\frac{\partial T}{\partial x}\bigg|_{x=x_0+\Delta x/2} \approx \frac{T(x_0+\Delta x)-T(x_0)}{\Delta x}$$

yields the following approximation of the heat flux through the face of the cube that dissects the x axis at $x = x_0 + \Delta x/2$:

$$\Delta y\Delta z\, q_{x_0+\Delta x/2} = \rho\left(x_0+\frac{\Delta x}{2},y_0,z_0\right)\Delta y\Delta z\frac{T(x_0+\Delta x,y_0,z_0)-T(x_0,y_0,z_0)}{\Delta x}$$

$$= \frac{\Delta T\left(x_0+\dfrac{\Delta x}{2},y_0,z_0\right)}{R\left(x_0+\dfrac{\Delta x}{2},y_0,z_0\right)} \tag{5.11}$$

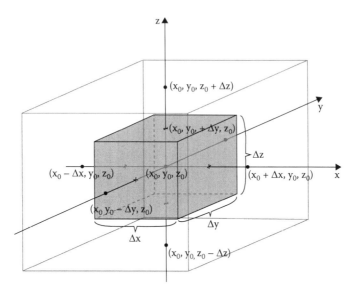

FIGURE 5.10   One element of the finite difference mesh in the 3D case.

whereby ΔT and R are defined as:

$$\Delta T\left(x_0 + \frac{\Delta x}{2}, y_0, z_0\right) = T\left(x_0 + \Delta x, y_0, z_0\right) - T\left(x_0, y_0, z_0\right) \tag{5.12a}$$

$$R\left(x_0 + \frac{\Delta x}{2}, y_0, z_0\right) = \frac{\Delta x}{\rho\left(x_0 + \frac{\Delta x}{2}, y_0, z_0\right)\Delta y \Delta z} \tag{5.12b}$$

Following the same procedure for the other faces of the cube allows the heat flux term in 5.1 to be approximated in the following way:

$$\oiint_S \vec{q} \mathrm{d}S \approx \Delta y \Delta z \left(q_{x_0+\Delta x/2} + q_{x_0-\Delta x/2}\right) + \Delta x \Delta z \left(q_{x_0+\Delta y/2} + q_{x_0-\Delta y/2}\right)$$

$$+ \Delta y \Delta x \left(q_{x_0+\Delta z/2} + q_{x_0-\Delta z/2}\right) \tag{5.13}$$

The approximation of the heat generation term gives:

$$\iiint_V Q \mathrm{d}V \approx \Delta x \Delta y \Delta z Q\left(x_0, y_0, z_0\right) \tag{5.14}$$

Combining 5.13 and 5.14, while observing that the medium within the cube is homogenous and has the thermal conductivity ρ, yields the following approximation of 5.1, in the absence of temporal changes in the temperature:

$$\Delta y \Delta z \left(\frac{T\left(x_0 + \Delta x, y_0, z_0\right) - T\left(x_0, y_0, z_0\right)}{\Delta x} + \frac{T\left(x_0 - \Delta x, y_0, z_0\right) - T\left(x_0, y_0, z_0\right)}{\Delta x} \right)$$

$$+ \Delta x \Delta z \left(\frac{T\left(x_0, y_0 + \Delta y, z_0\right) - T\left(x_0, y_0, z_0\right)}{\Delta y} + \frac{T\left(x_0, y_0 - \Delta y, z_0\right) - T\left(x_0, y_0, z_0\right)}{\Delta y} \right)$$

$$+ \Delta y \Delta x \left(\frac{T\left(x_0, y_0, z_0 + \Delta z\right) - T\left(x_0, y_0, z_0\right)}{\Delta z} + \frac{T\left(x_0, y_0, z_0 - \Delta z\right) - T\left(x_0, y_0, z_0\right)}{\Delta z} \right)$$

$$= \Delta x \Delta y \Delta z \frac{Q\left(x_0, y_0, z_0\right)}{\rho} \tag{5.15}$$

Comparing 5.15 with 5.8 leads to the conclusion that the approximation 5.15 yields an equivalent thermal resistance circuit (Figure 5.11) of the elementary unit cell from Figure 5.10.

In 2D the surface integrals reduce to line integrals while the volume integrals reduce to area integrals. Considering, therefore, the energy balance for the finite difference mesh element shown in Figure 5.12 yields the following approximation of Equation 5.1:

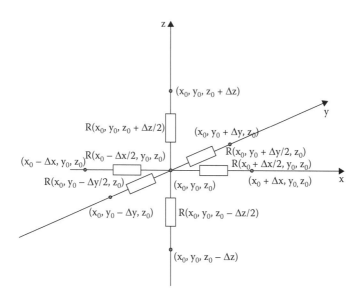

FIGURE 5.11 Equivalent thermal resistance network for a finite difference node from Figure 5.10.

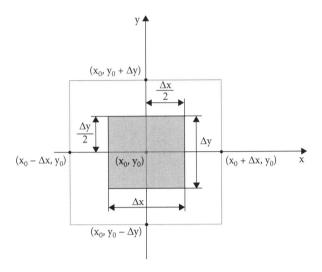

FIGURE 5.12 One element of the finite difference mesh in the 2D case.

$$\Delta y \left(\frac{T(x_0 + \Delta x, y_0) - T(x_0, y_0)}{\Delta x} + \frac{T(x_0 - \Delta x, y_0) - T(x_0, y_0)}{\Delta x} \right)$$

$$+ \Delta x \left(\frac{T(x_0, y_0 + \Delta y) - T(x_0, y_0)}{\Delta y} + \frac{T(x_0, y_0 - \Delta y)T(x_0, y_0)}{\Delta y} \right) = \Delta x \Delta y \frac{Q(x_0, y_0)}{\rho}$$

$$(5.16)$$

One of the important advantages of the energy balance method is that it can be easily extended to derive an approximation to 5.1 on an irregular finite difference mesh [19].

Now for comparison we will derive the finite difference approximation of Equation 5.2 using the Taylor series. We consider the 2D case again in steady state. Under this assumption 5.2 reduces to

$$\frac{\partial^2 T}{\partial^2 x} + \frac{\partial^2 T}{\partial^2 y} = \frac{Q}{\rho} \qquad (5.17)$$

We derive first the finite difference approximation to the second derivative of T with respect to x. A Taylor series expansion of T at $x = x_0 + \Delta x$ yields:

$$T(x_0 + \Delta x) = T(x_0) + \frac{\partial T}{\partial x}\bigg|_{x=x_0} \Delta x + \frac{1}{2} \frac{\partial^2 T}{\partial x^2}\bigg|_{x=x_0} \Delta x^2 + \frac{1}{6} \frac{\partial^3 T}{\partial x^3}\bigg|_{x=x_0} \Delta x^3$$

$$+ \frac{1}{24} \frac{\partial^4 T}{\partial x^4}\bigg|_{x=x_0} \Delta x^4 + \cdots \qquad (5.18a)$$

Similarly at $x = x_0 - \Delta x$ we obtain:

$$T\left(x_0 - \Delta x\right) = T\left(x_0\right) - \frac{\partial T}{\partial x}\bigg|_{x=x_0} \Delta x + \frac{1}{2}\frac{\partial^2 T}{\partial x^2}\bigg|_{x=x_0} \Delta x^2 - \frac{1}{6}\frac{\partial^3 T}{\partial x^3}\bigg|_{x=x_0} \Delta x^3$$

$$+ 24\frac{\partial^4 T}{\partial x^4}\bigg|_{x=x_0} \Delta x^4 + \cdots \qquad (5.18b)$$

Adding 5.18a and 5.18b together and dividing both sides by Δx^2 gives a formula that expresses the second derivative of temperature at $x = x_0$, in terms of the temperatures at $x = x_0$, $x = x_0 + \Delta x$, and $x = x_0 - \Delta x$:

$$\frac{\partial^2 T}{\partial x^2}\bigg|_{x=x_0} \approx \frac{T\left(x_0 + \Delta x\right) - 2T\left(x_0\right) + T\left(x_0 - \Delta x\right)}{\Delta x^2} - \frac{1}{12}\frac{\partial^4 T}{\partial x^4}\bigg|_{x=x_0} \Delta x^2 \qquad (5.19)$$

In 5.19, we included also the leading error term. Since the error term is proportional to Δx^2, the approximation is second order accurate. Similarly, we can approximate the second derivative with respect to y:

$$\frac{\partial^2 T}{\partial y^2}\bigg|_{y=y_0} \approx \frac{T\left(y_0 + \Delta y\right) - 2T\left(y_0\right) + T\left(y_0 - \Delta y\right)}{\Delta y^2} - \frac{1}{12}\frac{\partial^4 T}{\partial y^4}\bigg|_{y=y_0} \Delta y^2 \qquad (5.20)$$

Substituting 5.19 and 5.20 into 5.17 yields the following approximation to Equation 5.17 at the point (x_0, y_0):

$$\left(\frac{\partial^2 T}{\partial x^2} + \frac{\partial^2 T}{\partial y^2}\right)\bigg|_{(x=x_0, y=y_0)} \approx \frac{T\left(x_0 + \Delta x, y_0\right) - 2T\left(x_0, y_0\right) + T\left(x_0 - \Delta x, y_0\right)}{\Delta x^2}$$

$$+ \frac{T\left(x_0, y_0 + \Delta y\right) - 2T\left(x_0, y_0\right) + T\left(x_0, y_0 - \Delta y\right)}{\Delta y^2} - \frac{1}{12}\frac{\partial^4 T}{\partial y^4}\bigg|_{y=y_0} \Delta y^2$$

$$- \frac{1}{12}\frac{\partial^4 T}{\partial x^4}\bigg|_{x=x_0} \Delta x^2 = \frac{Q\left(x_0, y_0\right)}{\rho} \qquad (5.21)$$

Dividing 5.16 by $\Delta x \Delta y$ and neglecting the error terms in 5.21 shows that these two equations are equivalent. Hence, both methods lead to exactly the same approximations of the equations describing the heat flow. The first approach allows fairly easy identification of the main heat flow paths within the device and is much easier to apply if the medium is not continuous or if the mesh is irregular, whilst the latter approach immediately provides the error estimates. In this section we will rely mostly on the rigorous derivation of the finite difference approximations using the Taylor series while in the next section we will apply the energy balance equation method. Now we turn our attention to a set of simple illustrating examples that explain how the FDM is applied to study heat conduction.

We consider first an infinite copper rod and calculate the temperature distribution within its cross-section subject to a given temperature distribution on the boundary (Figure 5.13). Such a copper rod could represent approximately a heat sink of a photonic device. We assume that the temperature distribution on the boundary does not vary along the rod, that is, independent of the z spatial variable. We consider a problem where $W = H = 1$ cm and introduce a set of equidistant sampling points with $\Delta x = \Delta y = W/4$ (Figure 5.14). When $\Delta x = \Delta y$ in the absence of heat sources, Equation 5.21 reduces to a particularly simple form:

$$\frac{T\left(x_0 + \Delta x, y_0\right) + T\left(x_0 - \Delta x, y_0\right) + T\left(x_0, y_0 + \Delta y\right) + T\left(x_0, y_0 - \Delta y\right) - 4T\left(x_0, y_0\right)}{\Delta x^2}$$

$$+ O\left(\Delta x^2\right) = 0 \tag{5.22}$$

where $O(\Delta x^2)$ denotes terms proportional to the second and higher powers of Δx. The form of 5.22 shows that the finite difference approximation of the Laplace equation can be conveniently represented by the stencil shown in Figure 5.15, where the numbers positioned next to the nodal points give the finite difference weights associated with the nodal values of the temperature.

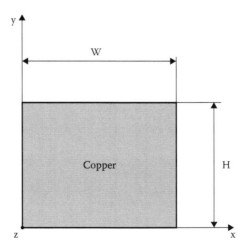

FIGURE 5.13   Cross-section of the copper rod.

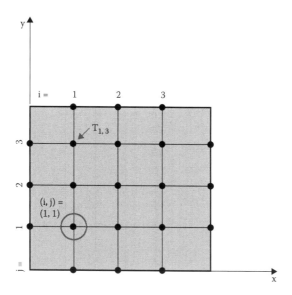

FIGURE 5.14 Selection of sampling points for the copper rod cross-section from Figure 5.13.

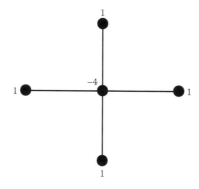

FIGURE 5.15 Finite difference stencil approximating Laplace Equation (5.11).

In the selected set of sampling points, the temperature at all the points that lay on the external boundary is known. We therefore have nine unknown values of the temperature. Writing Equation 5.22 for each nodal point with an unknown temperature yields a set of nine equations with nine unknowns, which can be solved using standard techniques of linear algebra. Figure 5.16 illustrates the process of setting up the equations at each nodal point.

It should be noted that if the finite difference stencil includes points laying on the boundary, the known values of the temperature are substituted and moved over with opposite sign to the right hand side of the equation. In order to avoid confusion, the nodes referring to the unknown values of the temperature have to be numbered consistently, usually starting from the bottom left corner (Figure 5.17).

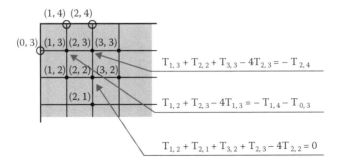

$$T_{1,3} + T_{2,2} + T_{3,3} - 4T_{2,3} = -T_{2,4}$$

$$T_{1,2} + T_{2,3} - 4T_{1,3} = -T_{1,4} - T_{0,3}$$

$$T_{1,2} + T_{2,1} + T_{3,2} + T_{2,3} - 4T_{2,2} = 0$$

FIGURE 5.16 Finite difference equations approximating Laplace Equation (5.11) at selected nodal points.

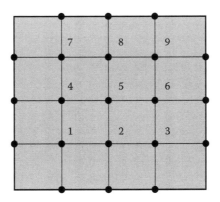

FIGURE 5.17 Numbering finite difference nodes for the rod problem from Figure 5.13.

Writing the finite difference equations at each node and assembling them together using matrix notation in such a way that the row number corresponds to the node number yields:

$$\begin{bmatrix} -4 & 1 & 0 & 1 & 0 & 0 & 0 & 0 & 0 \\ 1 & -4 & 1 & 0 & 1 & 0 & 0 & 0 & 0 \\ 0 & 1 & -4 & 0 & 0 & 1 & 0 & 0 & 0 \\ 1 & 0 & 0 & -4 & 1 & 0 & 1 & 0 & 0 \\ 0 & 1 & 0 & 1 & -4 & 1 & 0 & 1 & 0 \\ 0 & 0 & 1 & 0 & 1 & -4 & 0 & 0 & 1 \\ 0 & 0 & 0 & 1 & 0 & 0 & -4 & 1 & 0 \\ 0 & 0 & 0 & 0 & 1 & 0 & 1 & -4 & 1 \\ 0 & 0 & 0 & 0 & 0 & 1 & 0 & 1 & -4 \end{bmatrix} \begin{bmatrix} T_1 \\ T_2 \\ T_3 \\ T_4 \\ T_5 \\ T_6 \\ T_7 \\ T_8 \\ T_9 \end{bmatrix} = \begin{bmatrix} -T_{0,1} - T_{1,0} \\ -T_{2,0} \\ -T_{4,1} - T_{3,0} \\ -T_{0,2} \\ 0 \\ -T_{4,2} \\ -T_{0,3} - T_{1,4} \\ -T_{2,4} \\ -T_{4,3} - T_{3,4} \end{bmatrix} \qquad (5.23)$$

> **ALGORITHM 5.1 FINITE DIFFERENCE SOLUTION OF THE LAPLACE EQUATION SUBJECT TO A SET TEMPERATURE VALUE AT THE BOUNDARY**
>
> 1. Start
> 2. Select Δx, Δy, and the sampling nodal points
> 3. Given the values of the temperature at the boundary assemble the coefficient matrix and the augmented vector
> 4. Calculate the temperature at nodal points by solving the set of linear algebraic equations 5.23
> 5. Stop

The set of equations 5.23 can be written in a more compact way:

$$AT = \bar{b} \tag{5.24}$$

where A is the coefficient matrix, T is a vector containing all the unknown values of the temperature, and the vector $\bar{b}$ contains the right hand side of Equation 5.23. The solution of the set of algebraic equations yields the temperature distribution within the device.

Algorithm 5.1 summarizes all steps necessary to solve Equation 5.17 (in the absence of heat sources) using the FDM subject to a set value of the temperature at the boundary.

Next, we give a listing of the MATLAB software that implements Algorithm 5.1 and calculates the solution of 5.24 if the temperature on the boundary is constant and equal to 300 K. We use nine nodal points.

```
% program calculates temperature distribution
clear % clears variables
clear global % clears global variables
format short

Nx = 3;% number of points along x axis
Ny = 3;% number of points along y axis
N = Nx*Ny;%total number of points

for i = 1:N% setting off diagonals
a1(i) = 1;
a2(i) = 1;
a4(i) = 1;
a5(i) = 1;
end

for i = 1:Ny% setting zeros in off diagonals
a4((i-1)*Nx+1) = 0;
```

```
a2(i*Nx) = 0;
end
for i = 1:N% setting the main diagonal
a3(i) = -4;
end

for i = 1:N% setting the augmented vector
b(i) = -300;
end
b(1) = -600;
b(Nx) = -600;
b(N-Nx+1) = -600;
b(N) = -600;
for j = 2:Ny-1
for i = 2:Nx-1
       b((j-1)*Nx+i) = 0;
end
end

c = [-Nx,-1,0,1,Nx];%setting the matrix pattern
%forming directly the matrix in sparse format
A = spdiags([a1' a2' a3' a4' a5'], c, N, N);
% a1-a5 are transposed
x = A\b'% solving the set of equations
```

In agreement with the maximum principle, that is, that maximum (or minimum) of the solution of the Laplace equation is attained at the boundary, the program yields a constant value of the temperature in the entire domain (Table 5.2).

If the temperature distribution on the boundary is not constant, the part of the MATLAB code that sets the augmented vector b needs to be modified and a larger number of sampling points might be required in order to obtain a sufficiently accurate solution. This usually results in a set of linear algebraic equations with a large sparse coefficient matrix. Due to this reason, the presented MATLAB code is written using the sparse matrix format. The solution of the resulting set of algebraic Equation 5.24 can be performed using many numerical methods since the coefficient matrix A is symmetrical and has a penta-diagonal structure with the dominant main diagonal. The most often used techniques are the LU decomposition for a banded matrix, Gauss-Seidel method, successive over-relaxation (SOR) method, alternate direction implicit (ADI) method, and projection methods, for example, Arnoldi or CG method [26,27].

As for the second simple illustrating example, we consider the same rod as in the previous example, however, we set different boundary conditions (Figure 5.18).

TABLE 5.2

The Values of the Temperature Calculated by Solving Problem from Figure 5.13 at the Nodes Given in Figure 5.17

Number of the Node	1	2	3	4	5	6	7	8	9
Temperature [K]	300	300	300	300	300	300	300	300	300

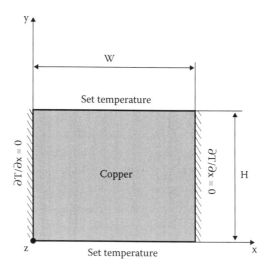

FIGURE 5.18 Cross-section of the copper rod with zero normal derivative condition set at the vertical boundaries.

Namely, on the vertical boundaries we force the normal derivative to be equal to zero. Such boundary condition implies that there is no heat flux through the boundary (cf. 5.3b). We consider again the same square box as in the previous example with $W = H = 1$ cm. We introduce a square finite difference mesh (Figure 5.19a). Hence, the only difference with the previous example is the vanishing normal derivative of the temperature that was imposed as the boundary condition on the vertical walls of the rod. We will therefore first consider how to handle such a boundary condition. As an illustrating example, let's derive a finite difference approximation to the Laplace equation at the node $(i,j) = (1,2)$, cf. Figure 5.19a:

$$\frac{T_{1,3} + T_{2,2} + T_{0,2} + T_{1,1} - 4T_{1,2}}{\Delta x^2} + O\left(h^2\right) = 0 \qquad (5.25)$$

The value of the temperature at the point $(i,j) = (0,2)$ is not available. However, we know the value of the temperature derivative

$$\left. \frac{\partial T}{\partial x} \right|_{0,2}$$

at the point $(i,j) = (0,2)$. In order to obtain an appropriate approximation to the Laplace equation that allows imposing a derivative boundary condition, we substitute $x_0 = x_{0,2}$ and $x_0 + \Delta x = x_{1,2}$ into 5.18a and obtain:

$$T_{1,2} = T_{0,2} + \left. \frac{\partial T}{\partial x} \right|_{0,2} \Delta x + \frac{1}{2} \left. \frac{\partial^2 T}{\partial x^2} \right|_{0,2} \Delta x^2 + \frac{1}{6} \left. \frac{\partial^3 T}{\partial x^3} \right|_{0,2} \Delta x^3 + \frac{1}{24} \left. \frac{\partial^4 T}{\partial x^4} \right|_{0,2} \Delta x^4 + \cdots \quad (5.26a)$$

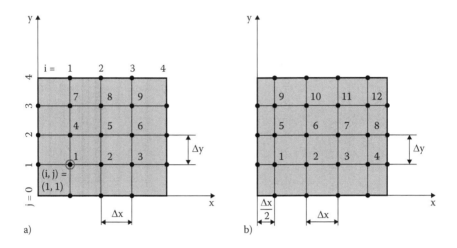

FIGURE 5.19 a) Numbering finite difference nodes for the rod problem from Figure 5.18. b) Shifting the finite difference nodes by $\Delta x/2$ for the rod problem from Figure 5.18.

Truncating Equation 5.26a after the second derivative term gives us the following approximation for $T_{0,2}$:

$$T_{0,2} \approx T_{1,2} - \frac{\partial T}{\partial x}\bigg|_{0,2} \Delta x - \frac{1}{2}\frac{\partial^2 T}{\partial x^2}\bigg|_{0,2} \Delta x^2 \qquad (5.26b)$$

which can be substituted into 5.25:

$$\frac{T_{1,3} + T_{2,2} - \dfrac{\partial T}{\partial x}\bigg|_{0,2} \Delta x + T_{1,1} - 3T_{1,2}}{\Delta x^2} - \frac{1}{2}\frac{\partial^2 T}{\partial x^2}\bigg|_{0,2} + O\left(\Delta x^2\right) = 0 \qquad (5.27)$$

Since the term

$$-\frac{1}{2}\frac{\partial^2 T}{\partial x^2}\bigg|_{0,2}$$

introduces an $O(\Delta x^0)$ error term, the overall accuracy of 5.27 degrades to zeroth order when compared with 5.22. In order to improve the finite difference approximation accuracy, one could use a central derivative approximation to the temperature derivative

$$\frac{\partial T}{\partial x}\bigg|_{0,2}$$

at the point $(i,j) = (0,2)$ [28]. Alternatively, one can shift the nodes by $\Delta x/2$ and expand T into the Taylor series at $x = x_0 + \Delta x/2$:

$$
T\left(x_0 + \frac{\Delta x}{2}\right) = T\left(x_0\right) + \left.\frac{\partial T}{\partial x}\right|_{x=x_0} \frac{\Delta x}{2} + \frac{1}{2}\left.\frac{\partial^2 T}{\partial x^2}\right|_{x=x_0} \frac{\Delta x^2}{4} + \frac{1}{6}\left.\frac{\partial^3 T}{\partial x^3}\right|_{x=x_0} \frac{\Delta x^3}{8}
$$

$$
+ \frac{1}{24}\left.\frac{\partial^4 T}{\partial x^4}\right|_{x=x_0} \frac{\Delta x^4}{16} + \cdots
\tag{5.28a}
$$

and similarly at $x = x_0 - \Delta x/2$

$$
T\left(x_0 - \frac{\Delta x}{2}\right) = T\left(x_0\right) - \left.\frac{\partial T}{\partial x}\right|_{x=x_0} \frac{\Delta x}{2} + \frac{1}{2}\left.\frac{\partial^2 T}{\partial x^2}\right|_{x=x_0} \frac{\Delta x^2}{4} - \frac{1}{6}\left.\frac{\partial^3 T}{\partial x^3}\right|_{x=x_0} \frac{\Delta x^3}{8}
$$

$$
+ \frac{1}{24}\left.\frac{\partial^4 T}{\partial x^4}\right|_{x=x_0} \frac{\Delta x^4}{16} + \cdots
\tag{5.28b}
$$

Subtracting 5.28b from 5.28a and rearranging terms yields an approximation of the first derivative:

$$
\left.\frac{\partial T}{\partial x}\right|_{x=x_0} = \frac{T\left(x_0 + \dfrac{\Delta x}{2}\right) - T\left(x_0 - \dfrac{\Delta x}{2}\right)}{\Delta x} - \frac{1}{24}\left.\frac{\partial^3 T}{\partial x^3}\right|_{x=x_0} \Delta x^2 + \cdots
\tag{5.29a}
$$

Neglecting the leading error term and the higher order terms yields a finite difference approximation of the first derivative:

$$
\left.\frac{\partial T}{\partial x}\right|_{x=x_0} \approx \frac{T\left(x_0 + \dfrac{\Delta x}{2}\right) - T\left(x_0 - \dfrac{\Delta x}{2}\right)}{\Delta x}
\tag{5.29b}
$$

From 5.29a it follows also that:

$$
\left.\frac{\partial^2 T}{\partial x^2}\right|_{x=x_0} \approx \frac{\left.\dfrac{\partial T}{\partial x}\right|_{x=x_0+\frac{\Delta x}{2}} - \left.\dfrac{\partial T}{\partial x}\right|_{x=x_0-\frac{\Delta x}{2}}}{\Delta x} - \frac{1}{24}\left.\frac{\partial^4 T}{\partial x^4}\right|_{x=x_0} \Delta x^2
\tag{5.30}
$$

It can be observed that 5.30 is consistent with 5.13 since using 5.29a to approximate first derivatives on the right hand side of 5.28 yields 5.13 again. It should be noted that the error terms introduced by the application of 5.29a, that is,

$$
\frac{1}{24}\left.\frac{\partial^3 T}{\partial x^3}\right|_{x=x_0+\Delta x/2} \Delta x^2 \quad \text{and} \quad \frac{1}{24}\left.\frac{\partial^3 T}{\partial x^3}\right|_{x=x_0-\Delta x/2} \Delta x^2
$$

combine to increase the coefficient in front of the leading error term in 5.30 to 1/12, which agrees with 5.13.

In order to apply the finite difference approximation 5.30 at the nodes neighbour-ing the vertical walls, we need to shift the sampling points by $\Delta x/2$ (Figure 5.19b).

Now, the zero derivative boundary condition can be directly imposed for either

$$\left.\frac{\partial T}{\partial x}\right|_{x_0 + \frac{\Delta x}{2}} \text{ or } \left.\frac{\partial T}{\partial x}\right|_{x_0 - \frac{\Delta x}{2}}$$

depending on whether the left or right vertical boundary is adjacent to the mesh node. The other first derivative in 5.30 is then approximated using 5.29a.

Here for the sake of simplicity, we use 5.27 to handle all nodal points adjacent to the vertical walls while for the other nodal points we use 5.25. In the example stud-ied, the second derivative of temperature with respect to x is equal zero so the zeroth order error term disappears anyway. However, it should be noted that adopting such an approach when the second derivative is not equal to zero would significantly reduce the accuracy of the results. Hence, in a general case one should use either 5.30 or the technique described in Strikwerda [28].

Combining all finite difference equations from all nine nodal points (Figure 5.19a) yields the following set of nine linear algebraic equations:

$$\begin{bmatrix} -3 & 1 & 0 & 1 & 0 & 0 & 0 & 0 & 0 \\ 1 & -4 & 1 & 0 & 1 & 0 & 0 & 0 & 0 \\ 0 & 1 & -3 & 0 & 0 & 1 & 0 & 0 & 0 \\ 1 & 0 & 0 & -3 & 1 & 0 & 1 & 0 & 0 \\ 0 & 1 & 0 & 1 & -4 & 1 & 0 & 1 & 0 \\ 0 & 0 & 1 & 0 & 1 & -3 & 0 & 0 & 1 \\ 0 & 0 & 0 & 1 & 0 & 0 & -3 & 1 & 0 \\ 0 & 0 & 0 & 0 & 1 & 0 & 1 & -4 & 1 \\ 0 & 0 & 0 & 0 & 0 & 1 & 0 & 1 & -3 \end{bmatrix} \begin{bmatrix} T_1 \\ T_2 \\ T_3 \\ T_4 \\ T_5 \\ T_6 \\ T_7 \\ T_8 \\ T_9 \end{bmatrix} = \begin{bmatrix} -T_{1,0} \\ -T_{2,0} \\ -T_{3,0} \\ 0 \\ 0 \\ 0 \\ -T_{1,4} \\ -T_{2,4} \\ -T_{3,4} \end{bmatrix} \quad (5.31)$$

The coefficient matrix has the same structure as in the previous example so the same numerical techniques can be used for the solution of 5.29. We impose the tem-perature at the top of the rod equal to 330 K while at the bottom equal to 300 K. For such cases, the temperature distribution can be calculated analytically, that is: $T(x,y) = 330 \text{ K} - (30 \text{ K/cm})*y$. The solution of the problem from Figure 5.18 results in algorithm 5.2.

The implementation of algorithm 5.2 requires only minor modifications to the code developed in the previous example. The calculated values of the temperature at the nodal points are given in Table 5.3.

It should be noted that even though the number of sampling nodes is small the numerically calculated solution gives the same numbers of the temperature (to 15 digital places) as the analytical method. This results from one property of the finite difference approximations used in this example, namely, that all polynomials up to the order 2 fulfil it exactly [29].

ALGORITHM 5.2   FINITE DIFFERENCE SOLUTION OF EQUATION 5.11 SUBJECT TO THE BOUNDARY CONDITIONS SET IN FIGURE 5.18

1. Start
2. Select Δx, Δy, and the sampling nodal points
3. Given the values of the temperature and the temperature derivative at the boundary assemble the coefficient matrix and the augmented vector
4. Calculate the temperature at nodal points by solving the set of linear algebraic equations 5.31
5. Stop

TABLE 5.3

The Values of the Temperature Calculated by Solving Problem from Figure 5.18 at the Nodes Given in Figure 5.19a

Number of the Node	Temperature [K]	
	Numerical	Analytical
1	307.5	307.5
2	307.5	307.5
3	307.5	307.5
4	315	315
5	315	315
6	315	315
7	322.5	322.5
8	322.5	322.5
9	322.5	322.5

The next problem we consider is depicted in Figure 5.20. We consider again the same copper rod as in the previous examples. However, on the top of the rod, we place a heat source, with the total heating power P_H. Such a heat source might represent an operating laser diode mounted on a copper heat sink (Figure 5.5).

If the heat flux is distributed homogenously over the entire cross-section of the heater, the y component of the heat flux $q_y = P_H/(WL)$. However, form 5.4b:

$$q_y = -\rho_c \frac{\partial T}{\partial y}$$

where ρ_c is the heat conductivity for copper. Consequently, we obtain the condition for the normal derivative adjacent to the heater:

$$\frac{\partial T}{\partial y} = \frac{q_y}{\rho_c} \qquad (5.32a)$$

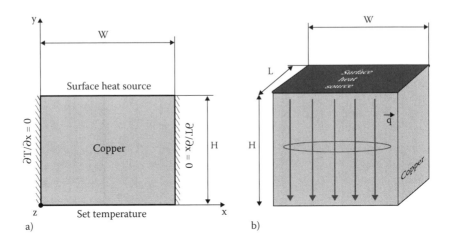

FIGURE 5.20 Copper rod with surface heat source.

or for a homogenously distributed heat source:

$$\frac{\partial T}{\partial y} = \frac{P_H}{\rho_c WL} \tag{5.32b}$$

The more general condition 5.32a can be used if the heat flux distribution is arbitrary. Note that the "+" sign in 5.32 follows from the fact that in the coordinate system introduced in Figure 5.20a the heat flows from the heater along the negative y direction.

Now we can use again 5.27 and substitute 5.32b to derive the finite difference approximation for 5.13 for all the nodes that are adjacent to the heater. We assume that the width of the heat source equals the rod width. If we introduce a set of sampling points (Figure 5.21), then at a node adjacent to the heater, we obtain:

$$\left.\left(\frac{\partial^2 T}{\partial x^2} + \frac{\partial^2 T}{\partial y^2}\right)\right|_{2,3} \approx \frac{\dfrac{P_H \Delta y}{\rho_c WL} + T_5 - 3T_8 + T_7 + T_9}{\Delta x^2} = 0 \tag{5.33a}$$

Similarly, for an upper corner node we obtain:

$$\left.\left(\frac{\partial^2 T}{\partial x^2} + \frac{\partial^2 T}{\partial y^2}\right)\right|_{1,3} \approx \frac{\dfrac{P_H \Delta y}{\rho_c WL} + T_4 - 2T_7 + T_8}{\Delta x^2} \tag{5.33b}$$

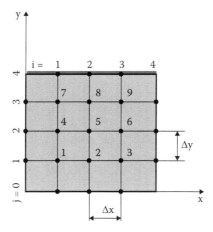

FIGURE 5.21 Numbering finite difference nodes for the rod problem from Figure 5.20.

and a lower corner node:

$$\left(\frac{\partial^2 T}{\partial x^2} + \frac{\partial^2 T}{\partial y^2} \right)\bigg|_{1,1} \approx \frac{-3T_1 + T_2 + T_4 + T_{1,0}}{\Delta x^2} \tag{5.33c}$$

Equations 5.15 and 5.37 can be used for all other nodal points, so we can write the finite difference equations in a matrix vector product form:

$$\begin{bmatrix} -3 & 1 & 0 & 1 & 0 & 0 & 0 & 0 & 0 \\ 1 & -4 & 1 & 0 & 1 & 0 & 0 & 0 & 0 \\ 0 & 1 & -3 & 0 & 0 & 1 & 0 & 0 & 0 \\ 1 & 0 & 0 & -3 & 1 & 0 & 1 & 0 & 0 \\ 0 & 1 & 0 & 1 & -4 & 1 & 0 & 1 & 0 \\ 0 & 0 & 1 & 0 & 1 & -3 & 0 & 0 & 1 \\ 0 & 0 & 0 & 1 & 0 & 0 & -2 & 1 & 0 \\ 0 & 0 & 0 & 0 & 1 & 0 & 1 & -3 & 1 \\ 0 & 0 & 0 & 0 & 0 & 1 & 0 & 1 & -2 \end{bmatrix} \begin{bmatrix} T_1 \\ T_2 \\ T_3 \\ T_4 \\ T_5 \\ T_6 \\ T_7 \\ T_8 \\ T_9 \end{bmatrix} = \begin{bmatrix} -T_{1,0} \\ -T_{2,0} \\ -T_{3,0} \\ 0 \\ 0 \\ 0 \\ b \\ b \\ b \end{bmatrix} \tag{5.34}$$

The coefficient b is defined as:

$$b = -\frac{P_H \Delta y}{\rho_c W L}$$

**ALGORITHM 5.3   FINITE DIFFERENCE SOLUTION
OF EQUATION 5.11 SUBJECT TO THE BOUNDARY
CONDITIONS SET IN FIGURE 5.20**

1. Start
2. Select Δx, Δy, and the sampling nodal points
3. Calculate the values of the normal derivative in the nodal points adjacent to the heater
4. Given the values of the temperature and the temperature derivative at the boundary assemble the coefficient matrix and the augmented vector
5. Calculate the temperature at nodal points by solving the set of linear algebraic equations 5.34
6. Stop

The coefficient matrix in 5.34 is symmetrical therefore the same procedures can be followed as in the previous examples. Algorithm 5.3 gives all the steps that need to be followed.

Again, the implementation of algorithm 5.3 requires only minor modifications to the MATLAB code used in the previous case. We impose the temperature at the bottom of the rod equal to 300 K. For such case, the temperature distribution can be calculated analytically:

$$T(x,y) = \frac{P_H \Delta y}{\rho_c WL} y + 300 \ K.$$

If we further assume that $P_H = 4$ W (a fairly typical value for a laser diode), $L = 1$cm, and $\rho_c = 400$ W/m K, the temperature distribution can be calculated both analytically and numerically. The calculated values of the temperature at the nodal points are given in Table 5.4.

Again, they do not differ from the analytical solution despite the zero order accuracy near the boundary. This results from the fact that the second derivative is intrinsically equal to zero in this example since the solution depends linearly on y and there is no variation with y.

The last example that we consider is a silica rod within which there is a heat source (Figure 5.22). This example is related to the study of heat flow in fibre lasers [13].

The boundary condition imposed in this case is a constant temperature equal to 300 K. We assume that the temperature along the rod is constant. Therefore, in order to calculate the temperature distribution in the transverse cross-section, it is necessary to solve the 2D Poisson equation:

$$\frac{\partial^2 T}{\partial^2 x} + \frac{\partial^2 T}{\partial^2 y} = -\frac{Q}{\rho} \tag{5.35}$$

The only major modification, when compared with the first problem considered, is the presence of the heat source distribution. Under the assumption that the source

TABLE 5.4

The Values of the Temperature Calculated by Solving Problem from Figure 5.20 at the Nodes Given in Figure 5.21

	Temperature [K]	
Number of the Node	Numerical	Analytical
1	302.5	302.5
2	302.5	302.5
3	302.5	302.5
4	305	305
5	305	305
6	305	305
7	307.5	307.5
8	307.5	307.5
9	307.5	307.5

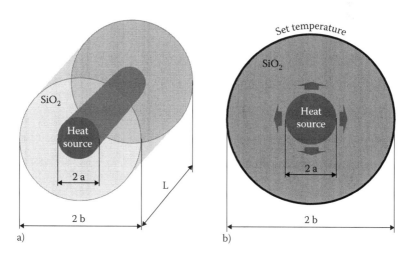

FIGURE 5.22 A silica glass fibre with a) a volume heat source and b) schematic diagram of the heat flow in the cross-section.

is homogenously distributed, the density of energy per time unit for the heat source power density $Q = P_{HS}/(L \pi a^2)$ where P_{HS} is the total power of the heat source, which is assumed equal to 200 W. We assume a rod length of 10 m and a radius of 400 μm. The radius of the heat source is 100 μm. The heat conductivity of silica is 1.38 W/mK (Table 5.1). We introduce a set of sampling points given in Figure 5.23. Note that, unlike previous examples, the distances between the nodal points are not constant. To avoid ambiguity when referring to the nodal points, in Figure 5.23 we give explicitly the nodal indexes on the left side of the outer boundary. The same convention is applied on the right side.

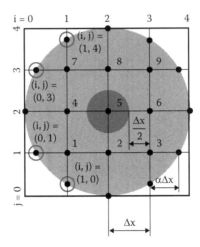

FIGURE 5.23 Numbering finite difference nodes for the problem from Figure 5.22.

Since the distribution of the nodal points in Figure 5.23 is not equidistant, it is necessary to derive more general finite difference approximations. For this purpose, let's first derive a finite difference representation of the second derivative w.r.t. x on three nodal points: $x_0 - \Delta x$, x_0, and $x_0 + a\Delta x$. For this purpose, we expand T at $x = x_0 + a\Delta x$:

$$T(x_0 + a\Delta x) = T(x_0) + \left.\frac{\partial T}{\partial x}\right|_{x=x_0} a\Delta x + \frac{1}{2}\left.\frac{\partial^2 T}{\partial x^2}\right|_{x=x_0} a^2\Delta x^2 + \frac{1}{6}\left.\frac{\partial^3 T}{\partial x^3}\right|_{x=x_0} a^3\Delta x^3$$

$$+ \frac{1}{24}\left.\frac{\partial^4 T}{\partial x^4}\right|_{x=x_0} a^4\Delta x^4 + \cdots \tag{5.36a}$$

and, similarly, at $x = x_0 - \Delta x$:

$$T(x_0 - \Delta x) = T(x_0) - \left.\frac{\partial T}{\partial x}\right|_{x=x_0} \Delta x + \frac{1}{2}\left.\frac{\partial^2 T}{\partial x^2}\right|_{x=x_0} \Delta x^2 - \frac{1}{6}\left.\frac{\partial^3 T}{\partial x^3}\right|_{x=x_0} \Delta x^3$$

$$+ 24\left.\frac{\partial^4 T}{\partial x^4}\right|_{x=x_0} \Delta x^4 + \cdots \tag{5.36b}$$

Eliminating the term containing the first derivative yields the following approximation to the second derivative:

$$\left.\frac{\partial^2 T}{\partial x^2}\right|_{x=x_0} \approx 2\frac{T(x_0 + a\Delta x) - (1+a)T(x_0) + aT(x_0 - \Delta x)}{(a+a^2)\Delta x^2} - \frac{1}{3}\frac{(1+a^2)}{(1+a)}\left.\frac{\partial^3 T}{\partial x^3}\right|_{x=x_0} \Delta x$$

$$\tag{5.37a}$$

Similarly, we can obtain:

$$\frac{\partial^2 T}{\partial y^2}\bigg|_{y=y_0} \approx 2\frac{T\left(y_0 + a\Delta y\right)-\left(1+a\right)T\left(y_0\right)+aT\left(y_0 - \Delta y\right)}{\left(a+a^2\right)\Delta y^2} - \frac{1}{3}\frac{\left(1+a^2\right)}{\left(1+a\right)}\frac{\partial^3 T}{\partial y^3}\bigg|_{y=y_0}\Delta y$$

(5.37b)

Applying 5.37 at the nodal point 3 in Figure 5.23 for instance, we obtain the following approximation to the left hand side of 5.35:

$$\left(\frac{\partial^2 T}{\partial x^2}+\frac{\partial^2 T}{\partial y^2}\right)\bigg|_{3,1} \approx \frac{2}{\left(a+a^2\right)\Delta x^2}\left(aT_{2,1}+T_{4,1}+aT_{3,2}+T_{3,0}-2\left(1+a\right)T_{3,1}\right) \text{ (5.38)}$$

Similarly, we can derive the finite difference equations for the nodal points 1, 7, and 9. For the other points, we can use 5.13, keeping in mind that we need to include the source term at the node 5. This procedure yields the following equations:

$$\begin{bmatrix} -2a_2 & a_1 & 0 & a_1 & 0 & 0 & 0 & 0 & 0 \\ 1 & -4 & 1 & 0 & 1 & 0 & 0 & 0 & 0 \\ 0 & a_1 & -2a_2 & 0 & 0 & a_1 & 0 & 0 & 0 \\ 1 & 0 & 0 & -4 & 1 & 0 & 1 & 0 & 0 \\ 0 & 1 & 0 & 1 & -4 & 1 & 0 & 1 & 0 \\ 0 & 0 & 1 & 0 & 1 & -4 & 0 & 0 & 1 \\ 0 & 0 & 0 & a_1 & 0 & 0 & -2a_2 & a_1 & 0 \\ 0 & 0 & 0 & 0 & 1 & 0 & 1 & -4 & 1 \\ 0 & 0 & 0 & 0 & 0 & a_1 & 0 & a_1 & -2a_2 \end{bmatrix}\begin{bmatrix} T_1 \\ T_2 \\ T_3 \\ T_4 \\ T_5 \\ T_6 \\ T_7 \\ T_8 \\ T_9 \end{bmatrix} = \begin{bmatrix} -a_3\left(T_{0,1}+T_{1,0}\right) \\ -T_{2,0} \\ -a_3\left(T_{4,1}+T_{3,0}\right) \\ -T_{0,2} \\ -\dfrac{P_{HS}\Delta x^2}{\rho_{silica}L\pi a^2} \\ -T_{4,2} \\ -a_3\left(T_{0,3}+T_{1,4}\right) \\ -T_{2,4} \\ -a_3\left(T_{4,3}+T_{3,4}\right) \end{bmatrix}$$

(5.39)

The coefficients a_i are given by:

$$a_1 = \frac{2}{\left(1+a\right)}, \quad a_2 = \frac{2}{a}, \text{ and } a_3 = \frac{2}{\left(a+a^2\right)}$$

The set of algebraic Equation 5.36 can be easily solved using the same techniques as in the first example. The solution of 5.35 for the problem from Figure 5.22 results in algorithm 5.4.

Following algorithm 5.4, the MATLAB code can be written by introducing only minor modifications to the code developed for problem 1. The calculated values of the temperature at the nodal points are given in Table 5.5.

**ALGORITHM 5.4   FINITE DIFFERENCE SOLUTION
OF EQUATION 5.11 SUBJECT TO THE BOUNDARY
CONDITIONS SET IN FIGURE 5.22**

1. Start
2. Select Δx, Δy, and the sampling nodal points
3. Calculate the values of the heat source power density in the nodes that overlap with the heat source spatial distribution
4. Given the values of the temperature at the boundary and the values of the heat source power density at the nodes that overlap with the heat source spatial distribution, assemble the coefficient matrix and the augmented vector
5. Calculate the temperature at nodal points by solving the set of linear algebraic equations 5.39
6. Stop

TABLE 5.5

The Values of the Temperature Calculated by Solving Problem from Figure 5.22 at the Nodes Given in Figure 5.23

	Temperature [K]	
Number of the Node	Numerical	Analytical
1	3.009048877586438e+002	3.007994043483792e+002
2	3.021409874245248e+002	3.015988086967583e+002
3	3.009048877586439e+002	3.007994043483792e+002
4	3.021409874245248e+002	3.015988086967583e+002
5	3.067541741808117e+002	3.043509140825884e+002
6	3.021409874245249e+002	3.015988086967583e+002
7	3.009048877586439e+002	3.007994043483792e+002
8	3.021409874245248e+002	3.015988086967583e+002
9	3.009048877586438e+002	3.007994043483792e+002

It can be observed that there is a significant difference between the results obtained from the analytical solution and the numerically calculated ones. This results from the fact that the number of finite difference mesh points is small. The difference between the analytical and numerical results can be reduced by refining the finite difference mesh, however, at the loss of the penta-diagonal structure of the coefficient matrix. One can also observe that there is a significant redundancy in the set of Equation 5.39. This follows from the symmetry of the considered structure. A quick inspection of the Table 5.4 reveals that there are only three independent values of the temperature. In fact due to the presence of symmetry planes, we could have only considered one octant of the circular cross-sections. This octant could for instance contain the nodal points 5, 7, and 8. The finite difference equations for these nodes can be derived using the fifth, seventh, and eighth equation from the set 5.39. The values of temperature at the nodal points: 2, 4, 6, and 9 that are present in these equations can be eliminated using the symmetry planes. For instance, $T_4 = T_8$, due to the symmetry plane passing through the nodal points 7, 5, and 3.

A better advantage of the symmetry can be taken if the Poisson Equation 5.17 is recast into the cylindrical coordinate system:

$$\frac{1}{r}\frac{\partial}{\partial r}\left(r\frac{\partial T}{\partial r}\right) + \frac{1}{r^2}\frac{\partial^2 T}{\partial^2 \theta} = -\frac{Q}{\rho} \qquad (5.40)$$

where $x = r\cos(\theta)$ and $y = r\sin(\theta)$. In fact the considered problem can be easily solved analytically in the cylindrical coordinate system [4,5]. Equation 5.40 can also be approximated using the FDM [25]. The cylindrical coordinate system has also two other advantages when compared with the rectangular one. Namely, it is easy in this coordinate system to impose the value of the normal derivative at the boundary for the considered problem, unlike the rectangular coordinate system [19,30,31], and this results in a penta-diagonal sparse matrix. The value of the normal derivative is imposed if the fibre laser is cooled by convection [13].

If the shape of the considered domain is not rectangular, the finite difference mesh can be adopted to handle such a problem by either varying the mesh size, introducing other types of finite difference stencils, for example, a triangular mesh, or using other coordinate systems. A description of these techniques can be found in books on heat flow [19], the finite difference solution of partial differential equations [25], and in standard textbooks on numerical modelling [32,33]. In some situations, the application of the FDM to a problem with an irregular boundary can result in the loss of the penta-diagonal pattern of the coefficient matrix. In such case, the computational domain can be complemented to a regular shape using either an immersed interface [30,34] or an embedded boundary method [35,36].

FINITE DIFFERENCE ANALYSIS OF HEAT FLOW IN INHOMOGENEOUS MEDIA

The numerical techniques presented in the section "Finite Difference Analysis of Heat Flow in Homogenous Media" fail if the thermal conductivity distribution has stepwise discontinuities within the considered domain. A simple example of such a problem is shown in Figure 5.24. The thermal conductivity of copper ρ_C is 400 W/mK while that of alumina is more than ten times smaller, that is, $\rho_A = 27$ W/mK (Table 5.1). Hence, there is a large stepwise discontinuity of thermal conductivity at the interface between two materials. This type of problem can be encountered, for instance, when studying the heat flow through the submount into the heat sink.

There are two fundamental ways in which the problem shown in Figure 5.24 can be solved numerically. The first of them consists in dividing the considered area into subdomains in which the heat conductivity is constant. The finite difference equations are written at a set of selected nodal points in each subdomain and then combined, consistently the solutions in each subdomain, by imposing the continuity of the heat flux and temperature distribution at the interface. The other approach

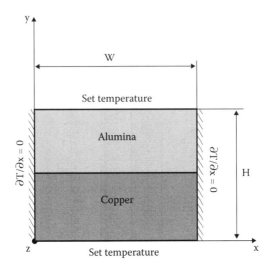

FIGURE 5.24 An example of a heat flow problem in an inhomogeneous domain with a step-wise thermal conductivity distribution.

consists in deriving a finite difference approximation that is accurate also when step-wise discontinuities of the heat conductivity are present and treat the entire problem in one domain. In this section we will compare both approaches. For this purpose, we will use the problem shown in Figure 5.24 as an example. We assume that $H_C = H_A = H/2$ and that $H = W = 1$ cm. We assume a constant temperature distribution at both horizontal boundaries. At the top horizontal boundary, the temperature is equal to 330 K, while at the bottom one it is set to 300 K. It is assumed also that there is no heat flow through the vertical boundaries.

We consider first the approach whereby we introduce two subdomains, each containing on homogenous region, and then impose the continuity of the temperature and of the heat flux across the interface between the two regions. We denote the subdomain made of copper as "subdomain 1" and the subdomain made of alumina as "subdomain 2." We introduce the sampling points in both subdomains separately. Figure 5.25 gives the distribution of the nodal points in each subdomain and provides the numbering of the nodes.

Then we treat all the nodal points that are within subdomain 1 using the techniques developed in the section "Finite Difference Analysis of Heat Flow in Homogenous Media." We write the finite difference equations for them using 5.15 and 5.21. We follow the same procedure for the nodal points that fall within the subdomain 2. However, the nodal points that lay on the interface are treated differently. At these points, we impose the continuity of the heat flux that is equivalent to imposing the continuity of the normal component of the heat flux vector:

$$\rho_C \left. \frac{\partial T}{\partial y} \right|_C = \rho_A \left. \frac{\partial T}{\partial y} \right|_A \tag{5.41}$$

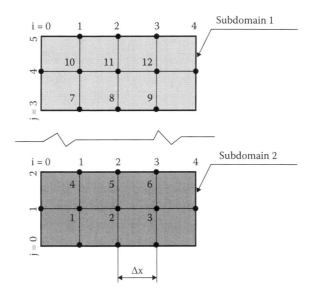

FIGURE 5.25   Numbering finite difference nodes for the problem from Figure 5.24 when splitting the computational domain in two subdomains.

where ρ_C and ρ_A is the heat conductivity for copper and alumina, respectively,

$$\frac{\partial T}{\partial y}\bigg|_C \quad \text{and} \quad \frac{\partial T}{\partial y}\bigg|_A$$

are the values of the normal derivative at the interface on the side of the copper and alumina, respectively. In order to evaluate this derivative on the alumina side, we use 5.25 and obtain:

$$\frac{\partial T}{\partial y}\bigg|_{y=y_0} = \frac{T(y_0 + \Delta y) - T(y_0)}{\Delta y} - \frac{1}{2}\frac{\partial^2 T}{\partial y^2}\bigg|_{y=y_0}\Delta y - \frac{1}{6}\frac{\partial^3 T}{\partial y^3}\bigg|_{y=y_0}\Delta y^2$$

$$- \frac{1}{24}\frac{\partial^4 T}{\partial y^4}\bigg|_{y=y_0}\Delta y^3 + \cdots \tag{5.42}$$

Neglecting the higher order terms, the following expression gives the finite difference approximation for the first derivative at $y = y_0$:

$$\frac{\partial T}{\partial y}\bigg|_{y=y_0} \approx \frac{T(y_0 + \Delta y) - T(y_0)}{\Delta y} \tag{5.43}$$

Similarly, we obtain the approximation for the first derivative on the copper side:

$$\frac{\partial T}{\partial y}\bigg|_{y=y_0} \approx \frac{T(y_0) - T(y_0 - \Delta y)}{\Delta y} \tag{5.44}$$

Substituting 5.43 and 5.44 into 5.41 yields the required equation:

$$\rho_C\left[T(y_0) - T(y_0 - \Delta y)\right] = \rho_A\left[T(y_0 + \Delta y) - T(y_0)\right] \tag{5.45}$$

We also impose the continuity of the temperature distribution. This is equivalent to equating the values of temperature at the points on the interface that coincide, that is, 4 and 7, 6 and 9, and so on. Assembling all equations together yields the following set of linear algebraic equations:

$$\begin{bmatrix} -3 & 1 & 0 & 1 & 0 & 0 & 0 & 0 & 0 & 0 & 0 & 0 \\ 1 & -4 & 1 & 0 & 1 & 0 & 0 & 0 & 0 & 0 & 0 & 0 \\ 0 & 1 & -3 & 0 & 0 & 1 & 0 & 0 & 0 & 0 & 0 & 0 \\ \rho_C & 0 & 0 & -\rho_C & 0 & 0 & -\rho_A & 0 & 0 & \rho_A & 0 & 0 \\ 0 & \rho_C & 0 & 0 & -\rho_C & 0 & 0 & -\rho_A & 0 & 0 & \rho_A & 0 \\ 0 & 0 & \rho_C & 0 & 0 & -\rho_C & 0 & 0 & -\rho_A & 0 & 0 & \rho_A \\ 0 & 0 & 0 & -1 & 0 & 0 & 1 & 0 & 0 & 0 & 0 & 0 \\ 0 & 0 & 0 & 0 & -1 & 0 & 0 & 1 & 0 & 0 & 0 & 0 \\ 0 & 0 & 0 & 0 & 0 & -1 & 0 & 0 & 1 & 0 & 0 & 0 \\ 0 & 0 & 0 & 0 & 0 & 0 & 1 & 0 & 0 & -3 & 1 & 0 \\ 0 & 0 & 0 & 0 & 0 & 0 & 0 & 1 & 0 & 1 & -4 & 1 \\ 0 & 0 & 0 & 0 & 0 & 0 & 0 & 0 & 1 & 0 & 1 & -3 \end{bmatrix} \begin{bmatrix} T_1 \\ T_2 \\ T_3 \\ T_4 \\ T_5 \\ T_6 \\ T_7 \\ T_8 \\ T_9 \\ T_{10} \\ T_{11} \\ T_{12} \end{bmatrix} = \begin{bmatrix} -T_{1,0} \\ -T_{2,0} \\ -T_{3,0} \\ 0 \\ 0 \\ 0 \\ 0 \\ 0 \\ 0 \\ -T_{1,4} \\ -T_{2,4} \\ -T_{3,4} \end{bmatrix}$$

$$\tag{5.46}$$

where $a_1 = \rho_C - \rho_A$. Following the same steps as in examples considered in the section "Finite Difference Analysis of Heat Flow in Homogenous Media," we obtain the solution of 5.46, which is shown in Table 5.6.

Now we will use the second approach. The distribution of the nodal points is given in Figure 5.26.

The nodes 4, 5, and 6 lie on the interface between the alumina and copper. In the section "Finite Difference Analysis of Heat Flow in Homogenous Media," we derived the finite difference equations that can be applied in all nodes, with the exception of those that lie on the interface between two regions with a different thermal conductivity. At these nodes, Equation 5.9 is not valid. The correct equation that applies at these nodal points can be derived directly from 5.4, in the absence of heat sources and temporal changes in the energy stored:

$$\rho\frac{\partial^2 T}{\partial x^2} + \frac{\partial}{\partial y}\left(\rho\frac{\partial T}{\partial y}\right) = 0 \tag{5.47}$$

whereby we took advantage of the fact that the interface between the two homogenous regions is parallel to the x-axis. For the approximation of the partial derivative with respect to x, we can use the energy balance method. Figure 5.27 shows one element of the finite difference mesh.

TABLE 5.6
Values of the Temperature Calculated by Solving Problem from Figure 5.24 at the Nodes Given in Figure 5.25

Number of the Node	Temperature	
	Numerical	Analytical
1	3.009484777517565e+002	3.009484777517565e+002
2	3.009484777517565e+002	3.009484777517565e+002
3	3.009484777517565e+002	3.009484777517565e+002
4	3.018969555035129e+002	3.018969555035129e+002
5	3.018969555035129e+002	3.018969555035129e+002
6	3.018969555035130e+002	3.018969555035129e+002
7	3.018969555035129e+002	3.018969555035129e+002
8	3.018969555035129e+002	3.018969555035129e+002
9	3.018969555035130e+002	3.018969555035129e+002
10	3.159484777517565e+002	3.159484777517565e+002
11	3.159484777517566e+002	3.159484777517565e+002
12	3.159484777517566e+002	3.159484777517565e+002

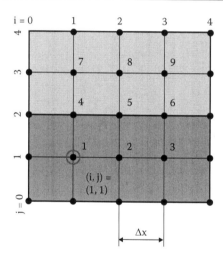

FIGURE 5.26 Numbering finite difference nodes for the problem from Figure 5.24 when calculating the temperature distribution directly in the entire domain.

Considering the energy balance [19] for the finite difference mesh element shown in Figure 5.27, that is, calculating a line integral of 5.47, yields the following approximation of Equation 5.47:

$$\frac{\Delta y}{2}\left(q_{x_0+\Delta x/2}+q_{x_0-\Delta x/2}\right)\bigg|_A + \frac{\Delta y}{2}\left(q_{x_0+\Delta x/2}+q_{x_0-\Delta x/2}\right)\bigg|_C$$
$$+\Delta x\, q_{x_0+\Delta y/2}+\Delta x\, q_{x_0-\Delta y/2}=0 \tag{5.48}$$

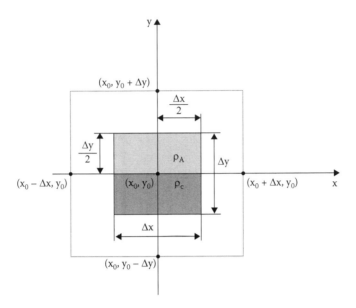

FIGURE 5.27   One element of the finite different mesh in an inhomogeneous medium.

In Equation 5.48 the heat flux through the vertical sides of the rectangle is split into two parts. One, with the subscript A, corresponds to the heat flux through alumina, while the second part, with the subscript C, corresponds to copper (Figure 5.27). Since the interface between the media is positioned half way between mesh points, both parts are multiplied by $\Delta y/2$. Following the same steps as in the case of 2D homogenous medium yields:

$$
\frac{\Delta y}{2}\rho_A\left(\frac{T\left(x_0+\Delta x,y_0\right)-T\left(x_0,y_0\right)}{\Delta x}+\frac{T\left(x_0-\Delta x,y_0\right)-T\left(x_0,y_0\right)}{\Delta x}\right)
$$
$$
+\frac{\Delta y}{2}\rho_C\left(\frac{T\left(x_0+\Delta x,y_0\right)-T\left(x_0,y_0\right)}{\Delta x}+\frac{T\left(x_0-\Delta x,y_0\right)-T\left(x_0,y_0\right)}{\Delta x}\right)
$$
$$
+\Delta x\left(\rho_A\frac{T\left(x_0,y_0+\Delta y\right)-T\left(x_0,y_0\right)}{\Delta y}+\rho_C\frac{T\left(x_0,y_0-\Delta y\right)-T\left(x_0,y_0\right)}{\Delta y}\right)=0 \quad (5.49)
$$

After recasting and dividing both sides of 5.49 by $\Delta x\Delta y$ one obtains:

$$
\frac{\left(\rho_A+\rho_C\right)}{2}\frac{T\left(x_0+\Delta x,y_0\right)-2T\left(x_0,y_0\right)+T\left(x_0-\Delta x,y_0\right)}{\Delta x^2}
$$
$$
+\frac{\rho_A T\left(x_0,y_0+\Delta y\right)-\left(\rho_A+\rho_C\right)T\left(x_0,y_0\right)+\rho_C T\left(x_0,y_0-\Delta y\right)}{\Delta y^2}=0 \quad (5.50)
$$

The formula 5.50 can be extended easily to any piecewise homogenous media if the boundaries between the regions, with a constant value of the conductivity, are parallel to the one of the coordinate axes [19,20]. The only limitation is that the finite difference nodal points must lie on the interfaces. An application of the Taylor series allows this limitation to be relaxed.

Now we can write and assemble the finite difference equations at all the nodes (Figure 5.26):

$$
\begin{bmatrix}
-3 & 1 & 0 & 1 & 0 & 0 & 0 & 0 & 0 \\
1 & -4 & 1 & 0 & 1 & 0 & 0 & 0 & 0 \\
0 & 1 & -3 & 0 & 0 & 1 & 0 & 0 & 0 \\
a_2 & 0 & 0 & -3 & 1 & 0 & a_1 & 0 & 0 \\
0 & a_2 & 0 & 1 & -4 & 1 & 0 & a_1 & 0 \\
0 & 0 & a_2 & 0 & 1 & -3 & 0 & 0 & a_1 \\
0 & 0 & 0 & 1 & 0 & 0 & -3 & 1 & 0 \\
0 & 0 & 0 & 0 & 1 & 0 & 1 & -4 & 1 \\
0 & 0 & 0 & 0 & 0 & 1 & 0 & 1 & -3
\end{bmatrix}
\begin{bmatrix}
T_1 \\ T_2 \\ T_3 \\ T_4 \\ T_5 \\ T_6 \\ T_7 \\ T_8 \\ T_9
\end{bmatrix}
=
\begin{bmatrix}
-T_{1,0} \\ -T_{2,0} \\ -T_{3,0} \\ 0 \\ 0 \\ 0 \\ -T_{1,4} \\ -T_{2,4} \\ -T_{3,4}
\end{bmatrix}
\tag{5.51}
$$

whereby

$$
a_1 = \frac{2\rho_A}{\rho_A + \rho_C} \quad \text{and} \quad a_2 = \frac{2\rho_C}{\rho_A + \rho_C}
$$

Following again the same steps as in the previous example, we obtain the solution of 5.51, which is shown in Table 5.7.

By comparing Tables 5.6 and 5.7, we can observe that both methods give the same results. The third column gives the values obtained from the analytical solution:

TABLE 5.7

The Values of the Temperature Calculated by Solving Problem from Figure 5.24 at the Nodes Given in Figure 5.26

	Temperature [K]	
Number of the Node	Numerical	Analytical
1	3.009484777517564e+002	3.009484777517565e+002
2	3.009484777517565e+002	3.009484777517565e+002
3	3.009484777517565e+002	3.009484777517565e+002
4	3.018969555035129e+002	3.018969555035129e+002
5	3.018969555035129e+002	3.018969555035129e+002
6	3.018969555035129e+002	3.018969555035129e+002
7	3.159484777517565e+002	3.159484777517565e+002
8	3.159484777517566e+002	3.159484777517565e+002
9	3.159484777517565e+002	3.159484777517565e+002

$$T(x,y) = \frac{60K}{H} \frac{1}{1+\rho_C/\rho_A} y + 300K \qquad\qquad 0 < y < \frac{H}{2}$$

$$T(x,y) = \frac{60K}{H} \frac{1}{1+\rho_A/\rho_C} y + 30K \frac{\rho_A - \rho_C}{\rho_A + \rho_C} + 300K \qquad \frac{H}{2} < y < H \quad (5.52)$$

The analytical solution and both sets of results obtained agree numerically on 15 decimal places. The calculations were performed using double precision arithmetic. This example shows that by splitting the considered region into subdomains one can use standard finite difference calculus, however, there is an increase in the total number of nodes due to nodal point duplication. When applying the global approach, the number of nodes is reduced, but there is a need to derive nonstandard finite difference approximations that can handle the discontinuities in the thermal conductivity distribution. These approximations may not be as flexible and accurate as the standard finite difference approximations.

A much more powerful approach that relies on dividing the problem into subdomains is obtained when using the overset grids [37]. In this approach, the problem is split into subdomains within which a very efficient finite difference approximation can be applied. All subdomains are made to overlap so that at the boundaries of each subdomain the missing nodal values of temperature are obtained by interpolation between the nodal values of the temperature in the adjacent subdomains. The overset grids allow an efficient analysis of problems with complicated irregular boundaries, and can also be applied very efficiently on multiprocessor computers using parallel processing techniques.

Finally, we note that, although, the energy balance method is fairly straightforward and easy to use when there are stepwise discontinuities in the distribution of thermal conductivity, it yields approximations with limited accuracy. A more rigorous approach that also allows for the derivation of the finite difference approximations in a medium with discontinuous thermal conductivity distribution is based on the extension of the Taylor series approach [38].

HEAT SOURCES, BOUNDARY CONDITIONS, AND THERMAL BOUNDARY RESISTANCE

Heat is generated in photonic devices when the transfer of the energy to the lattice occurs. In semiconductor optoelectronic devices there are many processes that result in such a transfer of energy. When electron hole pairs recombine radiatively, releasing a photon, no heat is generated. However, in the case of a nonradiative recombination, the excess energy is transferred to the lattice and phonons are generated. Another mechanism of heat generation in semiconductor devices involves free carrier absorption, whereby, a carrier that interacts with an incident photon is moved to a higher energy state. Then the carrier thermalizes to the bottom of the band while releasing phonons. Another significant source of heat in semiconductor optoelectronic devices is the Joule heat that accompanies the electron and hole flow through the semiconductor lattice. The heat can be also generated in semiconductor devices by thermoelectric processes, for example, Thompson heat. Interestingly, the thermoelectric effects can also be used to provide an internal cooling mechanism for semiconductor lasers [39]. It should be also noted that a semiconductor material lattice does not have to be in equilibrium with electrons and holes. In many physical phenomena related to semiconductors, the carrier temperatures can be temporarily, significantly higher than the lattice temperature. However, due to their much lower heat capacity, the carriers do not have a major direct impact on the lattice temperature. A more detailed description of the heat sources in semiconductor optoelectronic devices can be found in Piprek [40] and the references quoted therein.

In lanthanide ion–doped photonic devices, the main source of heat results from the interactions between the dopant ions and the host lattice. When the ionic level transition energy is comparable to the maximum phonon energy of the host lattice, then the phonon-assisted transitions become more probable than the radiative transitions. In such circumstances, a particular transition is effectively nonradiative, which results in heat generation, cf. Chapter 7.

In a number of circumstances, a nonnegligible contribution to a photonic device's operating temperature can originate from the thermal boundary resistances. These resistances can, for instance, originate from imperfect thermal contacts. However, in a semiconductor device that consists of many layers of various semiconductor materials, the phonons are reflected at the interfaces between materials with different properties, thus impeding the process of heat transfer. This results in a step wise change of temperature while crossing the semiconductor interface [41]. The treatment of the thermal boundary resistance can be found in standard textbooks on heat transfer modelling [19]. However, when considering the phonon reflections, it is essential to calculate the actual value of the resistance [41].

Finally, we note that the calculation of the temperature distribution in a photonic device depends on the boundary conditions assumed. In the case of optical fibres, it is assumed that the heat is extracted from the fibre through the heat convection [13]. The situation is more complicated in the case of laser diodes. Experimental results indicate that most of the heat is extracted from the device by heat conduction through the substrate into the heat sink. However, a small proportion of the generated heat is also dissipated through radiation and heat convection from the top surface [8]. Handling of these latter two processes is not straightforward. The simplest approach is to neglect them and to assume a Neumann boundary condition at the top surface. This however, was found inconsistent with the experimental data [42] and instead a Dirichlet boundary condition was found to predict more accurately the temperature distribution. We also note that it is possible to introduce an inherent singularity into the problem when discontinuities are present in the boundary condition. Such a situation can easily occur whilst modelling laser diodes whereby there can be an abrupt transition between a Dirichlet and Neumann boundary conditions. Such a discontinuity may introduce a singularity into the temperature distributions. The presence of the singularity will in turn prevent the higher order finite difference schemes from converging faster than the standard five-point stencil. A fairly detailed discussion of this topic can be found in Ames [43].

REFERENCES

1. Bejan, A., *Heat Transfer*. 1993, New York: Wiley.
2. Ochalski, T.J., et al., Thermoreflectance and micro-Raman measurements of the temperature distributions in broad contact laser diodes. *Optica Applicata*, 2005. 35(3): p. 479–484.
3. Siegal, B., Measurement of junction temperature confirms package thermal design. *Laser Focus World*, 2003. (11).
4. Collins, R.E., *Mathematical Methods for Physicists and Engineers*. 1999, New York: Dover Publications Inc.
5. Carslaw, H.S. and J.C. Jaeger, *Conduction of Heat in Solids*. 1959, London: Oxford University Press.
6. Moosburger, R., et al., Digital optical switch based on 'oversized' polymer rib waveguides. *Electronics Letters*, 1996. 32(6): p. 544–545.
7. Joyce, W.B. and R.W. Dixon, Thermal resistance of heterostructure lasers. *Journal of Applied Physics*, 1975. 46(2): p. 855–862.
8. Pipe, K.P. and R.J. Ram, Comprehensive heat exchange model for a semiconductor laser diode. *IEEE Photonics Technology Letters*, 2003. 15(4): p. 504–506.
9. Ronnie, T.J.W., Advances in High-Power Laser Diode Packaging, In *Semiconductor Laser Diode Technology and Applications*, D.S. Patil, Editor. 1999, www.intechopen. com.
10. Biwojno, K., et al., Thermal models for silicon-on insulator-based optical circuits. *Optica Applicata*, 2004. 34(2): p. 149–161.
11. Zhuo, C., et al., The thermal effect in a grazing-incidence slab laser with the novel composite cooling method. *Optical and Quantum Electronics*, 2009. 41(1): p. 27–38.
12. Vuksic, J.A., et al., Numerical Optimization of the single fundamental mode output from a surface modified vertical-cavity surface-emitting laser. *IEEE Journal of Quantum Electronics*, 2001. 37(1): p. 108–117.
13. Brown, D.C. and H.J. Hoffman, Thermal, stress, and thermo-optic effects in high average power double-clad silica fiber lasers. *IEEE Journal of Quantum Electronics*, 2001. 37(2): p. 207–217.
14. Nakwaski, W. and M. Osinski, On the thermal resistance of vertical-cavity surface-emitting lasers. *Optical and Quantum Electronics*, 1997. 29(9): p. 883–892.
15. Kotake, S. and K. Hijikata, *Numerical Simulations of Heat Transfer and Fluid Flow on a Personal Computer*. 1993, Amsterdam: Elsevier.
16. Shih, T.M., *Numerical Heat Transfer*. 1984, New York: Hemisphere Publishing Corporation.
17. Minkowycz, W.J., et al., *Handbook of Numerical Heat Transfer*. 2006, New York: John Wiley & Sons.
18. Bekker, A.A., *The Boundary Element Method in Engineering: A Complete Course*. 1992, London: McGraw-Hill.
19. Croft, D.R., Stone, J.A.R., *Heat Transfer Calculations Using Finite Difference Equations*. 1977, Sheffield: PAVIC Publications, Sheffield City Polytechnic.
20. Patankar, S.V., *Numerical Heat Transfer and Fluid Flow*. 1980, New York: Hemisphere Publishing Corporation.
21. Reddy, J.N., Gartling, D.K., *The Finite Element Method in Heat Transfer and Fluid Dynamics*. 2000: Taylor & Francis.
22. Szymanski, M., et al., Two-dimensional model of heat flow in broad-area laser diode mounted to a non-ideal heat sink. *Journal of Physics D-Applied Physics*, 2007. 40(3): p. 924–929.
23. Lee, C.C. and D.H. Chien, The effect of bonding wires on longitudinal temperature profiles of laser diodes. *Journal of Lightwave Technology*, 1996. 14(8): p. 1847–1852.

24. Mukherjee, J. and J.G. McInerney, Electrothermal analysis of CW high-power broad-area laser diodes: A comparison between 2-D and 3-D Modeling. *Ieee Journal of Selected Topics in Quantum Electronics*, 2007. 13(5): p. 1180–1187.

25. Smith, G.D., *Numerical Solution of Partial Differential Equations: Finite Difference Method*. 1988, Oxford: Oxford University Press.

26. Saad, Y., *Iterative Methods for Sparse Linear Systems*, 2nd edition. 2000, Philadelphia: SIAM.

27. Barret, R., et al., *Templates for the Solution of Linear Systems: Building Blocks for Iterative Methods*. 1994, Philadelphia: SIAM.

28. Strikwerda, J.C., *Finite Difference Schemes and Partial Differential Equations*. 2004, Philadelphia: SIAM.

29. Spiegel, M.R., *Calculus of Finite Differences and Difference Equations*. 1994: McGraw-Hill, Inc.

30. Gibou, F. and R. Fedkiw, A fourth order accurate discretisation for Laplace and heat equations on arbitrary domains with applications to the Stefan problem. *Journal of Computational Physics*, 2005. 202(2): p. 577–601.

31. Collatz, L., *The Numerical Treatment of Differential Equations*. 1960, Berlin: Springer-Verlag.

32. Pozirikidis, C., *Numerical Computation in Science and Engineering*. 1998, Oxford: Oxford University Press.

33. Rosloniec, S., *Fundamental Numerical Methods for Electrical Engineering*. 2008, Berlin: Springer.

34. LeVeque, R.J. and Z. Li, The immersed interface method for elliptic equations with discontinuous coefficients and singular sources. *SIAM Journal Numerical Analysis*, 1994. 31(4): p. 1019–1044.

35. Jomaa, Z. and C. Macaskill, The embedded finite difference method for the Poisson equation in a domain with irregular boundary and Dirichlet boundary conditions. *Journal of Computational Physics*, 2005. 202(1): p. 488–506.

36. Johansen, H. and P. Colella, A cartesian grid embedded boundary method for Poisson's equation on irregular domains. *Journal of Computational Physics*, 1998. 147(1): p. 60–85.

37. Sherer, S.E. and J.N. Scott, High-order compact finite-difference methods on general overset grids. *Journal of Computational Physics*, 2005. 210(2): p. 459–496.

38. Sujecki, S., Extended Taylor series and interpolation of physically meaningful functions. *Optical and Quantum Electronics*, 2013. 45(1): p. 53–66.

39. Pipe, K.P., R.J. Ram, and A. Shakouri, Internal cooling in semiconductor laser diode. *IEEE Photonics Technology Letters*, 2002. 14(4): p. 453–455.

40. Piprek, J., *Semiconductor Optoelectronic Devices Introduction to Physics and Simulation*. 2003, San Diego: Academic Press.

41. MacKenzie, R., et al., An investigation of thermal boundary resistance in 1.3 micrometer edge-emitting dilute nitride quantum well laser diodes. *Physica Status Solidi C*, 2008. 5(2): p. 485–489.

42. Szymanski, M., Two-dimensional model of heat flow in broad-area laser diode: Discussion of the upper boundary condition. *Microelectronics Journal*, 2007. 38(6–7): p. 771–776.

43. Ames, W.F., *Numerical Methods for Partial Differential Equations*. 1977, New York: Academic Press.

6 Flow of Current in Semiconductor Photonic Devices

Semiconductor photonic devices, for example, laser diodes (LDs), light emitting diodes (LEDs) [1], and photodiodes (PDs) [2,3] have many applications in telecom, medicine, and manufacturing industries. The advantages of these devices over alternative solutions include a compact device structure, low cost, high reliability, and ease of use. The modelling of the flow of current in LDs, LEDs, and PDs is essential for gaining an insight into the main physical processes that govern their operation, and also for the purpose of device optimisation and design.

In this chapter, we give an introduction to the modelling of the current flow in semiconductor devices. Chapter 8 brings this theory into the context of a semiconductor laser modelling. To ease the understanding of the presented material, we use the MATLAB programming environment to back up the theory with software implementation examples. We start this chapter with deriving the drift-diffusion equations. Then we present in detail the numerical solution of the drift diffusion equations in one-dimensional (1D) for an unbiased and biased p–n junction. In the last section, we discuss the application of the numerical techniques described in the sections "Potential Distribution in Unbiased p–n Junction" and "Potential and Quasi Fermi Level Distribution in Biased p–n Junction" in the analysis of photonic semiconductor devices. As in the previous chapter, there are many books available that discuss specifically the numerical modelling of carrier transport in semiconductor devices [4–6]. Here, we therefore focus only on selected aspects of the numerical solution of the drift-diffusion equations that are either essential in understanding when developing software tools or relevant to Chapter 8.

INTRODUCTION

Atoms in a semiconductor material are relatively closely spaced. The valence electrons of these atoms interact, and as a result of the interaction two bands of closely spaced energy levels form: the valence band and the conduction band. It is assumed in semiconductor theory that only the electrons of these two bands participate in the current flow, while the other electrons are bound to the atom cores. At 0 K the valence band is fully filled while the conduction band is empty; hence, current flow through a semiconductor material is not possible. However, at higher temperatures thermal generation processes can promote electrons from the valence band to the conduction band where the electrons become mobile. Thus, a semiconductor material gains an ability to conduct current. A promoted electron leaves an unoccupied energy state in the valence band (a hole) and, thus, also enables the valence band electrons to participate in the current flow.

In most currently available modelling and design tools, the drift-diffusion equations are used to study the current flow in semiconductor devices [7]. The drift-diffusion equations can be formally derived from the Boltzmann transport equation. However, they can also be obtained phenomenologically with the help of Maxwell's equations. Here we follow the latter approach. We derive the Poisson equation first and then carry out the derivation of the current continuity equations.

For a semiconductor material, Maxwell's equations have the following form:

$$\nabla \times \vec{E}\left(\vec{r},t\right) = -\frac{\partial \vec{B}\left(\vec{r},t\right)}{\partial t} \tag{6.1a}$$

$$\nabla \times \vec{H}\left(\vec{r},t\right) = \frac{\partial \vec{D}\left(\vec{r},t\right)}{\partial t} + \vec{J} \tag{6.1b}$$

Equation 6.1 is complemented by the divergence conditions:

$$\nabla \cdot \vec{D} = \rho \tag{6.2a}$$

$$\nabla \cdot \vec{B} = 0 \tag{6.2b}$$

In addition, the dependence between the electric field and the electric flux vectors is given by

$$\vec{D}\left(\vec{r},t\right) = \varepsilon_0 \int_{-\infty}^{\infty} \left(1 + \chi\ \left(\vec{r},t-\tau\right)\right)\vec{E}\left(\vec{r},\tau\right)d\tau \tag{6.3a}$$

where ε_0 is the electric permittivity of the free space and χ is the electric susceptibility. For static fields, Equation 6.3a reduces to:

$$\vec{D}\left(\vec{r}\right) = \varepsilon\ \vec{E}\left(\vec{r}\right) \tag{6.3b}$$

where ε is the static electric permittivity of the semiconductor material.

To derive the Poisson equation, we observe first that equation 6.1a under a static approximation reduces to:

$$\nabla \times \vec{E}\left(\vec{r}\right) = 0 \tag{6.4}$$

which implies that the electric field can be expressed as a gradient of a scalar potential ϕ:

$$\vec{E}(\vec{r}) = -\nabla\varphi \tag{6.5}$$

Substituting 6.5 into 6.2a and using 6.3b yields:

$$\nabla\cdot(\varepsilon\,\nabla\varphi) = -\rho \tag{6.6}$$

For a homogenous medium, Equation 6.6 reduces to the Poisson equation.

The application of 6.6 is in principle limited to the static case only. However, in Collins [5], it was shown that under several constraints, it is plausible to use 6.6 also when electromagnetic fields vary in time.

For a semiconductor material, the electric charge density ρ that appears on the right-hand side of 6.6 is typically expressed as a sum of four terms:

$$\rho = q(p - n - N_A + N_D) \tag{6.7}$$

where q is an elemental charge while n and p are the electron and hole concentrations, respectively. N_A and N_D, respectively, stand for the acceptor and donor concentrations.

To obtain the current continuity equations, we calculate the divergence of 6.1b:

$$\nabla\cdot\nabla\times\vec{H}(\vec{r},t) = \frac{\partial\left(\nabla\cdot\vec{D}(\vec{r},t)\right)}{\partial t} + \nabla\cdot\vec{J} = 0 \tag{6.8}$$

Using 6.2a in 6.8 yields the current continuity equation:

$$\nabla\cdot\vec{J} = \frac{\partial\rho}{\partial t} \tag{6.9}$$

By comparing 6.9 with 5.2, one can observe that 6.9 is another example of an equation that relates a flux with a rate of change. Equation 6.9 relates the electric charge flux (the current density) with the electric charge density variations and, in essence, mathematically expresses the charge conservation principle stating that, within a time interval, the amount of electric charge generated within a volume is equal to the net charge flux through the surface of the volume. Because both electrons and holes have to obey the particle conservation principle separately, we can write immediately the following equations:

$$\frac{\partial n}{\partial t} = \frac{1}{q}\nabla\cdot\vec{J}_n \tag{6.10a}$$

$$\frac{\partial p}{\partial t} = -\frac{1}{q}\nabla \cdot \vec{J}_p \qquad\qquad (6.10b)$$

where $\vec{J}_n$ and $\vec{J}_p$ stand for the electron and hole current density, respectively. The electron–hole pair generation and recombination processes can be included phenomenologically in 6.10 in a similar manner as the heat generation term was included in 5.2. This yields the final form of the current continuity equations for the electrons and holes:

$$\frac{\partial n}{\partial t} = \frac{1}{q}\nabla \cdot \vec{J}_n - R \qquad\qquad (6.11a)$$

$$\frac{\partial p}{\partial t} = -\frac{1}{q}\nabla \cdot \vec{J}_p - R \qquad\qquad (6.11b)$$

where R is the net electron hole pair recombination–generation rate. A positive value of R corresponds to recombination (i.e., reduces electron and hole concentrations in absence of other terms in 6.11) whereas a negative one corresponds to generation. Thus the electron–hole pair recombination–generation term in 6.11 plays the same role as the heat generation term in 5.2.

If complemented by equations relating the current densities with the carrier concentrations and electric potential, equations 6.11 and 6.6 form a self-consistent set that can be solved for the electric potential distributions and carrier concentrations (or quasi-Fermi levels). Suitable expressions for the current densities can be obtained from the Boltzmann equation [8,9]:

$$\vec{J}_n = q\mu_n n\, \nabla E_{Fn} - L_n\nabla T \qquad\qquad (6.12a)$$

$$\vec{J}_p = q\mu_p p\, \nabla E_{Fp} - L_p\nabla T \qquad\qquad (6.12b)$$

where μ_n and μ_p is the electron and hole mobility, respectively, and L_n and L_p are transport coefficients defined by integrals in k space [9]. E_{Fn} and E_{Fp} are quasi-Fermi levels for electrons and holes, respectively. For an isotropic material with a constant temperature distribution, equations 6.12a and 6.12b reduce to:

$$\vec{J}_n = -q\mu_n n\, \nabla \varphi_n \qquad\qquad (6.13a)$$

$$\vec{J}_p = -q\mu_p p\, \nabla \varphi_p \qquad\qquad (6.13b)$$

Equation 6.13 can be recast into the familiar drift-diffusion form [9]:

$$\vec{J}_n = q\mu_n n\vec{E} + qD_n\nabla n \qquad\qquad (6.14a)$$

$$\vec{J}_p = q\mu_p n\vec{E} - qD_p \nabla p \qquad (6.14b)$$

Equation 6.13 can be simply obtained from 6.14 by applying a Boltzmann statistics approximation [10]. However, this does not imply yet that the Boltzmann statistics limits the accuracy of the calculations when applying 6.14 or 6.13 because at this stage the formulae relating carrier concentrations with quasi-Fermi levels can be interpreted as a mathematical change of variables only (cf. [5], p. 136). The advantage of using 6.14 is that these equations can be derived phenomenologically (see, for instance, [11]) and, hence, also provide an immediate intuitive insight into the nature of the carrier transport.

Formulas 6.13 and 6.14 contain the electron and hole densities which can be related to the quasi-Fermi levels via the integrals:

$$n = \int_{E_c}^{\infty} \rho_c(E) f_n(E) dE \qquad (6.15a)$$

$$p = \int_{-\infty}^{E_v} \rho_v(E) f_p(E) dE \qquad (6.15b)$$

where ρ_c and ρ_v are the density of state functions for the conduction and valence band, respectively. E_c gives the energy of the conduction band edge and E_v the valence one. The f_p and f_n are the Fermi distribution functions:

$$f_n(E) = \frac{1}{1 + \exp\left(\dfrac{E - E_{Fn}}{kT}\right)} \qquad (6.16a)$$

$$f_p(E) = \frac{1}{1 + \exp\left(\dfrac{E_{Fp} - E}{kT}\right)} \qquad (6.16b)$$

Finally, typically in simulations, instead of the quasi-Fermi levels E_{Fn} and E_{Fp}, one uses the quasi-Fermi potentials ϕ_n and ϕ_p:

$$E_{Fn} = -q\phi_n$$
$$E_{Fp} = -q\phi_p$$

Thus solving 6.6 and 6.11 with the current densities given by 6.13 while using 6.15 to relate the carrier concentrations with the quasi-Fermi levels leads to a formulation of the problem whereby the unknown quantities are the electric potential and the quasi-Fermi potentials. However, other sets of unknowns can be also be selected to solve the drift-diffusion equations. One popular way of formulating the problem results in the carrier concentrations and the electric potential as the unknowns. Other approaches include the application of the Slootboom variables [12] and Mock stream potentials [13–15]. A thorough discussion of advantages resulting from the use of a particular set of variables can be found in the available literature [10,16] and hence we will not discuss this issue in more detail here. When modelling the current flow in semiconductor photonic devices, the preferred choice is the formulation of the problem in terms of the electric potential and quasi-Fermi potentials. We therefore use this formulation and discuss its application in detail in the next two sections.

When compared with the heat conduction equation, there are hardly any analytical solutions available for the drift-diffusion equations. A recent survey of approximate analytical solutions of the drift-diffusion equation, for a p–n junction can be found in Laux and Hess [17]. An exact analytical solution has been provided by Mock [14] (pp. 90–91) and Burgler et al. [18]. There are also several other semiconductor devices for which analytical solutions are available [19–22]. However, for all practical design purposes, the application of numerical methods is indispensable. There are two main techniques that have been applied for this purpose: the finite difference method (FDM) and the finite element method (FEM). Here we focus on the FDM. The description of the FEM application to the modelling of semiconductor devices can be found in Collins [5], Bank et al. [16], and Buturla et al. [23] and most recently in Bochev et al. [24]. Not to obscure the essential information in the sections "Potential Distribution in Unbiased p–n Junction" and "Potential and Quasi Fermi Level Distribution in Biased p–n Junction," we limit the derivations to a uniform FD mesh only.

POTENTIAL DISTRIBUTION IN UNBIASED P–N JUNCTION

The calculation of the electric potential distribution in an unbiased semiconductor device is often done as the first step when calculating the electric potential and carrier densities of a biased device. This is sometimes necessary as the convergence of the iterative algorithms used for solving Equations 6.6 and 6.11 strongly depends on the initial guess. Finding a good initial guess (i.e., such that would guarantee the convergence of the algorithm to the solution within few iterations) is particularly difficult when the bias voltage is large. Hence, one of the possible remedies for this problem consists in calculating the solutions of 6.6 and 6.11 at the zero bias first and then incrementing the bias voltage slowly up until the operating bias voltage is reached [25]. At each step, the solution obtained at the previous step is used as the initial guess.

If no bias is applied, then the spatial distribution of the quasi-Fermi potential is constant. Therefore, Equation 6.11 decouples from 6.6, and it is only necessary to solve 6.6 for the distribution of the electric potential. Once the distribution of the electric potential is known, the entire band diagram can be drawn.

To solve 6.6 one has to substitute 6.15 for n and p. This, however, introduces two new unknowns, namely, the conduction and valence band energies: E_c and E_v. Consequently, before we can proceed with the numerical solution of 6.6, it is necessary to express the values of the valence and conduction band energies that appear in

6.15 in terms of the electric potential and of the semiconductor material parameters: band gap energy and electron affinity. This reduces the number of unknowns to one only: the electric potential.

In a p-type or n-type semiconductor, E_c and E_v are constants and their values can be easily determined if the band gap energy and the electron affinity are known (Figure 6.1).

In a p–n junction on the other hand, E_c and E_v become functions of position and their distributions are not known until 6.6 is solved. However, both conduction and valence band energy can be related to the electric potential. Before this is accomplished, the unknown electric potential reference level has to be selected. Various values of the reference level can be found in the literature. In this section we align the reference level with the vacuum level in the n-type semiconductor that forms the p–n junction. With such choice of the reference level, the vacuum level equals $-q\phi$ (Figure 6.2). Once the reference level for the electric potential is selected, the position of the conduction band edge and the valence band edge can be expressed in terms of the electric potential, band gap energy E_g, and electron affinity X. This can be obtained by inspection from Figure 6.2:

$$E_v = -q\varphi - X - E_g \tag{6.17a}$$

$$E_c = -q\varphi - X \tag{6.17b}$$

With 6.17 and 6.14, the only unknown in Equation 6.6 is the electric potential. So, we are now in the position to discuss the numerical solution of 6.6. We consider in detail a solution of 6.6 in the 1D case. If we select x as the spatial variable, then the partial differential equation 6.6 reduces to the following ordinary differential equation:

$$\frac{d}{dx}\left(\varepsilon\frac{d\varphi}{dx}\right) = -\rho \tag{6.18}$$

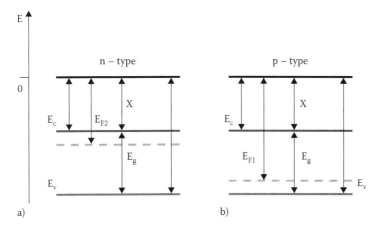

FIGURE 6.1 Energy level diagram for a) a n-type and b) a p-type semiconductor.

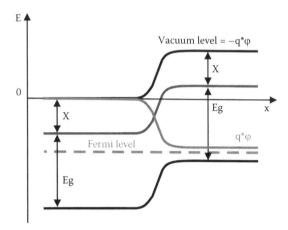

FIGURE 6.2 Energy level diagram for an unbiased p–n homojunction whereby the reference level of the electric potential was selected to coincide with the vacuum level of the n type semiconductor.

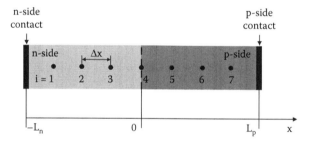

FIGURE 6.3 Schematic diagram showing an example distribution of the nodal mesh points in a p–n junction.

In the first step, we convert the solution of the ordinary differential equation 6.18 to a solution of a set of nonlinear algebraic equations. For this purpose we apply the FDM. We introduce a set of nodal points x_i (Figure 6.3) and use the approach essentially described in Chapter 5 when deriving finite difference approximations for modelling heat flow in inhomogeneous media. This gives the following equation at each nodal point x_i:

$$EQ_i(\varphi_{i-1}, \varphi_i, \varphi_{i+1}) = \frac{\varepsilon_L \varphi_{i-1} - (\varepsilon_L + \varepsilon_R)\varphi_i + \varepsilon_R \varphi_{i+1}}{\Delta x^2} + \rho(x_i) = 0 \qquad (6.19a)$$

In 6.19a, it is assumed that the dielectric discontinuity is coincident with an finite difference (FD) mesh point x_i. The value of ε_L is on the "left" side of the discontinuity, that is, on the side of the FD mesh point x_{i-1}, while ε_R is on the side of x_{i+1}. For a homogenous medium equation, 6.19a reduces to:

$$EQ_i(\varphi_{i-1}, \varphi_i, \varphi_{i+1}) = \frac{\varphi_{i-1} - 2\varphi_i + \varphi_{i+1}}{\Delta x^2} + \frac{\rho(x_i)}{\varepsilon} = 0 \qquad (6.19b)$$

At $x = -L_n$ we impose the boundary condition $\phi = 0$ whereas at $x = L_p$ (Figure 6.3) we set the electric potential equal to the built-in (or electric barrier) potential: $\phi = -V_{BARRIER}$. We note that such selection of the boundary conditions is consistent with the assumed electric potential reference level. Collating together Equation 6.19 at each FD nodal point, and including the boundary condition, one obtains a set of non-linear algebraic equations; whereby, the number of equations is equal to the number of unknowns, that is, the electric potential values at the FD nodal points: ϕ_i. This set of equations can be solved using Newton-Raphson method [26]:

$$\text{Jacobian} * \varphi^{NEW} = -EQ(\varphi^{OLD}) + \text{Jacobian} * \varphi^{OLD} \qquad (6.20)$$

whereby the vector φ^{NEW} is used to store the samples of φ_i obtained by the Newton-Raphson method; while, the vector φ^{OLD} stores the initial values for the samples φ_i while the Jacobian is given as:

$$\text{Jacobian} = \begin{bmatrix} \dfrac{\partial EQ_1}{\partial \varphi_1} & \cdots & \dfrac{\partial EQ_1}{\partial \varphi_i} & \cdots & \dfrac{\partial EQ_1}{\partial \varphi_N} \\ \vdots & & \vdots & & \vdots \\ \dfrac{\partial EQ_i}{\partial \varphi_1} & \cdots & \dfrac{\partial EQ_i}{\partial \varphi_i} & \cdots & \dfrac{\partial EQ_i}{\partial \varphi_N} \\ \vdots & & \vdots & & \vdots \\ \dfrac{\partial EQ_N}{\partial \varphi_1} & \cdots & \dfrac{\partial EQ_N}{\partial \varphi_i} & \cdots & \dfrac{\partial EQ_N}{\partial \varphi_N} \end{bmatrix} \qquad (6.21)$$

When φ^{NEW} is substituted for φ^{OLD}, an iteration procedure is established that converges to the solution of Equation 6.18, provided that the initial guess was selected sufficiently near to the solution [26]. This results in the Algorithm 6.1.

Before implementing Algorithm 6.1 in MATLAB, two comments are necessary. Firstly, the set of nonlinear algebraic equations that is obtained from 6.6, through applying the FDM (or other methods), may potentially have spurious solutions.

ALGORITHM 6.1   FINITE DIFFERENCE BASED CALCULATION OF ELECTRIC POTENTIAL DISTRIBUTION IN AN UNBIASED HOMOJUNCTION

1. Start
2. Select Δx and the sampling nodal points
3. Set the initial guess ϕ^{OLD}
4. Calculate ϕ^{NEW} from 6.20
5. If $|\phi^{OLD} - \phi^{NEW}| >$ residual $\phi^{OLD} = \phi^{NEW}$ go to 4
6. Stop

The best way of avoiding calculating the spurious solution, instead of the physically meaningful one, is the selection of the initial guess as near as possible to the physical solution. Secondly, the integrals 6.15 cannot be evaluated analytically (cf. [11]). This means that at each position x_i, they have to be evaluated numerically. This results in a large numerical overhead. Therefore, practical algorithms use analytical approximations for the integrals 6.15. Here we apply a Boltzmann statistics approximation, which is widely used for bulk semiconductor material regions in photonic device simulations. This choice does not limit the generality of the presented numerical approach, because at this stage the introduction of Boltzmann statistics can be understood as a simple mathematical change of the variables (Chapter 5 in Collins [5]). Under the Boltzmann statistics approximation, one obtains from 6.15 [11]:

$$p = N_v \exp\left(\frac{E_v - E_{Fp}}{kT}\right) \tag{6.22a}$$

$$n = N_c \exp\left(\frac{E_{Fn} - E_c}{kT}\right) \tag{6.22b}$$

where N_c and N_v are effective densities of states for the conduction and valence band, respectively. Substituting 6.17 to express the band energies in terms of material parameters in 6.22 leads to the following analytical expressions for the hole and electron concentrations:

$$p = N_v \exp\left(\frac{\left(-q\varphi - X - E_g\right) + q\varphi_p}{kT}\right) \tag{6.22c}$$

$$n = N_c \exp\left(\frac{-q\varphi_n - \left(-q\varphi - X\right)}{kT}\right) \tag{6.22d}$$

In 6.22c and 6.22d, both quasi-Fermi potentials are equal and constant so they can be effectively treated as a single simulation parameter. From Figure 6.2 it can be deduced that the numerical value of this parameter that is consistent with the assumed electric potential reference level can be obtained from:

$$q\varphi_n = q\varphi_p = E_c + kT \ln\left(\frac{n}{N_c}\right) \tag{6.23}$$

as applied to the n-type homogenous semiconductor material that forms the p–n junction. To evaluate 6.23 one needs to know the electron concentration in the n-type semiconductor. The carrier concentrations in a homogenous semiconductor can be conveniently obtained from the condition of the electric neutrality and the law of mass (which applies under the Boltzmann statistics approximation):

$$n + N_A = p + N_D \tag{6.24a}$$

$$np = n_i^2 \tag{6.24b}$$

where n_i is the intrinsic carrier concentration [11]. From 6.24 one can obtain the electron and hole concentrations expressed in terms of the dopant and intrinsic carrier concentrations:

$$n = \frac{N_D - N_A}{2} + \sqrt{\left(\frac{N_D - N_A}{2}\right)^2 + n_i^2} \tag{6.25a}$$

$$p = \frac{N_A - N_D}{2} + \sqrt{\left(\frac{N_A - N_D}{2}\right)^2 + n_i^2} \tag{6.25b}$$

When the donor concentration is high, it is convenient to approximate 6.25 with [11]:

$$n = N_D \tag{6.26a}$$

$$p = \frac{n_i^2}{N_D} \tag{6.26b}$$

Lastly we note that the built-in potential can be calculated from the difference between the Fermi potentials in the p-type and n-type homogenous semiconductor.

Thus applying Boltzmann statistics while implementing the Algorithm 6.1 results in the following MATLAB script:

```
% program calculates the electric potential in unbiased pn
% homojunction
% version 15.08.13,
clear % clears variables
clear global % clears global variables
format long e
% initial constants
I = sqrt(-1);% remember not to overwrite pi or i !
pi = 3.141592653589793e+000;
c = 2.998e+2;% speed of light in free space [mi/ps]
q = 1.602e-19;% elemental charge [C]
```

```
h_s = 6.62606885e-34;% Planck's constant [J*ps]
kB = 1.3807e-23;% Boltzman's constant [J/K]
T = 300.;% temperature [K]
kT = kB*T/q;% kT [eV]
UT = kB*T/q;% UT thermal potential [V]
eps0 = 8.854e-18;% vacum permittivity [F/mi]
m0= 9.11e-31;% electron mass [kg]

% input semiconductor parameters (GaAs)
mc = 0.067;% relative electron mass
mv = 0.37736;% relative hole mass
%(electrons) effective density of states [1/mi3]:
Nc = 2*sqrt((2*pi*mc*m0*kB*T/(h_s*h_s))^3)*1.e-18
%(holes) effective density of states [1/mi3]:
Nv = 2*sqrt((2*pi*mv*m0*kB*T/(h_s*h_s))^3)*1.e-18

Wg = 1.424;% bandgap [eV]
X = 4.07;% affinity [eV]
Wc = -4.07;% conduction band edge [eV] measured
% with respect to vacuum level
Wv = -5.494;% valence band edge [eV] measured
% with respect to vacuum level
eps = 13.18;% relative dielectric constant

% n side
ND = 1e+4;% donor concentration [1/mi3]
thick_n = 1.0;% n side thickness [mi]
% Fermi level at no bias measured with respect
% to vac level on n side [eV]:
WFeq_n = Wc+kT*log(ND/Nc)

% p side (the unchanged params are not redefined)
NA = 1e+4;% acceptor concentration [1/mi3]
thick_p = 1.0;% p side thickness [mi]
% Fermi level at no bias measured with respect
% to vac level on n side [eV]:
WFeq_p = Wv-kT*log(NA/Nv)
% Barrier potential (can be calculated directly
% since energy is in eV):
VBarrier = WFeq_n-WFeq_p

Up = (-X-Wg-WFeq_n);
Un = (X+WFeq_n);

% calculation parameters
thick = (thick_n+thick_p);% device length
N_points = 1000;% number of points on L
dx = thick/(N_points);% step dx [Li]

% vector initialisation
EQ = zeros(N_points,1);
```

```
Vi = zeros(N_points,1);
left_diag = zeros(N_points,1);
right_diag = zeros(N_points,1);
KN = zeros(N_points,1);
ue = ones(N_points+2,1);

% initial voltage distribution
for i = 1:N_points/2
    Vi(i) = 0.0;
    Vi(i+N_points/2) = -VBarrier;
end
ue(2:N_points+1) = Vi;
ue(1) = 0;
ue(N_points+2) = -VBarrier;

% setting up vectors for equations
for i = 1:N_points/2
    KN(i) = ND;
    KN(i+N_points/2) = -NA;
end
w = log(abs(KN(1)))


% setting off diagonals of Jacobian
for i = 1:N_points
    left_diag(i) = -1.0/(dx*dx);
    right_diag(i) = -1.0/(dx*dx);
end

for j = 1:10%main loop start
% setting main diagonal of Jacobian
main_diag = 2.0/(dx*dx)+(1/UT)*…
(Nv*exp((Up-Vi)/UT)+Nc*exp((Un+Vi)/UT))*q/(eps*eps0);
Jacobian = spdiags([left_diag main_diag right_diag],…
-1:1,N_points,N_points);

% calculating the function value at Vi
for i = 2:N_points+1
EQ(i-1) = -(q/(eps*eps0))*(Nv*exp((Up-ue(i))/UT)-…
Nc*exp((Un+ue(i))/UT)+…
KN(i-1))+(2.0*ue(i)-ue(i-1)-ue(i+1))/(dx*dx);
end

Right_side = -EQ+Jacobian*Vi;
Vi = Jacobian\Right_side;
res = ue(2:N_points+1)-Vi;
res'*res
ue(2:N_points+1) = Vi;

end %main loop stop
```

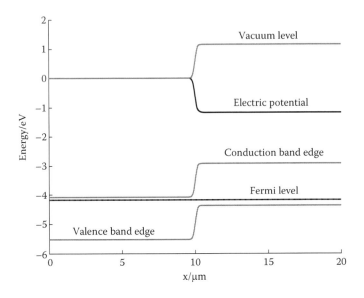

FIGURE 6.4 Calculated band diagram for GaAs homojunction with doping concentration of $10^{16}/cm^3$ on both sides. A thousand FD nodes were used while all other modelling parameters are listed in Table 6.1.

TABLE 6.1
Modelling Parameters for GaAs Homojunction at $T = 300$ K

Parameter	Value
Band gap energy	$E_g = 1.424$ eV
Electron affinity	X = 4.07 eV
Effective heavy hole mass	$m_v = 0.377/m_0$
Effective electron mass	$m_c = 0.067/m_0$
Intrinsic carrier concentration	$n_i = 1.74 \times 10^{12}$ 1/m^3
Effective density of states in Conduction Band	$N_c = 4.35 \times 10^{23}$ 1/m^3
Effective density of states in Valence Band	$N_v = 5.87 \times 10^{24}$ 1/m^3
Electron mobility	$\mu_e = 4000$ cm^2/(Vs)
Hole mobility	$\mu_h = 400$ cm^2/(Vs)
SRH recombination lifetime	$\tau = 10$ ps
Relative dielectric constant	$\varepsilon = 13.18$
n-side width	$L_n = 10$ μm
p-side width	$L_p = 10$ μm

We note that in the above code the check of the residual is not included because the convergence is fast and hence several iterations easily yield an accurate solution. The last important comment that should be mentioned here is that in developing the code implementing Algorithm 6.1 it is essential to follow-up consistently the units.

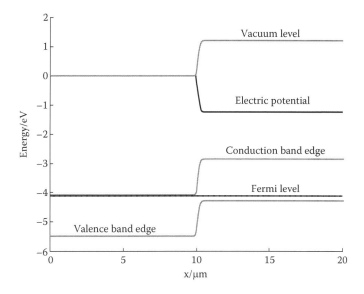

FIGURE 6.5 Calculated band diagram for GaAs homojunction with doping concentration of 10^{17}/cm^3 on the n side and 10^{16}/cm^3 on the p side. A thousand FD nodes were used while all other modelling parameters are listed in Table 6.1.

Figure 6.4 shows an example of an electric potential distribution calculated using the presented MATLAB script for a GaAs homojunction at $T = 300$ K. The modelling parameters are given in Table 6.1. The doping concentration on both the p- and n-sides is equal to 10^{16}/cm^3. For comparison we provide in Figure 6.5 the electric potential distribution when the doping concentration on the n-side is increased to 10^{17}/cm^3. As expected, the transition region on the n side becomes much narrower, the band bending is much steeper, and the junction becomes effectively nearly one-sided. In both simulations 1000 nodal points were used.

POTENTIAL AND QUASI-FERMI LEVEL DISTRIBUTION IN BIASED p–n JUNCTION

In this section we extend the FDM presented in the section "Potential Distribution in Unbiased p–n Junction" to the analysis of biased p–n junctions. Again to avoid obscuring the main message, we consider a p–n homojunction in the 1D case. The extension of the FDM presented here to multidimensional analysis is discussed in the section "Modelling of Current Flow in Photonic Semiconductor Devices." When a p–n junction is biased, the quasi-Fermi levels bend and become functions of position. As a result, Equation 6.6 cannot be decoupled from the current continuity equation 6.11. Nonetheless, essentially the same approach to the solution can be followed as in the case of an unbiased junction. Namely, converting first the differential equations into a set of nonlinear algebraic equations and then calculating the solution using the Newton-Raphson method.

If the spatial variable is aligned with x, in the 1D case, 6.6 and 6.11 reduce to the following equations:

$$\frac{d}{dx}\left(\varepsilon\frac{d\phi}{dx}\right)+q(p-n+N_D-N_A)=0 \qquad (6.27a)$$

$$\frac{dJ_{nx}}{dx}=R$$
$$\frac{dJ_{px}}{dx}=-R \qquad (6.27b)$$

where J_{nx} and J_{px} are the x components of the current density vector for the electrons and holes, respectively. The current densities can be expressed with the help of quasi-Fermi potentials:

$$J_{nx}=-q\mu_n n\frac{d\varphi_n}{dx}$$
$$J_{px}=-q\mu_p p\frac{d\varphi_p}{dx} \qquad (6.28)$$

or the carrier concentrations:

$$J_{nx}=q\mu_n nE_x+qD_n\frac{dn}{dx}$$
$$J_{px}=q\mu_p nE_x-qD_p\frac{dn}{dx} \qquad (6.29)$$

At this point it should be noted that, when studying the carrier transport across a heterojunction, depending on the junction doping profile, composition, and the bias voltage the inclusion of thermionic emission [4,27–29] or tunnelling current [30] may be necessary.

Again the integrals 6.15 need to be approximated analytically. When the device is biased, the Boltzmann approximation is only plausible under moderate values of the bias voltage. For large values of the bias voltage, other approximations have to be applied. A discussion of this topic can be found in standard textbooks [5,31]. A helpful guidance on the application of analytical approximations to Fermi-Dirac statistics in semiconductor device modelling can be found in Purbo et al. [32]. In quantum well regions for confined carriers, the integrals 6.15 can be evaluated analytically. The coupling between the confined and three dimensional (3D) carriers can be then obtained using the capture-escape model [33]. Here, for the sake of simplicity, we use the Boltzmann statistics.

Substituting 6.28a into 6.27b and using Boltzmann statistics yields:

$$\begin{vmatrix} \dfrac{d}{dx}\left(\varepsilon\dfrac{d\phi}{dx}\right)+q(p-n+N_D-N_A)=0=0 \\[2mm] \dfrac{d}{dx}\left\{\mu_n n\dfrac{d\varphi_n}{dx}\right\}=-R \\[2mm] \dfrac{d}{dx}\left\{\mu_p p\dfrac{d\varphi_p}{dx}\right\}=R \end{vmatrix} \tag{6.30}$$

where p and n are given by 6.18a and 6.18b. Before proceeding further, however, we need to express the carrier concentrations in terms of the potentials, so that there are three unknowns only, electric potential and quasi-Fermi potential. For this purpose, we need to define the reference level for the electric potential first. One could use again the vacuum level as in the previous section. However, in this section we will use another popular choice, which is the intrinsic semiconductor Fermi level E_i [5]. To specify E_i we substitute into the argument of 6.22a and 6.22b, $E_i - E_i$, and obtain:

$$p = n_i \exp\left(\frac{E_i - E_{Fp}}{kT}\right) \tag{6.31a}$$

$$n = n_i \exp\left(\frac{E_{Fn} - E_i}{kT}\right) \tag{6.31b}$$

where under Boltzmann statistics approximation, n_i is defined as:

$$n_i = \sqrt{N_v N_c}\,\exp\left(-\frac{E_g}{2kT}\right) \tag{6.32}$$

while the intrinsic energy level is obtained from the equation:

$$N_c \exp\left(\frac{E_c - E_i}{kT}\right) = N_v \exp\left(\frac{E_i - E_v}{kT}\right) \tag{6.33}$$

The calculation of the intrinsic energy level under Fermi statistics is explained in Collins [5]. Now we align the electric potential with the intrinsic energy level (Figure 6.6), that is:

$$E_i = -q\varphi \tag{6.34}$$

Equation 6.34 is valid for a homojunction. In a heterojunction, E_i becomes a piece-wise continuous function and one needs to select one value of E_i from a homogenous subregion that forms p–n junction and align with it the electric potential.

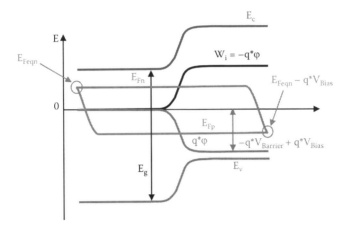

FIGURE 6.6 Energy level diagram for a biased p–n semiconductor homojunction.

Using 6.34 yields the following expressions for p and n:

$$p = n_i \exp\left(\frac{-q\varphi + q\varphi_p}{kT}\right) \tag{6.35a}$$

$$n = n_i \exp\left(\frac{-q\varphi_n + q\varphi}{kT}\right) \tag{6.35b}$$

Using 6.35 and 6.30 yields a set of three equations with three unknowns, which can be solved numerically. Now we focus on the approximation of the function values midway between two FD nodes. To avoid obscuring the problem by complicated formulae with a large number of independent parameters, we consider a homojunction with a homogenous distribution of carrier mobility and scale equations according to DeMari [34]. (Because of the current continuity, the presented approach can be readily extended to a heterojunction if only the recombination rate does not have discontinuities at the interfaces between the different semiconductors forming the heterojunction.) Thus substituting 6.35 into 6.30 yields, after scaling, the following equations:

$$\begin{cases} \dfrac{d^2 u}{d\tilde{x}^2} - \exp(u-v) + \exp(w-u) + \dfrac{(N_D^+ - N_A^-)}{n_i} = 0 = EQu \\[2ex] \dfrac{\mu_n n_i U_T}{L_i^2} \dfrac{d}{d\tilde{x}}\left\{\exp(u-v)\dfrac{dv}{d\tilde{x}}\right\} + R = 0 = EQv \\[2ex] \dfrac{\mu_p n_i U_T}{L_i^2} \dfrac{d}{d\tilde{x}}\left\{\exp(w-u)\dfrac{dw}{d\tilde{x}}\right\} - R = 0 = EQw \end{cases} \tag{6.36}$$

where the variables u, v, w, and $\tilde{x}$ are given by:

$$\tilde{x} = x / L_i; \quad L_i = \sqrt{\frac{\varepsilon U_T}{q n_i}}; \quad u = \frac{\varphi}{U_T}; \quad v = \frac{\varphi_n}{U_T}; \quad w = \frac{\varphi_p}{U_T}$$

The first step in the solution of 6.36 is the approximation of the set of nonlinear differential equations by a set of nonlinear algebraic equations. As in the section "Potential Distribution in Unbiased p–n Junction," this can be again achieved with the help of the FDM. The application of the FDM to the Poisson equation has already been described in detail in the Section "Potential Distribution in Unbiased p–n Junction." Here we therefore focus on the current continuity equations. For the sake of simplicity, we apply a uniform FD mesh only. If the discontinuities of ε, μ_n, and μ_p are coincident with the FD mesh point positions, then the following FD approximation is obtained for 6.36:

$$
\left[
\begin{array}{l}
\dfrac{u_{i+1} - 2u_i + u_{i-1}}{\Delta x^2} - \exp(u_i - v_i) + \exp(w_i - u_i) + \left.\dfrac{(N_D^+ - N_A^-)}{n_i}\right|_i = 0 = EQu_i \\[4mm]
\left\{\dfrac{\mu_n n_i U_T}{L_i^2} \dfrac{1}{\Delta x}\left[\exp(u_{i+1/2} - v_{i+1/2})\left.\dfrac{dv}{d\tilde{x}}\right|_{i+1/2} - \exp(u_{i-1/2} - v_{i-1/2})\left.\dfrac{dv}{d\tilde{x}}\right|_{i-1/2}\right]\right\} + R = 0 = EQv_i \\[4mm]
\left\{\dfrac{\mu_p n_i U_T}{L_i^2} \dfrac{1}{\Delta x}\left[\exp(w_{i+1/2} - u_{i+1/2})\left.\dfrac{dw}{d\tilde{x}}\right|_{i+1/2} - \exp(w_{i-1/2} - u_{i-1/2})\left.\dfrac{dw}{d\tilde{x}}\right|_{i-1/2}\right]\right\} - R = 0 = EQw_i
\end{array}
\right.
$$

$$(6.37)$$

where Δx is the distance between two FD nodes and the symbol $|_i$ means that the preceding expression is evaluated at the point $x = x_i$. In 6.37, the values of carrier concentrations and quasi-Fermi potential derivatives at the midpoint between two FD nodes are not readily available. Hence, there is a need to approximate these values. The calculation of suitable approximations is not a trivial task so we will now turn our attention to this problem. There are several ways known in the literature in which these approximate values can be obtained. For instance, one could simply use the arithmetic averaging of the exponential function argument [35] which for electron current continuity equation in 6.11, in the absence of recombination, would yield the following approximation:

$$
\exp\left[\frac{(u_{i+1} - v_{i+1} + u_i - v_i)}{2}\right]\frac{(v_{i+1} - v_i)}{\Delta x^2} -
$$

$$
\exp\left[\frac{(u_i - v_i + u_{i-1} - v_{i-1})}{2}\right]\frac{(v_i - v_{i-1})}{\Delta x^2} = EQv_i
$$

However, such finite difference discretisation scheme may produce spurious oscillations [16,36]. The approach that is widely recognised in the literature as correctly predicting the solution behaviour involves using the Bernoulli function. Its development is credited to Scharfetter and Gummel [37]. There are two ways known in the literature in which such an approximation can be derived. One of them relies on realising that 6.37 is equivalent to:

$$
\left|
\begin{aligned}
&\left. \frac{u_{i+1} - 2u_i + u_{i-1}}{\Delta x^2} - \exp\left(u_i - v_i\right) + \exp\left(w_i - u_i\right) + \frac{(N_D^+ - N_A^-)}{n_i} \right|_i = 0 = EQu_i \\
&\frac{1}{\Delta x}\left\{ J_{nx}\big|_{i+1/2} - J_{nx}\big|_{i-1/2} \right\} + R = 0 = EQv_i \\
&\frac{1}{\Delta x}\left\{ J_{px}\big|_{i+1/2} - J_{px}\big|_{i-1/2} \right\} - R = 0 = EQw_i
\end{aligned}
\right.
$$

So the problem reduces to calculating the optimal value of the current at the half distance between the FD nodes subject to known values of the potentials and carrier densities at the FD nodal points. This yields a differential equation [16,38]. For instance, for calculating the optimal value of J_{nx} at $x = x_{i+1/2}$, one can use 6.29. This results in the following differential equation:

$$
J_{nx}\big|_{i+1/2} = q\mu_n \frac{d\varphi}{dx} n + qD_n \frac{dn}{dx} \tag{6.38}
$$

which can be solved subject to the boundary conditions imposed by the known values of the electron concentration at $x = x_i$ and $x = x_{i+1}$. The unknowns in 6.38 are the electron concentration $n(x)$ and the value of J_{nx} at $x = x_{i+1/2}$ while the factor

$$
q\mu_n \frac{d\varphi}{dx}
$$

is assumed known and constant. Thus the problem 6.38 is essentially an inhomogeneous ordinary differential equation that can be readily solved for the electron distribution $n(x)$ and the value of J_{nx} at $x = x_{i+1/2}$. Once the solution is obtained, the value of J_{nx} at $x = x_{i+1/2}$ is expressed in terms of the values of n at the boundary points, that is, $n(x = x_i)$ and $n = (x = x_{i+1})$.

The other approach originates from [16] and is based on the application of the mean value theorem. This is the approach that is followed here in detail because it allows easy tracking of the truncation errors involved in the approximation and is directly applicable to 6.37.

The mean value theorem for integrals states that if a function $f(x)$ is continuous within the interval (a,b) then there exists such c in (a,b) that [39]:

$$f(x=c)(b-a) = \int_a^b f(x)\,dx \qquad (6.39)$$

One can observe that if 6.39 is true then also:

$$g(x=c_1)f(x=c_1)(b-a) = \int_a^b g(x)f(x)\,dx \qquad (6.40)$$

holds subject to continuity of the product of functions $f(x)$ and $g(x)$. However, point c_1 is in a general case different from c that appears in 6.39. Combining 6.39 and 6.40 yields:

$$g(x=c_1)\frac{f(x=c_1)}{f(x=c)}\int_a^b f(x)\,dx = \int_a^b g(x)f(x)\,dx \qquad (6.41)$$

Now for the functions $f(x)$ and $g(x)$ that stand in front of the integral on the left hand side of 6.41, we apply a Taylor series and expand $g(x)$ at a point c_2 that is positioned exactly in the middle of the interval (a,b) while $f(x)$ at point c:

$$\left(g(x=c_2)+g'(x=c_2)(c_1-c_2)+HOT\right)\left(1+\frac{f'(x=c)}{f(x=c)}(c_1-c)+HOT\right)\int_a^b f(x)\,dx$$

$$= \int_a^b g(x)f(x)\,dx \qquad (6.42)$$

In 6.42 the prime sign denotes the derivative while "HOT" stands for the higher order terms of the Taylor series. We observe that $|c_1-c_2| \le |b-a|/2 = h/2$ and $|c_1-c| \le |b-a| = h$, where $h = |a-b|$. Hence we finally obtain:

$$g(x=c_2)\int_a^b f(x)\,dx + O(h) \approx \int_a^b g(x)f(x)\,dx \qquad (6.43)$$

where the leading error term of the approximation is proportional to h. One could arrive more easily at 6.43 using the generalised mean value theorem [40]. However, the applicability of such an approximation would be limited to situations whereby the function $f(x)$ keeps the same sign within the interval (a,b).

Now we turn attention to the following identity [16]:

$$e^u \frac{d}{dx} e^{-u} = -\frac{du}{dx} \tag{6.44}$$

Integrating 6.18 between two FD nodal points x_i and x_{i+1} yields:

$$\int_{x_i}^{x_{i+1}} e^u \frac{d}{dx} e^{-u} dx = u_i - u_{i+1} \tag{6.45}$$

Now on the left hand side of 6.45, we identify e^u with $g(x)$ and $d/dx(e^{-u})$ with $f(x)$, and apply 6.43:

$$e^{u_{i+1/2}} \int_{x_i}^{x_{i+1}} \frac{d}{dx} e^{-u} dx = u_i - u_{i+1} \tag{6.46}$$

Carrying out the integration in 6.46 yields the following approximation for e^u at $x = x_{i+1/2}$:

$$e^{u_{i+1/2}} = \frac{u_i - u_{i+1}}{e^{-u_{i+1}} - e^{-u_i}} = B(\Delta u) e^{u_i} = B(-\Delta u) e^{u_i + 1} \tag{6.47}$$

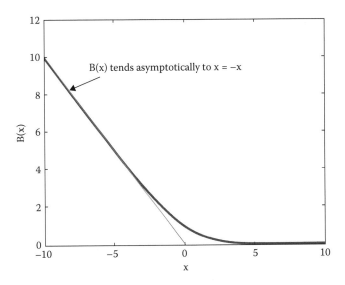

FIGURE 6.7   Bernoulli function $B(x)$.

where $\Delta u = u_i - u_{i+1}$ and $B(x)$ is the Bernoulli function (Figure 6.7):

$$B(x) = \frac{x}{\exp(x) - 1}$$

Using 6.47 to express the values of exponent function at the midpoint between two FD mesh points yields the following approximation for Equation 6.27 at a point x_i:

$$\left\{\begin{aligned}
&\left.\left|\frac{u_{i+1} - 2u_i + u_{i-1}}{\Delta x^2} - \exp(u_i - v_i) + \exp(w_i - u_i) + \frac{(N_D^+ - N_A^-)}{n_i}\right|\right|_i \\
&= 0 = EQu_i\left(u_{i+1}, u_i, u_{i-1}, v_i, w_i\right) \\
&B\left(u_i - v_i - u_{i+1} + v_{i+1}\right)\frac{(v_{i+1} - v_i)}{\Delta x^2} - B\left(u_i - v_i - u_{i-1} + v_{i-1}\right)\frac{(v_i - v_{i-1})}{\Delta x^2} \\
&\quad + \frac{L_i^2}{\mu_n n_i U_T}R\ \exp(v_i - u_i) = 0 = EQv_i\left(u_{i+1}, u_i, u_{i-1}, v_{i+1}, v_i, v_{i-1}\right) \\
&B\left(w_i - u_i - w_{i+1} + u_{i+1}\right)\frac{(w_{i+1} - w_i)}{\Delta x^2} - B\left(w_i - u_i - w_{i-1} + u_{i-1}\right)\frac{(w_i - w_{i-1})}{\Delta x^2} \\
&\quad - \frac{L_i^2}{\mu_n n_i U_T}R\ \exp(u_i - w_i) = 0 = EQw_i\left(u_{i+1}, u_i, u_{i-1}, w_{i+1}, w_i, w_{i-1}\right)
\end{aligned}\right. \tag{6.48}$$

We note that to obtain Equation 6.48 the first derivatives in 6.37 were approximated using the standard central finite differences:

$$\left.\frac{dv}{d\tilde{x}}\right|_{i+1/2} = \frac{1}{\Delta x}(v_{i+1} - v_i) \qquad \left.\frac{dv}{d\tilde{x}}\right|_{i-1/2} = \frac{1}{\Delta x}(v_i - v_{i-1}) \tag{6.49}$$

For obtaining a robust code, it is also important to approximate the Bernoulli function and its derivative for small values of the argument by:

$$B(x) \approx 1 - \frac{1}{2}x;$$

$$B'(x) \approx -\frac{1}{2} + \frac{1}{6}x$$

Again, writing equation 6.48 at all FD nodal points and imposing boundary conditions yields a set of nonlinear algebraic equations that approximates 6.36 at a set on FD nodal points. There are two approaches that have been used for solving the resulting set of equations. Both of them are based on calculating the solution of the set of nonlinear algebraic equations iteratively. One of the two algorithms (simultaneous solution method) updates simultaneously the values of the electric and quasi-Fermi potentials using the last available values of the unknowns while the other one calculates the updates separately for the electric potential and quasi-Fermi potential. The implementation of the first technique relies on the Newton-Raphson (N-R) method. The latter one (separate solution method) originates from Gummel [41] and has been further elaborated on by Slotboom [12]. The disadvantage of solving the equations separately is a relatively slow convergence of the resulting algorithm, especially, for the large values of the bias current, and when studying transient problems [15]. Furthermore, in some cases the separate solution method can fail to converge even if the initial guess is very near to the solution [42]. Some effort was invested into improving the performance of this method [43]; nonetheless, the convergence of the separate solution method for an arbitrary structure at a given bias cannot be guaranteed. The N-R algorithm, which forms the core of the approach that solves the equations simultaneously, on the other hand, has a guaranteed quadratic convergence in the vicinity of the solution [26]. However, even though N-R algorithm always moves in the direction of the largest gradient, convergence may not occur, because the N-R algorithm tends to overshoot significantly at large values of the bias voltage, especially during the initial iterations when the subsequent N-R updates are still relatively far from the solution [36]. This can result in a sequence of subsequent updates produced by N-R algorithm that can wonder off from the solution. Several strategies were elaborated to remedy this problem. One way of solving this problem is calculating an initial guess that is very near to the solution by applying approximate analytical techniques [44]. However, this makes the algorithm applicable only to selected structures, for which the approximate analytical solution is available. Another approach consist in calculating the solution at the zero bias first and then increasing the bias voltage using small steps up until the operating bias voltage is achieved. At each step the initial guess is set to the solution obtained at the previous bias voltage [25]. The disadvantage of this approach is that, especially at large values of the bias voltage, very small voltage increments must be applied to insure the convergence and hence the calculation time can become prohibitively large. An algorithm that is both robust and relatively efficient can be obtained by applying global N-R algorithms [26]. An optimisation of the backtracking strategy for a global N-R algorithm for drift-diffusion equations was performed by Rose and Bank [45,46]. An alternative strategy that provides a robust global algorithm for the solution of the drift-diffusion equations is based on the numerical path following (NPF) [47,48]. In this method, the calculations are initiated for a simple problem with a known solution. Then a parameter is introduced that defines a path, which allows tracking the solution from the initial problem to the actual one. Tracking of the solution is done in two steps. First, the solution is predicted along the vector that is tangential to the path. In the second step, the solution is corrected using the N-R method. For the sake of completeness, we should mention also techniques that are based on blending the separate solution methods with simultaneous solution methods [44,49,50]. The main difficulty with applying such techniques is the lack of clear guideline on when to stop the separate solution method and switch to simultaneous solution method.

Because all globally convergent numerical algorithms that are used for the solution of the drift-diffusion equation are based on N-R algorithm, we will now discuss in detail N-R algorithm application to the solution of equation 6.37 using FDM 6.48.

The application of the Newton-Raphson method to 6.48 after applying the Bernoulli function approximations for the current continuity equations yields:

$$\text{Jacobian} * V^{\text{NEW}} = -EQ(V^{\text{OLD}}) + \text{Jacobian} * V^{\text{OLD}} \qquad (6.50)$$

where the Jacobian matrix is given by:

$$\text{Jacobian} = \begin{vmatrix} \dfrac{\partial EQu}{\partial u} & \dfrac{\partial EQu}{\partial v} & \dfrac{\partial EQu}{\partial w} \\[2em] \dfrac{\partial EQv}{\partial u} & \dfrac{\partial EQv}{\partial v} & \dfrac{\partial EQv}{\partial w} \\[2em] \dfrac{\partial EQw}{\partial u} & \dfrac{\partial EQw}{\partial v} & \dfrac{\partial EQw}{\partial w} \end{vmatrix} \qquad (6.51)$$

and the vector V arranges together functions u, v, and w:

$$V = \begin{bmatrix} u \\ v \\ w \end{bmatrix} \qquad (6.52)$$

ALGORITHM 6.2   FINITE DIFFERENCE BASED CALCULATION OF ELECTRIC POTENTIAL DISTRIBUTION AND QUASI-FERMI LEVEL DISTRIBUTIONS IN A BIASED HOMOJUNCTION

1. Start
2. Select Δx and the sampling nodal points
3. Select the bias voltage
4. Select initial bias voltage (usually zero bias) and a bias voltage step
5. Set the initial guess V^{OLD} at initial bias voltage
6. Calculate V^{NEW} from 6.50
7. If $|V^{\text{OLD}} - V^{\text{NEW}}| >$ residual $V^{\text{OLD}} = V^{\text{NEW}}$, go to 6
8. If bias voltage not reached, increase bias voltage by the bias voltage step and go to 6 while using V^{NEW} as initial guess
9. Stop

The Jacobian consists of nine submatrices. The submatrix

$$\frac{\partial EQu}{\partial u}$$

for instance has the following form:

$$\frac{\partial EQu}{\partial u} = \begin{vmatrix} \dfrac{\partial EQu_1}{\partial u_1} & \dfrac{\partial EQu_1}{\partial u_2} & 0 & \cdots & \cdots & \cdots & 0 \\ 0 & \ddots & & & & & \vdots \\ \cdots & 0 & \dfrac{\partial EQu_i}{\partial u_{i-1}} & \dfrac{\partial EQu_i}{\partial u_i} & \dfrac{\partial EQu_i}{\partial u_{i+1}} & 0 & \cdots \\ \vdots & & & & \ddots & & 0 \\ 0 & \cdots & \cdots & \cdots & 0 & \dfrac{\partial EQu_N}{\partial u_{N-1}} & \dfrac{\partial EQu_N}{\partial u_N} \end{vmatrix} \quad (6.53)$$

We impose the following boundary conditions on the quasi-Fermi levels (Figure 6.6):

$$E_{Fn}(x=-L_L) = E_{Feqn} - W_i(x=-L_L) = kT \ln\left(\frac{|N_d(x=-L_L)|}{n_i(x=-L_L)}\right) \quad (6.54a)$$

$$E_{Fp}(x=-L_L) = E_{Fn}(x=-L_L) \quad (6.54b)$$

$$E_{Fn}(x=L_P) = kT \ln\left(\frac{|N_d(x=-L_L)|}{n_i(x=-L_L)}\right) - qV_{Bias} \quad (6.54c)$$

$$E_{Fp}(x=L_P) = E_{Fn}(x=L_P) \quad (6.54d)$$

Finally, the implementation of the Newton-Raphson method to the solution of 6.37 yields the Algorithm 6.2.

The Algorithm 6.2 can be implemented using MATLAB script following the same steps as in the case of the Algorithm 6.1. This results in the following code:

```
% program calculates the band structure of biased
% diode,Bernoulli
% SRH recombination and N-R backtracking
% Si like diode from IEEE TED 1999 pp396-412
clear % clears variables
clear global % clears global variables
```

```
% initial constants
I = sqrt(-1);% remember not to overwrite pi or i !
pi = 3.141592653589793e+000;
c = 2.998e+2;% speed of light in free space [mi/ps]
q = 1.602e-19;% elemental charge [C]
h_s = 6.62606885e-34;% Planck's constant [J*ps]
kB = 1.3807e-23;% Boltzman's constant [J/K]
T = 300.;% temperature [K]
kT = kB*T/q;% kT [eV]
UT = kB*T/q;% UT thermal potential [V]
eps0 = 8.854e-18;% vacum permittivity [F/mi]
m0 = 9.11e-31;% electron mass [kg]

% input diode parameters (Si)
Wg_n = 1.12;% bandgap [eV]
eps_n = 11.7;% relative dielectric constant
miu_n = 0.1;%electron mobility [mi^2/(V*ps)]
miu_p = 0.1;%hole mobility [mi^2/(V*ps)]

% n side first
ND = 1e+4;% effective dopant concentration [1/mi3]
thick_n = 5.0;% thickness [mi]

% p side
NA = ND;% effective dopant concentration [1/mi3]
thick_p = 5.0;% thickness [mi]

% calculating basic parameters for n side and p side:
Nc_n = 2.e+7;%electron effective density of states [1/mi3]
Nv_n = 2.e+7;%hole effective density of states [1/mi3]
ni_n = 7.82e-3;% intrinsic carrier concentration [1/mi3]

% Barrier potential calculated directly since energy
% is expressed in eV
VBarrier = kT*log(abs(ND)/ni_n)+kT*log(abs(NA)/ni_n)

% recombination coefficient
tau = 10;%SRH lifetime [ps]
B_n = eps_n*eps0/(tau*ni_n*q*miu_n);
B_p = eps_n*eps0/(tau*ni_n*q*miu_p);

% normalising relevant parameters
Nc_n = Nc_n/ni_n
Nv_n = Nv_n/ni_n
ND = ND/ni_n
NA = NA/ni_n
Wg_n = Wg_n/UT
Li = sqrt(eps_n*eps0*UT/(q*ni_n))%Debye length
VBarrier = VBarrier/UT
```

```
% calculation parameters
thick = (thick_n+thick_p)/Li;% device length [Li]
N_po = 100;% number of points on L
dx = thick/(N_po);% step dz [Li]
x_pos = dx*(1:N_po)';

% setting doping profile
for i = 1:N_po/2
    KN(i) = ND;
    KN(i+N_po/2) = -NA;
end
w = log(abs(KN(1)))%boundary condition for Fermi levels

% bias voltage
VBias = 0.6;%bias voltage [V]
dVBias = 0.01;%bias voltage step
%normalisation of bias voltage and of the bias voltage step
VBias = VBias/UT;
dVBias = dVBias/UT;

% initial voltage and quasi-Fermi level distribution
for i = 1:N_po/2
    ui(i) = 0.0;
    ui(i+N_po/2) = -(VBarrier-VBias);
    vi(i) = -w;
    vi(i+N_po/2) = -(w-VBias);
    wi(i) = -w;
    wi(i+N_po/2) = -(w-VBias);
end

ue(2:N_po+1) = ui;
ve(2:N_po+1) = vi;
we(2:N_po+1) = wi;

%external loop stepping up bias voltage
for jbias = 1:2
% setting boundary conditions
ue(1) = 0;
ue(N_po+2) = -(VBarrier-VBias);
ve(1) = -w;
ve(N_po+2) = -(w-VBias);
we(1) = -w;
we(N_po+2) = -(w-VBias);

for j = 1:4000%main loop start
% setting auxiliary vectors
for i = 1:N_po
a1(i) = ue(i+1)-ve(i+1)-ue(i+2)+ve(i+2);
a2(i) = ue(i+1)-ve(i+1)-ue(i)+ve(i);
b1(i) = we(i+1)-ue(i+1)-we(i+2)+ue(i+2);
b2(i) = we(i+1)-ue(i+1)-we(i)+ue(i);
```

```
A3(i) = (ve(i+2)-ve(i+1))/(dx*dx);
A4(i) = (ve(i+1)-ve(i))/(dx*dx);
B3(i) = (we(i+2)-we(i+1))/(dx*dx);
B4(i) = (we(i+1)-we(i))/(dx*dx);
end

for i = 1:N_po
a3(i) = ui(i)-vi(i);
b3(i) = wi(i)-ui(i);
A1(i) = Bern(a1(i));
A2(i) = Bern(a2(i));
AP1(i) = Bernp(a1(i));
AP2(i) = Bernp(a2(i));
B1(i) = Bern(b1(i));
B2(i) = Bern(b2(i));
BP1(i) = Bernp(b1(i));
BP2(i) = Bernp(b2(i));
E1(i) = exp(a3(i));
E2(i) = exp(b3(i));
end

DEN = E1+E2+2;
NUM1 = E2-1./E1;
NUM2 = E1-1./E2;

% setting off diagonals of Jacobian for dEQudu
for i = 1:N_po
    left_diag(i) = 1.0/(dx*dx);
    right_diag(i) = 1.0/(dx*dx);
end

% setting Jacobian elements start VVVVVVVVVVV
% setting dEQudu, dEQudv, dEQudw
main_diag = -2.0/(dx*dx)-E1-E2;
dEQudu = spdiags([left_diag' main_diag' right_diag'],…
-1:1, N_po, N_po);
dEQudv = spdiags([E1'],0,N_po, N_po);
dEQudw = spdiags([E2'],0,N_po, N_po);

% setting dEQvdu, dEQvdv, dEQvdw
left_diag = AP2.*A4;
left_diag = [left_diag(2:N_po) 0];%shifting for spdiags
main_diag = AP1.*A3-AP2.*A4+B_n*((-E2+1./E1).*DEN-NUM1.*…
(E1+E2))./(DEN.*DEN);
right_diag = -AP1.*A3;
right_diag = [0 right_diag(1:N_po-1)];%shifting for spdiags
dEQvdu = spdiags([left_diag' main_diag' right_diag'],…
-1:1, N_po, N_po);

left_diag = -AP2.*A4+A2/(dx*dx);
left_diag = [left_diag(2:N_po) 0];
```

```
main_diag = -AP1.*A3+AP2.*A4-A1/(dx*dx)-A2/(dx*dx)+…
    B_n*((-DEN./E1)+NUM1.*E1)./(DEN.*DEN);
right_diag = AP1.*A3+A1/(dx*dx);
right_diag = [0 right_diag(1:N_po-1)];
dEQvdv = spdiags([left_diag' main_diag' right_diag'],…
-1:1, N_po, N_po);

dEQvdw = spdiags([(B_n*(E2.*DEN-NUM1.*(E1+E2))./…
(DEN.*DEN))'],0,N_po, N_po);

% setting dEQwdu, dEQwdv, dEQwdw
left_diag = -BP2.*B4;
left_diag = [left_diag(2:N_po) 0];
main_diag = -BP1.*B3+BP2.*B4-B_p*((E1-1./E2).*DEN-NUM2.*…
(E1+E2))./(DEN.*DEN);
right_diag = BP1.*B3;
right_diag = [0 right_diag(1:N_po-1)];
dEQwdu = spdiags([left_diag' main_diag' right_diag'],…
-1:1, N_po, N_po);

left_diag = BP2.*B4+B2/(dx*dx);
left_diag = [left_diag(2:N_po) 0];
main_diag = BP1.*B3-BP2.*B4-B1/(dx*dx)-B2/(dx*dx)-…
B_p*((DEN./E2)-NUM2.*E2)./(DEN.*DEN);
right_diag = -BP1.*B3+B1/(dx*dx);
right_diag = [0 right_diag(1:N_po-1)];
dEQwdw = spdiags([left_diag' main_diag' right_diag'],…
-1:1, N_po, N_po);

dEQwdv = spdiags([(-B_p*((DEN.*E1)+NUM2.*E1)./…
(DEN.*DEN))'], 0, N_po, N_po);

% setting the Jacobian
Jacobian = [dEQudu dEQudv dEQudw;dEQvdu dEQvdv…
dEQvdw;dEQwdu dEQwdv dEQwdw];

% setting Jacobian elements finish VVVVVVVVVV

% setting dEQu, dEQv and EQw
for i = 2:N_po+1
    EQu(i-1)=-(2.0*ue(i)-ue(i-1)-ue(i+1))/(dx*dx)-…
    exp(ue(i)-ve(i))+exp(we(i)-ue(i))+KN(i-1);
end
EQv = A1.*A3-A2.*A4+B_n*(E2-1./E1)./DEN;
EQw = B1.*B3-B2.*B4-B_p*(E1-1./E2)./DEN;

% setting F
EQ = [EQu EQv EQw];
FF1 = EQ*EQ';
```

```
%creating unknown vector
Vi = [ui vi wi];

dVi = -Jacobian\EQ';
Vi = Vi+dVi';
ui = Vi(1:N_po);vi = Vi(N_po+1:2*N_po);
wi = Vi(2*N_po+1:3*N_po);
ue(2:N_po+1) = ui;ve(2:N_po+1) = vi;we(2:N_po+1) = wi;
residual = dVi'*dVi;
[j residual VBias*UT]
if residual < 1.e-5,break,end%residual check

% backtracking start *********************
for i = 1:N_po
a1(i) = ue(i+1)-ve(i+1)-ue(i+2)+ve(i+2);
a2(i) = ue(i+1)-ve(i+1)-ue(i)+ve(i);
b1(i) = we(i+1)-ue(i+1)-we(i+2)+ue(i+2);
b2(i) = we(i+1)-ue(i+1)-we(i)+ue(i);
A3(i) = (ve(i+2)-ve(i+1))/(dx*dx);
A4(i) = (ve(i+1)-ve(i))/(dx*dx);
B3(i) = (we(i+2)-we(i+1))/(dx*dx);
B4(i) = (we(i+1)-we(i))/(dx*dx);
end

for i = 1:N_po
a3(i) = ui(i)-vi(i);
b3(i) = wi(i)-ui(i);
A1(i) = Bern(a1(i));
A2(i) = Bern(a2(i));
AP1(i) = Bernp(a1(i));
AP2(i) = Bernp(a2(i));
B1(i) = Bern(b1(i));
B2(i) = Bern(b2(i));
BP1(i) = Bernp(b1(i));
BP2(i) = Bernp(b2(i));
E1(i) = exp(a3(i));
E2(i) = exp(b3(i));
end

DEN = E1+E2+2;
NUM1 = E2-1./E1;
NUM2 = E1-1./E2;

% setting dEQu, dEQv and EQw
for i = 2:N_po+1
    EQu(i-1)=-(2.0*ue(i)-ue(i-1)-ue(i+1))/(dx*dx)-…
    exp(ue(i)-ve(i))+exp(we(i)-ue(i))+KN(i-1);
end
EQv = A1.*A3-A2.*A4+B_n*(E2-1./E1)./DEN;
EQw = B1.*B3-B2.*B4-B_p*(E1-1./E2)./DEN;
```

```
% setting F
EQ = [EQu EQv EQw];
FF2 = EQ*EQ';

lambda = 0.5;%N-R backtracking parameter
if FF2 > FF1
   Vi = Vi-lambda*dVi';
   ui=Vi(1:N_po);vi=Vi(N_po+1:2*N_po);
   wi=Vi(2*N_po+1:3*N_po);
   ue(2:N_po+1)=ui;ve(2:N_po+1)=vi;we(2:N_po+1)=wi;
end
% backtracking end ************************


end%main loop stop

[j dVi'*dVi VBias*UT]
VBias = VBias+dVBias;


end%external loop stop

plot(x_pos*Li,ui*UT,x_pos*Li,-ui*UT,x_pos*Li,-vi*UT,…
x_pos*Li,-wi*UT)
xlabel('x [micrometer]','FontSize',24)
ylabel('E/UT','FontSize',24)
```

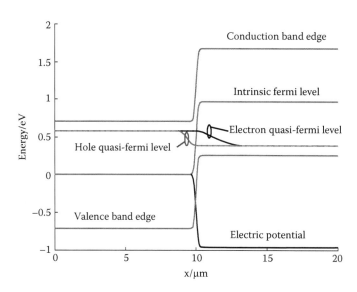

FIGURE 6.8   Calculated band diagram for GaAs homojunction with the doping concentration of 10^{16}/cm^3 on both the p side and n side at the bias voltage of 0.2 V. A thousand FD nodes were used while all other modelling parameters are listed in Table 6.1.

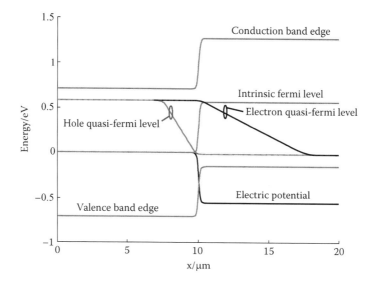

FIGURE 6.9 Calculated band diagram for GaAs homojunction with doping concentration of 10^{16}/cm^3 on both the p side and n side at the bias voltage of 0.6 V. A thousand FD nodes were used while all other modelling parameters are listed in Table 6.1.

In the directory, the main MATLAB program must be accompanied by two files storing the Bernoulli function and its derivative, that is:

Bern.m

```
% calculates Bernoulli function
function wynik = Bern(x)
if abs(x) < 0.01
       wynik = 1-x/2+x*x/12-x*x*x*x/720;
else
       wynik = x/(exp(x)-1);
end
```

Bernp.m

```
% calculates derivative of Bernoulli function
function wynik = Bernp(x)
if abs(x) < 0.01
       wynik = -1/2+x/6-x*x*x/180;
else
       wynik = (exp(x)-1-x*exp(x))/(exp(x)-1)^2;
end
```

We used a fairly simple computer implementation of the Bernoulli function, which we found sufficiently robust for the test calculations. A more elaborate approach is described in Collins [5]. The presented software includes Shockley-Read-Hall

(SRH) recombination only, which for a single thermal recombination level at the midgap under Boltzmann statistics is given by the formula:

$$R = \frac{pn - n_i^2}{\tau_n \left(n + n_i\right) + \tau_p \left(p + n_i\right)} \tag{6.55}$$

In simulations we assumed that the SRH lifetime $\tau_p = \tau_n = \tau = 10$ ps. We set the initial quasi-Fermi level distribution to be equal to E_{Feqn} on the n-side and $E_{\mathrm{Feqn}} - qV_{\mathrm{Bias}}$ on the p-side (cf. Figure 6.6) while the initial distribution of the electric potential is equal to zero on the n-side and $-(V_{\mathrm{Barrier}} - V_{\mathrm{Bias}})$ on the p-side. An external loop has been added that allows for starting the calculations at zero bias (or a selected bias) and following the solution up to the operating point by increasing the bias voltage by a selected bias voltage step.

The software includes also a simple N-R backtracking routine, which is essential at higher bias voltage values. Here we applied a very simple backtracking algorithm. The parameter "lambda" sets the amount of backtracking that is applied. The backtracking algorithm consists in moving back from V^{NEW} obtained from 6.50 by the amount equal to lambda*($V^{\mathrm{NEW}} - V^{OLD}$). Hence, the value of lambda = 0 corresponds to no backtracking while lambda = 1 effects a return to the unknown vector from the previous iteration. The price to pay for backtracking with a fixed backtracking parameter is a slower convergence rate. A more efficient algorithm is obtained if the amount of backtracking can be varied during the calculations, thus allowing for longer steps when getting nearer to the solution [5,46].

For the "Si-like" test diode structure from [17], the presented software allows bias voltages of 0.55 V to be reached without backtracking. At the bias voltage of 0.6 V, an application of N-R backtracking is necessary. In this case, setting lambda = 0.5 was found sufficient.

Figures 6.8 and 6.9 show calculated example band diagrams for GaAs homojunction with doping concentration of $10^{16}/\mathrm{cm}^3$ on both the p side and n side at the bias voltage of 0.2 and 0.6 V, respectively. In both simulations we used 1000 finite difference nodes. No backtracking was necessary when obtaining the results shown in Figures 6.8 and 6.9.

MODELLING OF CURRENT FLOW IN PHOTONIC SEMICONDUCTOR DEVICES

When the finite difference mesh is regular, then the methods described in the previous sections are fairly straightforward to extend to two dimensional (2D) and 3D analysis. In short, the principle remains the same; namely, first the partial differential equations are converted into a set of nonlinear algebraic equations and then the algebraic equations are solved using the N-R method. As an illustrating example, let's consider the 2D analysis of a symmetrical p–n junction. In this case, Equations 6.6 and 6.11 yield:

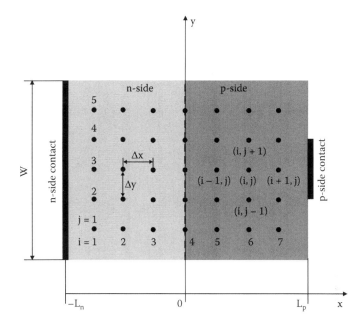

FIGURE 6.10 Schematic diagram showing an example distribution of the nodal mesh points in a p–n junction.

$$\begin{vmatrix} \dfrac{d}{dx}\left(\varepsilon\dfrac{d\varphi}{dx}\right)+\dfrac{d}{dy}\left(\varepsilon\dfrac{d\varphi}{dy}\right)+q(p-n+N_D-N_A)=0=0 \\[4mm] \dfrac{d}{dx}\left\{\mu_n n\dfrac{d\varphi_n}{dx}\right\}+\dfrac{d}{dy}\left\{\mu_n n\dfrac{d\varphi_n}{dy}\right\}=-R \\[4mm] \dfrac{d}{dx}\left\{\mu_p p\dfrac{d\varphi_p}{dx}\right\}+\dfrac{d}{dy}\left\{\mu_p p\dfrac{d\varphi_p}{dy}\right\}=R \end{vmatrix} \qquad (6.56)$$

Applying again DeMari's scaling for a symmetrical p–n junction yields:

$$\begin{vmatrix} \dfrac{d^2u}{d\tilde{x}^2}+\dfrac{d^2u}{d\tilde{y}^2}-\exp(u-v)+\exp(w-u)+\dfrac{(N_D^+-N_A^-)}{n_i}=0=EQu \\[4mm] \dfrac{\mu_n n_i U_T}{L_i^2}\left(\dfrac{d}{d\tilde{x}}\left\{\exp(u-v)\dfrac{dv}{d\tilde{x}}\right\}+\dfrac{d}{d\tilde{y}}\left\{\exp(u-v)\dfrac{dv}{d\tilde{y}}\right\}\right)+R=0=EQv \\[4mm] \dfrac{\mu_p n_i U_T}{L_i^2}\left(\dfrac{d}{d\tilde{x}}\left\{\exp(w-u)\dfrac{dw}{d\tilde{x}}\right\}+\dfrac{d}{d\tilde{y}}\left\{\exp(w-u)\dfrac{dw}{d\tilde{y}}\right\}\right)-R=0=EQw \end{vmatrix} \qquad (6.57)$$

Applying the Scharfeter-Gummel FD discretisation for 6.57 on a 2D regular square mesh yields:

$$
\left\{
\begin{aligned}
&\left. \frac{u_{i+1,j}+u_{i-1,j}+u_{i,j+1}+u_{i,j}-4u_{i,j-1}}{h^2} - \exp\left(u_{i,j}-v_{i,j}\right) + \exp\left(w_{i,j}-u_{i,j}\right) + \frac{(N_{\mathrm{D}}^{+}-N_{\mathrm{A}}^{-})}{n_i} \right|_i \\
&= 0 = EQu_i\left(u_{i+1,j},u_{i,j},u_{i-1,j},u_{i,j+1},u_{i,j-1},v_{i,j},w_{i,j}\right) \\[4pt]
&B\left(u_{i,j}-v_{i,j}-u_{i+1,j}+v_{i+1,j}\right)\frac{\left(v_{i+1,j}-v_{i,j}\right)}{h^2} - B\left(u_{i,j}-v_{i,j}-u_{i-1,j}+v_{i-1,j}\right)\frac{\left(v_{i,j}-v_{i-1,j}\right)}{h^2} \\[4pt]
&+B\left(u_{i,j}-v_{i,j}-u_{i,j+1}+v_{i,j+1}\right)\frac{\left(v_{i,j+1}-v_{i,j}\right)}{h^2} - B\left(u_{i,j}-v_{i,j}-u_{i,j-1}+v_{i,j-1}\right)\frac{\left(v_{i,j}-v_{i,j-1}\right)}{h^2} \\[4pt]
&+\frac{L_i^2}{\mu_n n_i U_T} R\ \exp\left(v_{i,j}-u_{i,j}\right)=0= \\[4pt]
&= EQv_i\left(u_{i,j},u_{i+1,j},u_{i-1,j},u_{i,j+1},u_{i,j-1},v_{i,j},v_{i+1,j},v_{i-1,j},v_{i,j+1},v_{i,j-1},\right) \\[4pt]
&B\left(w_{i,j}-u_{i,j}-w_{i+1,j}+u_{i+1,j}\right)\frac{\left(w_{i+1,j}-w_{i,j}\right)}{h^2} - B\left(w_{i,j}-u_{i,j}-w_{i-1,j}+u_{i-1,j}\right)\frac{\left(w_{i,j}-w_{i-1,j}\right)}{h^2} \\[4pt]
&+B\left(w_{i,j}-u_{i,j}-w_{i,j+1}+u_{i,j+1}\right)\frac{\left(w_{i,j+1}-w_{i,j}\right)}{h^2} - B\left(w_{i,j}-u_{i,j}-w_{i,j-1}+u_{i,j-1}\right)\frac{\left(w_{i,j}-w_{i,j-1}\right)}{h^2} \\[4pt]
&-\frac{L_i^2}{\mu_n n_i U_T} R\ \exp\left(u_{i,j}-w_{i,j}\right)=0= \\[4pt]
&= EQw_i\left(u_{i,j},u_{i+1,j},u_{i-1,j},u_{i,j+1},u_{i,j-1},w_{i,j},w_{i+1,j},w_{i-1,j},w_{i,j+1},w_{i,j-1},\right)
\end{aligned}
\right.
$$

$$(6.58)$$

where $h = \Delta x = \Delta y$ (Figure 6.10). Collating equation 6.58 together, while incorporating the boundary conditions, yields again 6.50; whereby, the Jacobian matrix has nine submatrices. The solution therefore follows in the same way as in the section "Potential and Quasi Fermi Level Distribution in Biased p–n Junction." The only practical difference is that the resulting Jacobian matrix dimensions in the 2D case are much larger than in the 1D case and the application of sparse matrix solvers becomes essential. Here projection methods are particularly helpful [2,3]. An alternative approach relies on the application of the block iterative methods. However, a block iterative technique should be applied carefully because they may reduce the rate of convergence for the N-R algorithm or in some cases even cause the N-R algorithm to diverge [5,15].

The extension of the semiconductor device model to the multidimensional case is explained in detail in Collins [5] and in Ronnie [35], both for the finite elements and finite differences. The in-depth discussion of the application of Equation 6.27 to LDs (i.e., devices that include heterojunctions) is given in Lee and Chien [4] and Piprek [6].

Multidimensional models usually require substantial computer resources. To reduce the resulting demand in terms of the computer memory and CPU time, a nonuniform FD (or finite element [24]) mesh can be applied so that the FD nodes are used more efficiently. In the FDM, a particularly useful technique consists in using the finite box method with terminating lines at an arbitrary point [5]. Many helpful comments on the development of a drift-diffusion equations solver can be found also in Buturla et al. [23].

For the sake of completeness, we note that when solving 6.11 in the time domain implicit L-stable and A-stable [51] techniques have to be applied due to the large stiffness of the resulting set of ordinary differential equations [25,52]. The Crank-Nicholson, highly popular in BPM computations, scheme is for instance not applicable due to the lack of L-stability, which results in the presence of unphysical ringing in the simulation results [25]. Further details of the application of drift diffusion analysis in photonic semiconductor devices, including the coupling with the thermal solver, will be discussed in the context of LD modelling in Chapter 7.

Finally, we give a short description of two approximate techniques that are most popular in photonics. These techniques reduce the computational overhead when modelling electric current flow in photonic devices. One of them is based on the ambipolar approximation within the active layer [53,54]. Under the ambipolar approximation, Equation 6.27 is reduced to the solution of:

$$D\frac{\partial^2 N}{\partial x^2} + D\frac{\partial^2 N}{\partial z^2} = -\frac{J(x,z)}{qd} + R \qquad (6.59)$$

where N is the carrier density within the active layer that expands in the plane x–z, D is a diffusion constant, $J(x,z)$ is the density distribution of the current that is injected into the active layer, d is the active layer thickness, q is an elemental charge, and R is the recombination rate. The main problem in applying 6.60 is the fact that the current density J needs to be known a priori before 6.60 can be solved. In principle, one needs to solve 6.27 to calculate the current density J. However, this would effectively eliminate the benefits of applying 6.60. An approximate technique that has been used for this purpose consists in solving the Laplace equation to calculate the electric potential distribution within the device. Then the current density is obtained from the Ohm's law [55]:

$$\vec{J} = \rho E = -\rho \nabla \Psi \qquad (6.60)$$

where ρ is the electric conductivity. Alternatively, the injected current distribution can be evaluated analytically [56]. A discussion of the ambipolar approximation when applied to heterojunctions is given in Nakwaski and Osinski [57].

REFERENCES

1. Joyce, W.B. and R.W. Dixon, Thermal resistance of heterostructure lasers. *Journal of Applied Physics*, 1975. 46(2): p. 855–862.
2. Saad, Y., *Iterative Methods for Sparse Linear Systems*, 2nd ed. 2000, Philadelphia: SIAM.

3. Barret, R., et al., *Templates for the Solution of Linear Systems: Building Blocks for Iterative Methods*. 1994, Philadelphia: SIAM.

4. Lee, C.C. and D.H. Chien, The effect of bonding wires on longitudinal temperature profiles of laser diodes. *Journal of Lightwave Technology*, 1996. 14(8): p. 1847–1852.

5. Collins, R.E., *Mathematical Methods for Physicists and Engineers*. 1999, New York: Dover Publications Inc.

6. Piprek, J., *Semiconductor Optoelectronic Devices Introduction to Physics and Simulation*. 2003, San Diego: Academic Press.

7. Palankovski, V. and S. Selberherr, Rigorous modeling of high-speed semiconductor devices. *Microelectronics Reliability*, 2004. 44(6): p. 889–897.

8. Marshak, A.H., Modeling semiconductor-devices with position-dependent material parameters. *IEEE Transactions on Electron Devices*, 1989. 36(9): p. 1764–1772.

9. Marshak, A.H. and C.M. Vanvliet, Electrical-current and carrier density in degenerate materials with nonuniform band-structure. *Proceedings of the IEEE*, 1984. 72(2): p. 148–164.

10. Siegal, B., Measurement of junction temperature confirms package thermal design. Laser Focus World, 2003(11).

11. Ochalski, T.J., et al., Thermoreflectance and micro-Raman measurements of the temperature distributions in broad contact laser diodes. *Optica Applicata*, 2005. 35(3): p. 479–484.

12. Slotboom, J.W., Computer-aided 2-dimensional analysis of bipolar transistors. *IEEE Transactions on Electron Devices*, 1973. ED20(8): p. 669–679.

13. Mock, M.S., 2-dimensional mathematical-model of insulated-gate field-effect transistor. *Solid-State Electronics*, 1973. 16(5): p. 601–609.

14. Mock, M.S., *Analysis of Mathematical Models of Semiconductor Devices*. 1983, Dublin: Bole.

15. Rafferty, C.S., M.R. Pinto, and R.W. Dutton, Iterative methods in semiconductor-device simulation. *IEEE Transactions on Electron Devices*, 1985. 32(10): p. 2018–2027.

16. Bank, R.E., D.J. Rose, and W. Fichtner, Numerical-methods for semiconductor-device simulation. *IEEE Transactions on Electron Devices*, 1983. 30(9): p. 1031–1041.

17. Laux, S.E. and K. Hess, Revisiting the analytic theory of p–n junction impedance: Improvements guided by computer simulation leading to a new equivalent circuit. *IEEE Transactions on Electron Devices*, 1999. 46(2): p. 396–412.

18. Burgler, J.F., et al., A new discretization scheme for the semiconductor current continuity equations. *IEEE Transactions on Computer-Aided Design of Integrated Circuits and Systems*, 1989. 8(5): p. 479–489.

19. Engl, W.L., H.K. Dirks, and B. Meinerzhagen, Device modeling. *Proceedings of the IEEE*, 1983. 71(1): p. 10–33.

20. Guo, J.Y. and C.Y. Wu, A new 2D analytic threshold-voltage model for fully depleted short-channel soi mosfets. *IEEE Transactions on Electron Devices*, 1993. 40(9): p. 1653–1661.

21. Meel, K., R. Gopal, and D. Bhatnagar, Three-dimensional analytic modelling of front and back gate threshold voltages for small geometry fully depleted SOI MOSFET's. *Solid-State Electronics*, 2011. 62(1): p. 174–184.

22. Rao, R., et al., Unified analytical threshold voltage model for non-uniformly doped dual metal gate fully depleted silicon-on-insulator MOSFETs. *Solid-State Electronics*, 2009. 53(3): p. 256–265.

23. Buturla, E.M., et al., Finite-element analysis of semiconductor devices: The FIELDAY program (Reprinted from IBM Journal of Research and Development, vol 25, 1981). *IBM Journal of Research and Development*, 2000. 44(1–2): p. 142–156.

24. Bochev, P., K. Peterson, and X. Gao, A new Control Volume Finite Element Method for the stable and accurate solution of the drift-diffusion equations on general unstructured grids. *Computer Methods in Applied Mechanics and Engineering*, 2013. 254: p. 126–145.

25. Pinto, M.R., et al., PISCES-IIB. 1985, Stanford Electronics Laboratories: Stanford CA 94305.

26. Press, W.H., Teukolsky, S.A., Vetterling, W.T., Flannery, B.P., *Numerical Recipes in C++: the Art of Scientific Computing*. 2002, Cambridge: Cambridge University Press.

27. Horio, K. and H. Yanai, Numerical modeling of heterojunctions including the thermionic emission mechanism at the heterojunction interface. *IEEE Transactions on Electron Devices*, 1990. 37(4): p. 1093–1098.

28. Yang, K.H., J.R. East, and G.I. Haddad, Numerical modeling of abrupt heterojunctions using a thermionic-field emission boundary-condition. *Solid-State Electronics*, 1993. 36(3): p. 321–330.

29. Grupen, M. and K. Hess, Simulation of carrier transport and nonlinearities in quantum-well laser diodes. *IEEE Journal of Quantum Electronics*, 1998. 34(1): p. 120–140.

30. Piprek, J., Semiconductor optoelectronic devices. 2003, San Diego: Elsevier.

31. Coldren, L.A. and S.W. Corzine, *Diode Lasers and Photonic Integrated Circuits*. 1995, New York: Wiley.

32. Purbo, O.W., D.T. Cassidy, and S.H. Chisholm, Numerical-model for degenerate and heterostructure semiconductor-devices. *Journal of Applied Physics*, 1989. 66(10): p. 5078–5082.

33. Esquivias, I., et al., Carrier dynamics and microwave characteristics of GaAs-based quantum-well lasers. *IEEE Journal of Quantum Electronics*, 1999. 35(4): p. 635–646.

34. De Mari, A., An accurate numerical steady-state one-dimensional solution of the p–n junction. *Solid-State Electronics*, 1968. 11(1): p. 33–58.

35. Ronnie, T.J.W., *Advances in High-Power Laser Diode Packaging, in Semiconductor Laser Diode Technology and Applications*, D.S. Patil, Editor. 1999, www.intechopen. com.

36. Polak, S.J., et al., Semiconductor-device modeling from the numerical point-of-view. *International Journal for Numerical Methods in Engineering*, 1987. 24(4): p. 763–838.

37. Scharfeter, D.l. and H.K. Gummel, Large-signal analysis of a silicon read diode oscillator. *IEEE Transactions on Electron Devices*, 1969. ED16(1): p. 64–77.

38. Frensley, W.R . Scharfetter-Gummel discretisation scheme for drift-diffusion equations. 2004; Available from: http://www.utdallas.edu/~frensley/minitech/ScharfGum.pdf.

39. Stewart, J., *Single Variable Calculus*. 2012, Pacific Grove, Calif.: Brooks/Cole Cengage Learning.

40. Malik, S.C. and S. Arora, *Mathematical Analysis*. 2010, New Delhi: New Age International Publishers.

41. Gummel, H.K., A self-consistent iterative scheme for one-dimensional steady state transistor calculations. *IEEE Transactions. on Electron Devices*, 1964. 11(10): p. 455–465.

42. Mock, M.S., Convergence of Gummels numerical algorithm. *Solid-State Electronics*, 1972. 15(1): p. 1–4.

43. Chin, S.P. and C.Y. Wu, A new methodology for 2-dimensional numerical-simulation of semiconductor-devices. *IEEE Transactions on Computer-Aided Design of Integrated Circuits and Systems*, 1992. 11(12): p. 1508–1521.

44. Perng, R.K., P.S. Lin, and C.Y. Wu, A new methodology for developing a fast 2-dimensional mosfet device simulator. *Solid-State Electronics*, 1991. 34(6): p. 635–647.

45. Bank, R.E. and D.J. Rose, Global approximate newton methods. *Numerische Mathematik*, 1981. 37(2): p. 279–295.

46. Bank, R.E. and D.J. Rose, Parameter selection for newton-like methods applicable to non-linear partial-differential equations. *SIAM Journal on Numerical Analysis*, 1980. 17(6): p. 806–822.

47. Bulashevich, K.A., et al., Simulation of visible and ultra-violet group-III nitride light emitting diodes. *Journal of Computational Physics*, 2006. 213(1): p. 214–238.

48. El Boukili, A. and A. Marrocco, Arclength continuation methods and applications to 2D drift-diffusion semiconductor equations. 1995, Institut National de Recherche en Informatique et en Automatique. p. 1–100.

49. Perng, R.K. and C.Y. Wu, A new algorithm for steady-state 2-D numerical-simulation of mosfets. *Solid-State Electronics*, 1990. 33(2): p. 287–293.

50. Akcasu, O.E., Convergence properties of newton method for the solution of the semiconductor transport-equations and hybrid solution techniques for multidimensional simulation of vlsi devices. *Solid-State Electronics*, 1984. 27(4): p. 319–328.

51. Smith, G.D., *Numerical Solution of Partial Differential Equations: Finite Difference Method*. 1988, Oxford: Oxford University Press.

52. Bank, R.E., et al., Transient simulation of silicon devices and circuits. *IEEE Transactions on Electron Devices*, 1985. 32(10): p. 1992–2007.

53. Mukherjee, J. and J.G. McInerney, Electrothermal analysis of CW high-power broad-area laser diodes: A comparison between 2-D and 3-D Modeling. *IEEE Journal of Selected Topics in Quantum Electronics*, 2007. 13(5): p. 1180–1187.

54. Szymanski, M., Two-dimensional model of heat flow in broad-area laser diode: Discussion of the upper boundary condition. *Microelectronics Journal*, 2007. 38(6–7): p. 771–776.

55. Szymanski, M., et al., Two-dimensional model of heat flow in broad-area laser diode mounted to a non-ideal heat sink. *Journal of Physics D-Applied Physics*, 2007. 40(3): p. 924–929.

56. Nakwaski, W. and M. Osinski, On the thermal resistance of vertical-cavity surface-emitting lasers. *Optical and Quantum Electronics*, 1997. 29(9): p. 883–892.

57. Nakwaski, W. and M. Osinski, Thermal-properties of etched-well surface-emitting semiconductor-lasers. *IEEE Journal of Quantum Electronics*, 1991. 27(6): p. 1391–1401.

7 Fibre Amplifiers and Lasers

Lanthanide ion–doped fibre optic devices have numerous applications. Commercially available fibre lasers cover wavelengths spanning from the visible range up to 2 μm, while mid-infrared fibre lasers are being developed in the research labs. When compared with other types of lasers, fibre lasers are characterized by a high quality of output beam and ease of beam delivery. Apart from fibre lasers, optical fibre amplifiers are of large practical importance. These devices play a prominent role in long haul telecom systems as signal regenerators.

In this chapter, we discuss the modelling of lanthanide ion–doped fibre amplifiers and lasers. We start with a general description of the theory of the light interaction with atoms in section one and then in the second section we continue with discussing the models of electronic levels in lanthanide ions. In the third section, we derive the equations describing the process of light amplification in a lanthanide ion–doped medium and introduce numerical techniques suitable for handling the resulting equations in the context of optical amplifiers. In the fourth and fifth sections, we discuss the extension of the presented theories to the modelling of fibre lasers and time domain analysis. In the sixth section, we discuss experimental methods used for extracting the modelling parameters. It is assumed that the reader is familiar with the basic principles of lanthanide doped fibre lasers and amplifiers [1–3].

PHOTONS AND ATOMS

Light interacts with atoms through three basic processes: spontaneous emission, stimulated emission, and absorption (Figure 7.1). During the process of spontaneous absorption, an atom makes a transition from a higher energy level to a lower one while releasing the energy difference in the form of a photon of energy $h\nu$ where ν is the photon frequency, while h is Planck's constant. The probability density of the spontaneous emission into an optical mode per atom, within a unit time, is $P = 1/\tau_{sp}$, where the spontaneous lifetime τ_{sp} is a parameter that is usually extracted from measurements. When a photon flux φ [photons/cm^{2}s] is incident onto an atom, the probability density of stimulated emission or absorption is $W = \varphi\, \sigma(\nu)$, where the function of photon frequency $\sigma(\nu)$ has units of area, and is hence referred to as the transition cross section. Similarly, as in the case of the spontaneous lifetime, the transition cross-section is extracted typically from measurements. A more detailed account of the photon-atom interaction can be found in Saleh and Teich [4].

Once the probability density of the stimulated emission is known, the equation describing the process of light amplification through stimulated emission can be

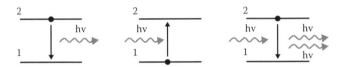

FIGURE 7.1   Interactions between atomic energy levels and a photon of energy $h\nu$.

derived. For this purpose we consider a fictitious cylinder with a cross-section A (Figure 7.2).

A photon flux φ is incident from the left side upon a section of the cylinder with a thickness dz. As a result of the interaction with the medium, via the process of stimulated emission, on the right side of the cylinder section the photon flux will increase by dφ. From the particle conservation principle, it follows that the number of photons generated per unit time within the cylinder section dz is equal to the net number of photons leaving the unit cylinder in the unit time. The number of atoms within the cylinder section is equal to the product NA dz, where N is the number of atoms per unit volume of the medium that are available for the stimulated emission. Consequently, the number of photons generated within the unit time in the cylinder section is equal to the product of the number of active atoms within the cylinder section and the probability of the stimulated emission W. Because the considered section of the cylinder is very short when compared with the transverse dimensions, we can neglect the photon leakage through the perimeter of the cylinder, and, hence, the net number of photons, leaving the cylinder per unit time, can be obtained from the difference between the photon flux entering and leaving the cylinder through the surface A, that is, dφ A. Thereafter, equating the number of photons generated in the cylinder to the net number of photons leaving the cylinder yields the equation:

$$N \ W \ A \ \mathrm{d}z = \mathrm{d}\phi \ A \tag{7.1}$$

Substituting into 7.1 $W = \varphi \ \sigma(\nu)$, and recasting, leads to the following equation that describes the process of the photon flux amplification by stimulated emission:

$$\frac{\mathrm{d}\phi(z)}{\mathrm{d}z} = N \ \sigma(\nu)\phi(z) \tag{7.2}$$

In 7.2 the dependence of the photon flux on z and of the transition cross-section on frequency is explicitly indicated. If we consider two energy levels with the energy difference equal to the photon energy (Figure 7.1), then the number of atoms per unit volume available for stimulated emission is equal to the number of atoms per unit volume that are in the energy level 2, which we denote by N_2. By following the same steps as in the case of the stimulated emission, we can observe that the stimulated absorption process is described by Equation 7.2 if N is interpreted as the number of atoms per unit volume that are available for the stimulated absorption. Again, in the case of the two level system considered in Figure 7.1, the number of atoms per unit

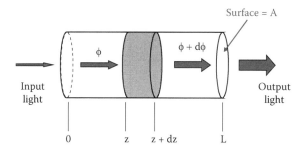

FIGURE 7.2 Photon flux amplification via stimulated emission process.

volume that are available for the stimulated absorption is equal to the number of atoms per unit volume that are in the energy level 1, which we denote by N_1. Finally, we can conclude that a photon flux that interacts with two energy levels (Figure 7.1) will experience the net gain $\gamma(v) = (N_2 - N_1)\,\sigma(v)$ according to the equation:

$$\frac{\mathrm{d}\phi(z)}{\mathrm{d}z} = \left(N_2 - N_1\right)\sigma(v)\phi(z) = \gamma(v)\phi(z) \tag{7.3}$$

Equation 7.3 shows that a net gain can only be achieved if $N_2 > N_1$. However, in the thermal equilibrium, the opposite condition is fulfilled because the level populations follow the Boltzman distribution. An inversion of the level population is therefore necessary.

Before 7.3 can be used for the calculation of the photon density distribution, one needs to know the spatial distribution of the volume concentrations of atoms in the levels 1 and 2. These atom concentrations are typically calculated using the rate equations approach. We have already discussed the rate equations approach with reference to laser diodes in the previous chapter. It should therefore be a fairly straightforward matter to introduce this method again in another context. For this purpose we will use an illustrating example that is borrowed from Saleh and Teich [4]. The considered energy level system consists of two specified energy levels (Figure 7.3).

The upper level is pumped at the rate R_2 while the lower level system is depleted at the rate R_1. The stimulated emission and absorption processes involve levels 1 and 2. The upper level is also depleted through the spontaneous emission to the lower lying level 1 with the time constant τ_{21} and to all lower lying levels with the time constant τ_{20}. Both processes can be combined and characterized by one time constant τ_2. The lower energy level is also depleted through the process of spontaneous emission to all lower lying levels with the time constant τ_1. The rate equations for changes to the atom concentrations in the levels 1 and 2 can thus be phenomenologically obtained. The rate of change of the atom concentration in level 2, $\mathrm{d}N_2/\mathrm{d}t$, is therefore proportional to the external pumping rate R_1, and the stimulated absorption rate (these terms come therefore with the positive sign in front of them) is inversely proportional to the stimulated emission and the combined spontaneous emission rate (which come with a negative sign in front):

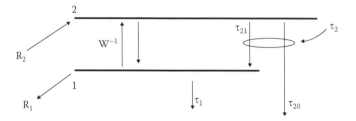

FIGURE 7.3 Schematic diagram of a two level system pumped at the rate R_2 into the upper energy level and depleted with the rate R_1 from the lover energy level.

$$\frac{dN_2}{dt} = R_2 + N_1 W - N_2 W - \frac{N_2}{\tau_2} \tag{7.4a}$$

Similarly, we can obtain the rate of the change of the atom concentration in level 1, dN_1/dt

$$\frac{dN_1}{dt} = \frac{N_2}{\tau_{21}} + N_2 W - R_1 - \frac{N_1}{\tau_1} - N_1 W \tag{7.4b}$$

where again the terms that feed into the level 1 appear with a positive sign in front of them while the terms that deplete the level 1 have a "minus" sign in front of them. If the rates R_1 and R_2 and all other constants are known, Equation 7.4 can be solved, thus providing the N_1 and N_2 concentrations for 7.3. The solution of 7.4 will strongly depend on all the parameters involved and the way the upper level is pumped and the lower level depleted. These parameters in turn depend on the number and other properties of all other energy levels involved. It is known from the theory of atom–photon interactions that in an optically pumped system the least number of the energy levels needed to achieve an inversion of population is 3. The so-called three-level system is the simplest energy level system that is used for the realization of the optical amplifiers and lasers. Figure 7.4 shows the schematic diagram of the three-level system.

The pump with the photon energy matching the energy difference between levels 1 and 3 feeds level 3 at the rate R. From level 3 the atoms quickly decay to level 2 by both radiative and nonradiative processes (therefore this transition is marked with a dash line arrow). The key issue for a three-level system is that the decay from level 3 to 2 is much faster than the spontaneous emission and nonradiative decay rates from level 2 to 1. If this condition is fulfilled, an inversion of population between levels 1 and 2 can be obtained and a lasing action at the corresponding wavelength can take place.

The three-level system has a major drawback, which results from the fact that the inversion of population is achieved with respect to the ground state. In the thermal equilibrium, the ground state is heavily populated and hence a strong pump is needed to achieve the inversion of population. To reduce the required pump laser power, a four-level system has been proposed (Figure 7.5). In the four-level system

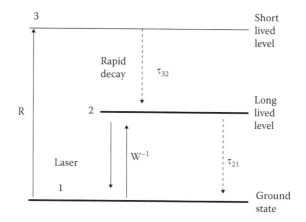

FIGURE 7.4 Schematic diagram of a three-level system pumped at rate R.

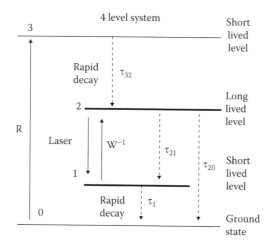

FIGURE 7.5 Schematic diagram of a four-level system pumped at rate R.

the inversion of population is achieved between levels 2 and 1 (Figure 7.5), that is, with respect to a higher lying level and not the ground state. The inversion of population can therefore be achieved with a lower pump power. However, the four-level system is in general more difficult to realize in practice. It requires, apart from the fast depopulation of the level 4, a method for an efficient removal of atoms from the lower lasing level 1 so that the inversion of population can be easily achieved and maintained. In the next section, we will show how a three-level system is realized using a silica glass host doped with erbium ions.

Finally, we use Figure 7.2 to derive phenomenologically the relationship between the photon flux density and the photon density of an optical waveguide mode. The rigorous derivation of this relationship can be found in Chapter 4 of Ebeling [5].

First, we consider photons contained in a section of a cylinder of length L and cross-section surface A that are travelling at the optical waveguide mode group velocity v_g (Figure 7.2). The total number of photons contained in this cylinder is $N_p L A$ where N_p is the photon density. Alternatively, we can calculate the total number of photons in the cylinder using the photon flux density φ. Thus we obtain $\varphi A L / v_g$. Equating both expressions for the total number of photons yields the following equation:

$$N_p L A = \phi \, A L \big/ v_g \qquad\qquad (7.5)$$

From 7.5 it follows immediately that:

$$N_p = \frac{\phi}{v_g} \qquad\qquad (7.6)$$

Equation 7.6 gives the required relationship between the photon flux and photon density.

SILICA GLASS–DOPED WITH ERBIUM IONS

Erbium–doped silica glass fibres find an application as optical amplifiers in long haul optical telecom systems that operate at the wavelength of approximately 1.55 μm. Erbium is incorporated into a silica glass host matrix in the form of trivalent ions, Er^{3+}. The atomic number of erbium is 68 while the electron configuration of an erbium atom is $1s^2\ 2s^2p^6\ 3s^2p^6d^{10}\ 4s^2p^6d^{10}f^{12}\ 5s^2p^6\ 6s^2$. When forming a trivalent ion, an erbium atom loses two electrons from the 6s shell and an electron from the 4f shell, and its electron configuration becomes: $1s^2\ 2s^2p^6\ 3s^2p^6d^{10}\ 4s^2p^6d^{10}f^{11}\ 5s^2p^6$. The 4f shell is the only shell that is not complete. All other shells are complete and there-fore not optically active [2]. Hence, in the ground state configuration the electrons from 4f shell are the ones that participate in the energy exchange with an external optical field. The 4f shell requires 14 electrons to be complete. Therefore, in Er^{3+} three electrons are missing from making this shell complete (these three 4f electrons should not be confused with the three electrons missing from the complete erbium atom configuration). It is known from the atomic spectroscopy theory that in the ground state configuration a 4f shell with three missing electrons results in 364 terms that correspond to various arrangements of 4f electrons [6]. The energy levels of lan-thanide ions are designated using the Russell-Saunders term notation [7]. In Er^{3+} ion the term symbols of the three lowest energy levels corresponding to 4f shell electrons are $^4I_{11/2}$, $^4I_{13/2}$, and $^4I_{15/2}$ [2]. We note that the derivation of these term symbols and the identification of particular energy levels of Er^{3+} photoluminescence spectra is in principle not trivial and can be found in the literature on atomic spectroscopy [7].

Figure 7.6 shows a schematic diagram of the three lowest lying energy levels of the erbium trivalent ion.

The energy difference between the ground state and the $^4I_{11/2}$ level corresponds to photon wavelength of about 980 nm. Low-cost, high power laser diodes at this wavelength are readily available in the AlGaAs material system. Therefore, efficient optical pumping can be easily realized in practice. Furthermore, the energy difference between the ground state and the $^4I_{13/2}$ level corresponds to the photon wavelength of about 1550 nm, which is of primary importance for long haul optical telecom systems. The other important fact is that the time constant τ_{32} is more than two orders of magnitude smaller than τ_{21}. This makes the Er^{3+} ion in a silica glass host, an example, of a classical three-level system; in which, an inversion of population can be achieved with a sufficiently strong pump (that is readily available). The reason for such large discrepancy between time constants is the difference in the physical process that facilitates both transitions. The transition between the level $^4I_{11/2}$ and $^4I_{13/2}$ is nonradiative. The energy difference for this transition is transferred to the host glass material via multiphonon emission, thus increasing the temperature of the host material (quantum defect). The transition between the level $^4I_{13/2}$ and $^4I_{15/2}$ on the other hand is radiative. The energy difference in this case is transferred to photons via the process of spontaneous emission. A comparative study of both these processes in the silica glass host is given in Desurvire [8].

Following the methodology outlined in the section "Photons and Atoms," we can write the rate equations for the energy level system shown in Figure 7.6 as:

$$\frac{dN_3}{dt} = -\frac{N_3}{\tau_{32}} + \left(N_1 - N_3\right)\phi_p\sigma_p$$

$$\frac{dN_2}{dt} = -\frac{N_2}{\tau_{21}} + \frac{N_3}{\tau_{32}} - \left(N_2 - N_1\right)\phi_s\sigma_s$$

$$\frac{dN_1}{dt} = \frac{N_2}{\tau_{21}} + \left(N_2 - N_1\right)\phi_s\sigma_s - \left(N_1 - N_3\right)\phi_p\sigma_p \qquad (7.7)$$

whereby N_1, N_2, and N_3 give the atom volume densities for the levels $^4I_{15/2}$, $^4I_{13/2}$, and $^4I_{11/2}$, respectively. The pump photon flux density and the corresponding transitions cross-section are denoted by φ_p and σ_p, while in the case of the signal wavelength (~1550 nm) they are denoted by φ_s and σ_s. The constants τ_{32} and τ_{21} characterise the photoluminescence decay time from the levels $^4I_{11/2}$ and $^4I_{13/2}$, respectively.

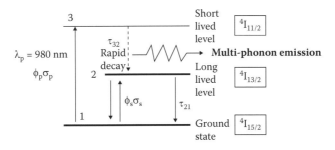

FIGURE 7.6 Schematic diagram of a three-level system of 980 nm pumped Er^{3+}.

First, let's consider the solution of 7.7 subject to constant pump and signal photon flux density. In such a case, the level populations do not vary with time, and the time derivatives on the left hand side of 7.7 are equal to zero:

$$0 = -\frac{N_3}{\tau_{32}} + \left(N_1 - N_3\right)\phi_p\sigma_p$$

$$0 = -\frac{N_2}{\tau_{21}} + \frac{N_3}{\tau_{32}} - \left(N_2 - N_1\right)\phi_s\sigma_s$$

$$0 = \frac{N_2}{\tau_{21}} + \left(N_2 - N_1\right)\phi_s\sigma_s - \left(N_1 - N_3\right)\phi_p\sigma_p \tag{7.8}$$

Equation 7.8 are linearly dependent; therefore, by solving them one cannot obtain unique values for the atomic populations in the three energy levels subject to given values of the signal and pump photon flux. Another equation needs to be added to create a system of equations with a unique solution. This equation follows from the observation that the sum of atom densities in all energy levels should be equal to the total Er^{3+} ion concentration in the silica glass host. When only three levels are considered, it follows that $N_1 + N_2 + N_3 = N$, where N is the total erbium dopant concentration. By replacing one of Equation 7.8, one obtains the following set of equations:

$$\left.\begin{array}{l} \dfrac{N_3}{\tau_{32}} + \left(N_1 - N_3\right)\phi_p\sigma_p = 0 \\[2mm] -\dfrac{N_2}{\tau_{21}} + \dfrac{N_3}{\tau_{32}} - \left(N_2 - N_1\right)\phi_s\sigma_s = 0 \\[2mm] N_1 + N_2 + N_3 = N \end{array}\right\} \tag{7.9}$$

that has a unique solution. It is noted that due to upconversion processes the last equation in 7.9 is only valid for moderate pump power levels and erbium dopant concentrations (this issue will be discussed in more detail in the last section of this chapter).

The set of algebraic Equation 7.9 tends to be numerically ill conditioned; therefore, an approximate analytical solution that is accurate within the typical parameter range is very useful. In the remaining part of this section, we will therefore derive an approximate solution following [2].

First, using the first equation we eliminate N_3 from the other two equations:

$$\left.\begin{array}{l} -\dfrac{N_2}{\tau_{21}} + \dfrac{1}{\tau_{32}} N_1 \dfrac{\phi_p\sigma_p}{\frac{1}{\tau_{32}} + \phi_p\sigma_p} - \left(N_2 - N_1\right)\phi_s\sigma_s = 0 \\[4mm] N_1 + N_2 + N_1 \dfrac{\phi_p\sigma_p}{\frac{1}{\tau_{32}} + \phi_p\sigma_p} = N \end{array}\right\} \tag{7.10}$$

Thus, we reduced the number of equations to two. Next we calculate the limit of 7.10 with τ_{32} tending to zero:

$$\left. \begin{array}{l} -\dfrac{N_2}{\tau_{21}} + N_1 \phi_p \sigma_p - \left(N_2 - N_1\right) \phi_s \sigma_s = 0 \\[3mm] N_1 + N_2 = N \end{array} \right\}$$

(7.11)

This is justified because τ_{32} is very small when compared with other parameters at typical operating conditions. From Equation 7.11 we can easily obtain N_1 and N_2:

$$N_1 = N \frac{1/_{\tau_{21}} + \phi_s \sigma_s}{1/_{\tau_{21}} + 2 \; \phi_s \sigma_s + \phi_p \sigma_p}$$

$$N_2 = N \frac{\phi_p \sigma_p + \phi_s \sigma_s}{1/_{\tau_{21}} + 2 \; \phi_s \sigma_s + \phi_p \sigma_p}$$

(7.12)

Because, τ_{32} is very small, we can safely assume that the $^4I_{11/2}$ level depopulates very quickly, and hence $N_3 = 0$. Therefore, we can obtain the following approximations for the populations' differences:

$$N_3 - N_1 \approx -N_1 = -N \frac{1/_{\tau_{21}} + \phi_s \sigma_s}{1/_{\tau_{21}} + 2 \; \phi_s \sigma_s + \phi_p \sigma_p}$$

$$N_2 - N_1 = N \frac{- 1/_{\tau_{21}} + \phi_p \sigma_p}{1/_{\tau_{21}} + 2 \; \phi_s \sigma_s + \phi_p \sigma_p}$$

(7.13)

Equation 7.13 can be used to calculate the level populations subject to given values of the pump and signal photon flux. These, however, are not known for an optical fibre amplifier and have to be calculated self-consistently subject to known incident pump and signal powers. We will discuss this problem in the next section.

FIBRE AMPLIFIER MODELLING

In this section we present models of fibre amplifiers and discuss the details of their software implementation. There are two basic configurations that are used for light amplification in fibre amplifiers: the copropagating pump configuration (Figure 7.7a) and the counterpropagating pump one (Figure 7.7b). We focus particularly on the co-propagating pump configuration. In this case the signal and pump beams are combined together at one end of the fibre section and launched into the lanthanide ion doped fibre of length L. During the propagation along the fibre, the pump transfers

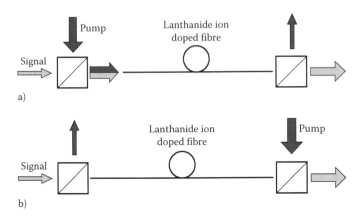

FIGURE 7.7 Schematic diagram of lanthanide ion–doped fibre amplifier in a) copropagating and b) counterpropagating pump configuration.

its energy to the signal, which results in signal amplification. In this section we first derive the equations, then describe optical beam propagation in a lanthanide ion doped medium, and then we solve these equations for the case of an erbium ion doped amplifier. In the second subsection, we discuss the handling of the amplified spontaneous emission.

COPROPAGATING AND COUNTERPROPAGATING PUMP FIBRE AMPLIFIER MODELS

In this subsection, we consider fibre amplifier models that allow for the calculation of the pump and the amplified signal power distribution along the fibre. To calculate the photon flux distribution along the fibre, first, we need to derive an equation that describes the optical beam evolution in the lanthanide ion–doped medium. We consider a step index circular fibre doped with lanthanide ions, take the advantage of the small refractive index difference between the core and the cladding of a typical core-cladding glass fibre, and consequently apply the scalar approximation. Furthermore, we assume that the gain resulting from doping the silica glass with lanthanide ions introduces only a small perturbation $\Delta n(x,y)$ to the refractive index distribution $n(x,y)$ of the fibre [3,9], and that the pump and the signal propagate in the fundamental mode of the unperturbed waveguide. Thereafter, we can represent the field distribution in the product form: $\Psi(x,y,z) = A(z)*F(x,y)*e^{j(\omega t - \beta z)}$, where $F(x,y)$ is the fundamental mode distribution of the unperturbed waveguide, $A(z)$ is a slowly changing function of z, and β is the fundamental mode propagation constant. To define $\Psi(x,y,z)$ uniquely, we impose on $F(x,y)$ the condition: $\iint F^*(x,y)F(x,y)\,dxdy = I_N$, where "*" stands for a complex conjugate and I_N is a normalization constant that can assume an arbitrary value, for example, 1. We substitute $\Psi(x,y,z)$ into the scalar wave Equation 2.1 and obtain:

$$-2j\beta\frac{dA(z)}{dz}F(x,y)+k_0^2\left(2n\Delta n\right)A(z)F(x,y)=0 \tag{7.14}$$

where k_0 stands for the wave number. In 7.14 we neglected the second derivative of the slowly varying function A, observed that for the fundamental mode distribution $F(x,y)$:

$$\frac{\partial^2 F}{\partial x^2} + \frac{\partial^2 F}{\partial y^2} + \left(k_0^2 n^2 - \beta^2 \right) F = 0,$$

and neglected Δn^2 when compared with $2n\Delta n$. In the next step, we multiply 7.14 by F^* and integrate the equation over the transverse plane. Because the integral of $F^*(x,y)$ $F(x,y)$ is normalized to I_N, we obtain:

$$I_N \frac{dA(z)}{dz} = -jk_0 I_N \overline{\Gamma} A(z) \tag{7.15}$$

where $\overline{\Gamma}$ is defined in the following way:

$$\overline{\Gamma} = \frac{k_0}{\beta} \iint F^*(x,y) n\Delta n F(x,y) dx dy \bigg/ I_N$$

Next we identify the imaginary Δn_I and real Δn_R part of the refractive index perturbation: $\Delta n = \Delta n_R + j\,\Delta n_I$ and consistently split $\overline{\Gamma}$ into its real and imaginary parts:

$$\overline{\Gamma}_R = \frac{k_0}{\beta} \iint F^*(x,y) n\Delta n_R F(x,y) dx dy \bigg/ I_N$$

$$\overline{\Gamma}_I = \frac{k_0}{\beta} \iint F^*(x,y) n\Delta n_I F(x,y) dx dy \bigg/ I_N$$

which allows recasting 7.15 into:

$$I_N \frac{dA(z)}{dz} = -jk_0 I_N \left(\overline{\Gamma}_R + j\overline{\Gamma}_I \right) A(z) \tag{7.16}$$

From Table 3.3 we obtain the relationships between Ψ and the photon flux φ and the total power P:

$$P = \iint \Psi^* \Psi dx dy = A^* A\, I_N \tag{7.17a}$$

$$\phi = \Psi^* \Psi / h\nu = A^* A\, F^* F / h\nu = P\, F^* F / (I_N h\nu) \tag{7.17b}$$

By multiplying 7.16 by A^* and its complex conjugate by A and adding both equations together, we obtain the equation for the evolution of the power guided by the waveguide:

$$\frac{dP(z)}{dz} = 2k_0 \bar{\Gamma}_I P(z) \tag{7.18}$$

If the scalar field depends locally on z according to $\exp(-jk_0(n+\Delta n_R+j\Delta n_I))$ while the power varies according to $\exp(g_m z)$, where g_m stands for the material gain, then it follows from 7.17b that $\Delta n_I = g_m/(2k_0)$. Therefore, we can recast 7.18 into:

$$\frac{dP(z)}{dz} = \hat{\Gamma} P(z) \tag{7.19a}$$

where

$$\hat{\Gamma} = \frac{k_0}{\beta} \iint F^*(x,y)\, n\, g_m F(x,y)\, dxdy \Big/ I_N \tag{7.19b}$$

In the case of an erbium-doped amplifier, whereby pump wave is interacting with one signal wave, from 7.19a one obtains a set of two equations:

$$\frac{dP_s(z)}{dz} = \hat{\Gamma}_s(z) P_s(z) \tag{7.20a}$$

$$\frac{dP_p(z)}{dz} = \hat{\Gamma}_p(z) P_p(z) \tag{7.20b}$$

where

$$\hat{\Gamma}_s(z) = \frac{k_0}{\beta} \iint F_s^*(x,y)\, n(x,y,z)\sigma_s \big(N_2(x,y,z) - N_1(x,y,z)\big) F_s(x,y)\, dxdy \Big/ I_N \tag{7.21a}$$

$$\hat{\Gamma}_p(z) = \frac{k_0}{\beta} \iint F_p^*(x,y)\, n\,(x,y,z)\sigma_p \big(N_3(x,y,z) - N_1(x,y,z)\big) F_p(x,y)\, dxdy \tag{7.21b}$$

In 7.20 and 7.21 we state all the function arguments explicitly to ease the understanding of the following discussion. To link 7.20 and 7.21 with 7.13 one needs a relationship that links the photon flux density with power. This can be obtained directly from 7.17b:

$$\phi_s = P_s \big(F_s^* F_s\big) \Big/ \big(I_N h \nu_s\big) \tag{7.22a}$$

$$\phi_p = P_p \left(F_p^* F_p \right) \Big/ \left(I_N h \nu_p \right) \tag{7.22b}$$

Thus, Equation 7.20 together with 7.21 and 7.22 allow for the calculation of the longitudinal distribution of signal and pump power within a fibre amplifier. To form a complete set, that can be solved self-consistently, Equation 7.20 has to be complemented by a set of equations that allows calculating the ionic level populations N_1 and N_2. In the case of erbium doping and the pumping wavelength of 980 nm, Equation 7.13 can be used for this purpose.

If we consider erbium ion–doped silica glass fibre amplifier as a specific example, then Equations 7.13 and 7.20 form a complete set that can be solved self-consistently. To obtain the numerical solutions of 7.13 and 7.20, we divide the fibre into M sections (Figure 7.8).

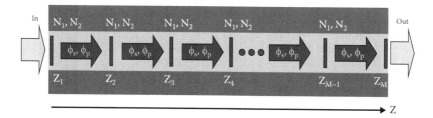

FIGURE 7.8 Longitudinal discretization.

ALGORITHM 7.1 1D ERBIUM-DOPED FIBRE AMPLIFIER MODEL FOR A COPROPAGATING PUMP CASE

1. Start
2. Set the values of all parameters
3. Set the longitudinal step Δz
4. Set the dimensions of the transverse computational window
5. Set the values of the incident pump and signal powers
6. Calculate the fundamental mode distribution for pump and signal wavelength
7. Calculate the incident pump and signal photon flux 7.22
8. Initialize z: $z = 0$
9. For the current value of the longitudinal position z calculate the transverse distribution of N_1 and N_2 using 7.13
10. Calculate the transverse distribution of gain coefficients 7.21
11. Calculate pump and signal powers at $z + \Delta z$ from 7.20
12. Update the value of z: $z = z + \Delta z$
13. If $z = L$ go to 15
14. Go to 9
15. Stop

In the computer memory, we only store the level populations and the values of the photon flux density at the discrete positions: z_1, z_2, and so on. We consider first an optical fibre amplifier whereby the signal wave is copropagating with the pump wave (Figure 7.7a). Furthermore, we assume that fibre ends are antireflection coated, and hence the residual wave reflection at the fibre–air interface is negligibly small. Under these assumptions the pump and signal power distributions within the fibre amplifier can be calculated starting from the known values of the incident pump and signal power at one end of the fibre, that is, at $z = 0$ (Figure 7.7a). If the first derivatives of the pump and signal powers with respect to z are approximated by a forward finite difference, Algorithm 7.1 is obtained.

The MATLAB implementation of Algorithm 7.1 is given next.

```
% erbium doped fibre amplifier co-propagating pump
% using the full calculation of transverse photon flux
% distribution

clear % clears variables
clear global % clears global variables
format long e
% initial constants
i = sqrt(-1);% remember not to overwrite pi or i !
pi = 3.141592653589793e+000;

fiber_scalar_beta_s

scalar_fibre_plot1D

FF_s = f.*f;

fiber_scalar_beta_p

scalar_fibre_plot1D

FF_p = f.*f;

subplot(3,1,1)

r = r*1e-4;% converting microns to cm for radius
a = a*1e-4;% and fibre radius

plot(r,FF_s,'b',r,FF_p,'r')

xlabel('r/micrometers','Fontsize',12)
ylabel('Intensity/a.u.','Fontsize',12)

h = 6.626e-34;%[Js]
v0 = 3e10;%[cm/s]
q = 1.609e-19;%[C]
sig_p = 2.7e-21;% pump absorption cross section [cm2]
sig_s = 7e-21;% signal emission cross section [cm2]
```

```
N = 0.7e19;% erbium ion concentration [1/cm3]
L = 1500;% amplifier length [cm]
R =.5e-4;% doping radius [cm]
if R > a
'doping radius larger than fibre core radius'
end
r_mask = r < R;% radius mask for integral calculation
dr = r(2)-r(1);% sampling step for integral calculation
A = pi*R*R;

gamma_s = sum(r_mask.*r.*FF_s)/(sum(r.*FF_s));
gamma_p = sum(r_mask.*r.*FF_p)/(sum(r.*FF_p));

lamp = 0.98e-4;% pump wavelength[cm]
lams = 1.53e-4;% signal wavelength[cm]
freqp = v0/lamp;%[1/s]
freqs = v0/lams;%[1/s]
tau21 = 9e-3;%[s]
Pp_ini = 0.05;%pump power [W]
Ps_ini = 1.e-6;%signal power [W]

% iteration loop
N_of_zsecs = 10000% number of z sections
dL = L/N_of_zsecs;% z step
z_pos = dL*(1:1:N_of_zsecs);

Ps = Ps_ini*ones(1,N_of_zsecs);
%initial signal wave relative intensity
Pp = Pp_ini*ones(1,N_of_zsecs);
%initial pump wave relative intensity

for k = 1:N_of_zsecs-1
Fi_s = Ps(k)*FF_s/((h*freqs)*2*pi*dr*sum(r.*FF_s));
Fi_p = Pp(k)*FF_p/((h*freqp)*2*pi*dr*sum(r.*FF_p));
mian = sig_p*Fi_p+2*sig_s*Fi_s+1/tau21;
N2mN1 = (sig_p*Fi_p-1/tau21).*r_mask*N./mian;
N3mN1 = -(sig_s*Fi_s+1/tau21).*r_mask*N./mian;
Gam_s = nc*sum(sig_s*N2mN1.*r.*FF_s)/(nin*sum(r.*FF_s));
Gam_p = nc*sum(sig_p*N3mN1.*r.*FF_p)/(nin*sum(r.*FF_p));
Ps(k+1) = Ps(k)+dL*Gam_s*Ps(k);
Pp(k+1) = Pp(k)+dL*Gam_p*Pp(k);
end
```

The program uses three external files: fiber_scalar_beta_s.m, fiber_scalar_beta_p.m, and scalar_fibre_plot1D.m. The listings for fiber_scalar_beta_s.m and scalar_fibre_plot1D.m are given in the following. The file fiber_scalar_beta_p.m calculates the propagation constant at the pump wavelength and can be easily obtained by changing the wavelength in fiber_scalar_beta_s.m to the pump wavelength. More detailed description of the propagation constant calculation in optical fibres is given in the section "Optical Fibres."

The content of the file fiber_scalar_beta_s.m:

```
% calculating beta for signal wavelength

lam = 1.53;%wavelength [micrometers]
nf = 1.45;
% refractive index in core IMAGINARY PART MUST BE ZERO!
nc = 1.44;% refractive index in cladding

a = 3.45;% waveguide radius [micrometers]

% calculation parameters
k = 2.0*pi/lam;% wavenumber
V = a*k*sqrt(nf^2-nc^2);% normalised frequency V
mode_order = 0;% LP mode order

% Calculation of initial interval locating B for bisection
% method
b = (0.00199999:0.001:0.9999999);
% Dispersion equation
u = sqrt(1.0-b)*V;
w = sqrt(b)*V;
DE = u.*besselj(mode_order+1,u)./besselj(mode_order,u)-…
w.*besselk(mode_order+1,w)./besselk(mode_order,w);

% precise location of B
length = size(b);
for j = length(2):-1:2 % size produces a 2 el. vector with
% first element being vert dim and second long dim hence
% length(2) not length
   if DE(j)/DE(j-1)<0
      break
   end
end


btop = b(j);
bbot = b(j-1);
jbrowse = j-1;
utop = sqrt(1.0-btop)*V;
wtop = sqrt(btop)*V;
ubot = sqrt(1.0-bbot)*V;
wbot = sqrt(bbot)*V;

% bisection method for root finding starting from the initial
% interval calculated above
DEtop = utop*besselj(mode_order+1,utop)/besselj…
(mode_order,utop)-wtop*besselk(mode_order+1,wtop)…
/besselk(mode_order,wtop);
DEbot = ubot*besselj(mode_order+1,ubot)/besselj…
(mode_order,ubot)-wbot*besselk(mode_order+1,wbot)…
```

```
/besselk(mode_order,wbot);
jmax = 100; % maximum number of bisections
tolerance = 0.000000000001; % tolerance for finding B

for j = 1:jmax
    bcen = (btop-bbot)/2+bbot;
    ucen = sqrt(1.0-bcen)*V;
    wcen = sqrt(bcen)*V;
    DEcen = ucen*besselj(mode_order+1,ucen)/besselj...
    (mode_order,ucen)-wcen*besselk(mode_order+1,wcen)...
    /besselk(mode_order,wcen);

    if abs(DEcen) < 1.d-15%added to avoid division by nearly zero
        break
    end

    if DEtop/DEcen > 0
        btop = bcen;
        DEtop = DEcen;
        else
        bbot = bcen;
        DEbot = DEcen;
    end

    delta = btop-bbot;
    if delta < tolerance
        break
    end
end

%end of bisection method loop

% checks for tolerance
if j = =jmax
    disp('maximum number of bisections exceeded before...
    reaching tolerance')
end

if jbrowse = =1
% disp('no zero found')
    out = 0;
else
    out = bcen;
end
% variable 'out' stores the relative propagation constant B

nin = abs(sqrt(out*(nf*nf-nc*nc)+nc*nc));
% calculation of effective index
'V B neff beta'
[V out nin nin*k]% plotting results in one line
```

The content of the file scalar_fibre_plot1D.m:

```
% program calculates 1D scalar potential distribution in a
% circ fibre using scalar approach

total_rad = 4*a;

u = sqrt(nf*nf*k*k-nin*nin*k*k);
v = sqrt(nin*nin*k*k-nc*nc*k*k);

B = 1;% B can be set to 1 while A needs to be calculated to
% impose%continuity
A = besselk(0,v*a)/besselj(0,u*a);

r1 = 0:0.00005:a;
f1 = A*besselj(0,u*r1);

r2 = a:0.00005:total_rad;
f2 = B*besselk(0,v*r2);

r = [r1 r2];
f = [f1 f2];

%normalisation for convenient plotting only

f = f/sqrt(f*f');
```

In steps 9 and 10 of Algorithm 7.1, there is a need to calculate the distributions of N_1, N_2, and of the gain coefficient within the entire transverse computational window. The rectangle numerical integration rule was used to calculate the transverse integrals. This requires an introduction of a sufficiently dense transverse grid. The storage and use of the photon flux density and population transverse distributions has its implications for the computer memory needed and the calculation time. However, if the fibre is doped with a constant dopant density within a small radius R around the centre of the fibre, one can assume that the pump and signal photon flux densities and N_1 and N_2 are nearly constant within the doped area. Thus the gain coefficient distributions 7.21 can be also assumed flat (i.e., independent of x and y) within the doped region. Under these assumptions the gain distribution becomes independent of x and y within the lanthanide ion–doped area A and is equal to zero outside. Because the refractive index within the doped area is constant and equal to n_c, it can also be extracted from under the integration sign in 7.19b, and hence 7.19a reduces to the following equation:

$$\frac{dP(z)}{dz} = \frac{n_c k_0}{\beta} \Gamma g_m(z) P(z) \approx \Gamma g_m(z) P(z) \tag{7.23}$$

where Γ is the confinement factor:

$$\Gamma = \iint_A F^*(x,y)F(x,y)\,dxdy \tag{7.24}$$

We note that for a circular step index fibre, if the doped area A is a circle with radius R, the calculation of the confinement factor can be simplified by taking advantage of the circular symmetry of the fundamental mode light intensity distribution:

$$\Gamma = \int_{r=0}^{r=R} F^*(r)F(r)rdr \bigg/ \int_{r=0}^{r=\infty} F^*(r)F(r)rdr$$

where r is the radial coordinate: $r = \sqrt{x^2 + y^2}$. Again, if the pump is only interacting with a single signal wave, then the equations for the pump power P_p and signal power P_s can be obtained from 7.23:

$$\frac{dP_s(z)}{dz} = \Gamma_s \sigma_s \left(N_2(z) - N_1(z) \right) P_s(z) \tag{7.25a}$$

$$\frac{dP_p(z)}{dz} = \Gamma_p \sigma_p \left(N_3(z) - N_1(z) \right) P_p(z) \tag{7.25b}$$

whereby N_1, N_2, and N_3 are, consistently with the assumptions made, functions of longitudinal position only. To complete the derivation a method has to be established that links 7.25 with 7.13. The standard approach, that is, well established in the literature, relies on calculating the constant values of the photon flux densities in the core from:

$$\varphi_{s,const} = \Gamma_s P_s / (Ah\nu_s) \tag{7.26a}$$

$$\varphi_{p,const} = \Gamma_p P_p / (Ah\nu_p) \tag{7.26b}$$

In the case of erbium ions pumped at 980 nm, the populations N_1, N_2, and N_3 are calculated directly from 7.13 with the values of φ_s and of φ_p substituted for φ_s and φ_p. To complete the derivation we note that 7.26 can be obtained by integrating 7.22. Thus, integrating first the right-hand side (RHS) of 7.22 over the entire transverse plane gives:

$$RHS = P_s / (h\nu_s)$$

Similarly for the left hand side (LHS) we can write:

$$LHS = \iint \phi_s \, dxdy = \iint \phi_s \, dxdy \, \frac{\displaystyle\iint_A \phi_s \, dxdy}{\displaystyle\iint_A \phi_s \, dxdy} = \Gamma \iint_A \phi_s \, dxdy \approx \Gamma \phi_s \left(x = 0, y = 0 \right) A$$

The last approximation is obviously, plausibly accurate if the doping area surface A is small and the photon flux density distribution is fairly flat within the doping area. Finally, equating LHS and RHS yields Equation 7.26a. Similarly we can obtain 7.26b.

Equation 7.25 is used in many lanthanide-doped optical fibre devices for the calculation of the pump and signal power distributions. Similarly, to 7.20 they need to be combined with equations that allow for the calculation of the ionic level populations.

In the case of erbium ion doping, Equation 7.26 can be used to combine 7.13 and 7.22 into a set of two ordinary differential equations:

$$\frac{dP_s}{dz} = \frac{\dfrac{\sigma_p P_p \Gamma_p}{Ah\nu_p} - \dfrac{1}{\tau_{21}}}{\dfrac{\sigma_p P_p \Gamma_p}{Ah\nu_p} + 2\dfrac{\sigma_s P_s \Gamma_s}{Ah\nu_s} + \dfrac{1}{\tau_{21}}} N \, \Gamma_s P_s \sigma_s$$

$$\frac{dP_p}{dz} = -\frac{\dfrac{\sigma_s P_s \Gamma_s}{Ah\nu_s} + \dfrac{1}{\tau_{21}}}{\dfrac{\sigma_p P_p \Gamma_p}{Ah\nu_p} + 2\dfrac{\sigma_s P_s \Gamma_s}{Ah\nu_s} + \dfrac{1}{\tau_{21}}} N \, \Gamma_p P_p \sigma_p \qquad (7.27)$$

Equation 7.27 forms a set of two ordinary differential equations and hence can be solved numerically in a much more efficient way than Equation 7.20 combined with 7.13. Furthermore, 7.27 is equally applicable to step index fibres as to photonic crystal fibres, if only the dopants are distributed near the fibre centre only, and the scalar approximation is applicable.

Additionally, we note that the real part of the refractive index perturbation is related through the Kramers-Kroning equations with the imaginary part, that is, not present in any way in 7.23 and 7.27. This is the consequence of the assumption made at the beginning, namely, that the light propagates in the fundamental mode of the unperturbed waveguide. In other words it is assumed that the real part of the refractive index perturbation does not modify the transverse field distribution of the fundamental fibre mode. It should be noted that this approximation is usually well justified for classical core–clad fibres but should be used with care in the case of microstructured fibres.

For the sake of completeness, we check if 7.27 fulfils the energy conservation principle. This can be shown by proving that 7.27 preserves the total number of photons per unit time. To this end we divide both sides of 7.27 by the respective photon energies and add them together:

> **ALGORITHM 7.2   1D ERBIUM DOPED FIBRE AMPLIFIER MODEL FOR A COPROPAGATING PUMP CASE**
>
> 1. Start
> 2. Set the values of all parameters, including the confinement factor 7.24
> 3. Set the longitudinal step Δz
> 4. Set the values of the incident pump and signal powers
> 5. Calculate photon flux density for pump and signal from 7.26
> 6. Initialise z: $z = 0$
> 7. For the current value of the longitudinal position z calculate the level populations using 7.13
> 8. Calculate pump and signal powers at $z + \Delta z$ from 7.27
> 9. Update the value of z: $z = z + \Delta z$
> 10. If $z = L$ go to 12
> 11. Go to 7
> 12. Stop

$$\frac{d\left(\dfrac{P_s}{h\nu_s} + \dfrac{P_p}{h\nu_p}\right)}{dz} = \frac{-\dfrac{1}{\tau_{21}}}{\dfrac{\sigma_p P_p \Gamma_p}{A h\nu_p} + 2\dfrac{\sigma_s P_s \Gamma_s}{A h\nu_s} + \dfrac{1}{\tau_{21}}} N\left(\Gamma_s \frac{P_s}{h\nu_s}\sigma_s + \Gamma_p \frac{P_p}{h\nu_p}\sigma_p\right) \qquad (7.28)$$

The right hand side of 7.28 tends to zero if τ_{21} tends to infinity, which is equivalent to neglecting the spontaneous emission. Thus, Equation 7.27 fulfils the energy conservation principle only if the spontaneous emission is neglected in 7.13.

The fact that Equation 7.27 does not fulfil the energy conservation principle may seem surprising. However, it should be noted that the lack of energy conservation inherent in Equation 7.27 is not an artefact introduced while constructing the mathematical model but an accurate reflection of the physical processes involved in the light amplification by lanthanide ions. It results from the fact that the spontaneous emission that accompanies the light amplification process has not been included in Equation 7.20 and consequently also in 7.25.

The numerical solution of Equation 7.27 for an optical fibre in which the signal and pump waves copropagate and the residual reflections are neglected results in Algorithm 7.2.

The MATLAB implementation of the Algorithm 7.2 is given next. Again, as for Algorithm 7.1, a simple explicit Euler method was used to integrate numerically the ordinary differential equations.

```
% erbium doped fibre amplifier co-propagating pump

clear
format long e
```

```
pi = 3.141592653589793e+000;

h = 6.626e-34;%[Js]
v0 = 3e10;%[cm/s]
q = 1.609e-19;%[C]
sig_p = 2.7e-21;% pump absorption cross section [cm2]
sig_s = 7e-21;% signal emission cross section [cm2]
N = 0.7e19;% erbium ion concentration [1/cm3]
L = 1500;% amplifier length [cm]
gamma_p = 9.349790276669653e-001;
%confinement factor at pump wavelength
gamma_s = 8.191686995681956e-001;
%confinement factor at signal wavelength
R = 3.4e-4;% doping radius [cm]
A = pi*R*R;

lamp = 0.98e-4;% pump wavelength[cm]
lams = 1.53e-4;% signal wavelength[cm]
freqp = v0/lamp;%[1/s]
freqs = v0/lams;%[1/s]
tau21 = 9e-3;%[s]
Pp_ini = 0.05;%pump power [W]
Ps_ini = 1.e-6;%signal power [W]

% iteration loop
N_of_zsecs = 1000% number of z sections
dL = L/N_of_zsecs;% z step
z_pos = dL*(1:1:N_of_zsecs);

Ps = Ps_ini*ones(1,N_of_zsecs);
%initial signal wave relative intensity
Pp = Pp_ini*ones(1,N_of_zsecs);
%initial pump wave relative intensity

for k = 1:N_of_zsecs-1
Ss = sig_s*Ps(k)*gamma_s/(A*h*freqs);
Sp = sig_p*Pp(k)*gamma_p/(A*h*freqp);
mian = Sp+2*Ss+1/tau21;
Ps(k+1) = Ps(k)+dL*(Sp-1/tau21)*sig_s*Ps(k)*N*gamma_s/mian;
Pp(k+1) = Pp(k)-dL*(Ss+1/tau21)*sig_p*Pp(k)*N*gamma_p/mian;
end
```

The comparison of the MATLAB implementation of the Algorithms 7.1 and 7.2 shows immediately the benefits of using the simplified approach apart from the obvious reduction in the computer memory needed and the calculation time. To gain insight into the accuracy of the simplified Algorithm 7.1, when compared with 7.2, we consider an erbium ion doped silica glass fibre amplifier that is made of 15 metres of uniformly doped fibre, with the doping radius R varying between 0.5 and 3.45 µm. The pump operates at the wavelength of 980 nm while the signal's operating wavelength is 1530 nm. The pump and signal transition cross sections are 2.7×10^{-21} cm^2

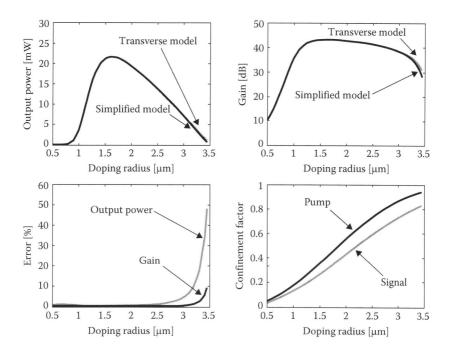

FIGURE 7.9 Comparison of the amplifier output power and gain calculated by Algorithms 7.1 and 7.2.

and 7×10^{-21} cm², respectively [2]. We assume that the refractive index of the core is 1.45 while in the cladding the refractive index is 1.44 at both wavelengths. The step-index circular fibre radius is 3.45 μm, which ensures that the fibre is single mode at 1.53 μm. The confinement factors for the pump and the signal are 0.64 and 0.4, respectively. The erbium ion concentration is 0.7×10^{19}/cm³. The spontaneous lifetime τ_{21} is assumed equal to 9 ms [2], which is a typical value observed in experiments. The input signal power equals 0.1 μW while the pump power is 50 mW. From the product of the dopant concentration and the cross-sections, we can obtain the upper limit for the signal and pump gain, which is 0.049/cm for the signal wave and 0.0189/cm for the pump wave. The corresponding imaginary parts of the refractive index perturbations Δn are 0.39×10^{-6} and -0.15×10^{-6} for the signal and pump wave, respectively. These values are much smaller than the value of the refractive index contrast for a typical step index single mode silica fibre, thus confirming Δn is a minor perturbation to the refractive index. In the case of photonic crystal fibres co-dopants are added to cancel out the effect of the erbium ion induced refractive index perturbation on the guiding properties of the fibre.

Figure 7.9 shows the dependence of the output signal power P_{out} and amplifier gain defined as 10 log (P_{out}/P_{in}) on the doping radius where P_{in} is the input signal power. Additionally, the dependence of the relative error for both the output signal power and the gain, and of the confinement factors, on the doping radius is given.

The transverse discretization step used in these calculations is 0.0001 μm while the differential equations were integrated with a longitudinal step of 0.1500 cm. The relative error is obtained by calculating the absolute value of the relative difference between the results obtained using the Algorithms 7.1 and 7.2. As expected the error increases with the doping radius. For the output signal power, it exceeds 10% for doping radius larger than 3 μm, which corresponds to the signal confinement factor values larger than 0.7. For small values of the confinement factor (and the doping radius), the difference between the results obtained using Algorithms 7.1 and 7.2 is small. However, there is a small growth of the error for very small values of the confinement factor, which results from the errors introduced by the transverse discretization.

In the case of the counterpropagating pump configuration (Figure 7.7b), Equation 7.27, in the absence of residual reflections at the fibre end facets, have to be solved subject to the following boundary conditions:

$$P_p(z = L) = P_{pump}$$

$$P_s(z = 0) = P_{signal}$$

where P_{pump} is the pump power and P_{signal} is the incident signal power. The value of pump power at $z = 0$ is therefore not known and the problem is essentially a two point boundary value problem, which is usually solved using the shooting method. We discuss this technique in more detail in section "Fibre Laser Modelling."

AMPLIFIED SPONTANEOUS EMISSION

The light amplification in optical fibres is accompanied by the process of amplified spontaneous emission Amplified spontaneous emission has a large impact on the operation of the optical fibre amplifiers [10].

In this subsection, we discuss how the amplified spontaneous emission can be included in the fibre amplifier model. As an illustrating example, we use an Yb³⁺ doped silica glass fibre amplifier. We apply a two level ytterbium ion model: $^2F_{5/2}$ and $^2F_{7/2}$ with pumping at 915 nm, for example [11]. Considering a comb of signal wavelengths λ_k with a constant wavelength spacing $\Delta\lambda$, the two level representation of the energy level structure results in a set of two equations [11]:

$$\begin{bmatrix} a_{11} & a_{12} \\ 1 & 1 \end{bmatrix} * \begin{bmatrix} N_1 \\ N_2 \end{bmatrix} = \begin{bmatrix} 0 \\ N \end{bmatrix} \tag{7.29}$$

that allows for calculation of the ionic level populations. In 7.29 the coefficients a_{xx} are as follows:

$$a_{11} = \frac{\Gamma_p \lambda_p \sigma_a(\lambda_p)}{Ahc} \left[P_p^+ + P_p^- \right] + \frac{\Gamma_s}{Ahc} \sum_{k=1}^{K} \lambda_k \sigma_a(\lambda_k) \left[P^+(\lambda_k) + P^-(\lambda_k) \right]$$

$$a_{12} = -\frac{\Gamma_p \lambda_p \sigma_e (\lambda_p)}{Ahc}\left[P_p^+ + P_p^-\right] - \frac{\Gamma_s}{Ahc}\sum_{k=1}^{K}\lambda_k \sigma_e (\lambda_k)\left[P^+(\lambda_k) + P^-(\lambda_k)\right] - \frac{1}{\tau}$$

where τ is the lifetime of level $^2F_{5/2}$. $P_p^{\pm}$ and $P^{\pm}(\lambda_k)$ are the optical powers for the pump and signal waves, respectively. The signs "+" and "−" refer to forward and backward travelling waves. A is the doping cross-section, h is Planck's constant, c is the speed of light in free space, $\Gamma_{s/p}$ is the confinement factor for signal/pump, and $\sigma_{a/e}$ is the absorption/emission cross-section for the $^2F_{5/2}$–$^2F_{7/2}$ transition. It is noted that, again, the approximate formulae 7.26 were adopted to relate the power with the photon flux.

Equation 7.28 is complemented by a set of ordinary differential equations that describe the evolution of the pump and signal power:

$$\frac{dP_p^{\pm}}{dz} = \mp\Gamma_p\left[\sigma_a(\lambda_p)N_2 - \sigma_e(\lambda_p)N_1\right]P_p^{\pm} \mp \alpha\ P_p^{\pm} \tag{7.30a}$$

$$\frac{dP_s^{\pm}}{dz} = \mp\Gamma_s\left[\sigma_a(\lambda_s)N_2 - \sigma_e(\lambda_s)N_1\right]P_s^{\pm} \mp \alpha\ P_s^{\pm} \tag{7.30b}$$

and by the following equations that account for the build-up of power in the waves that are sourced by the process of the amplified spontaneous emission:

$$\frac{dP^{\pm}(\lambda)}{dz} = \mp\Gamma_s\left[\sigma_a(\lambda)N_2 - \sigma_e(\lambda)N_1\right]P^{\pm}(\lambda)\mp \alpha\ P^{\pm}(\lambda) + 2\sigma_e(\lambda)N_2\frac{hc^2}{\lambda^3}\Delta\lambda$$

$$\tag{7.30c}$$

When compared with Equation 7.25, Equation 7.30a and b contain a term that accounts phenomenologically for the fibre attenuation, while Equation 7.30c additionally contains a term accounting for the process of the spontaneous emission (ASE) [2]:

$$2\sigma_e(\lambda)N_2\frac{hc^2}{\lambda^3}\Delta\lambda$$

In Equation 7.30 both forward and backward propagating waves have been included. Therefore, they are applicable also in the case of non-negligible residual reflections at the fibre ends. To find the solution of 7.29 and 7.30, we need to set the values of the incident pump and signal waves. The boundary conditions for the forward and backward propagating ASE waves are:

$$P^+(\lambda, z = 0) = 0$$

$$P^-(\lambda, z = L) = 0$$

Thus we obtain again a two point boundary value problem, the solution of which will be discussed in more detail in the section "Fibre Laser Modelling." Examples of fibre amplifier modelling that includes the effect of the spontaneous emission can be found for instance in Pedersen et al. [12], Sorbello et al. [13], and Gorjan [14].

Finally, we note that in the case of lanthanide doped planar waveguide amplifiers a single mode propagation assumption is not always applicable and a direct coupling of the Maxwell's equations with the rate equations has to be applied [15]. This requires much shorter longitudinal discretization step, and hence affects adversely the computational efficiency of the model.

FIBRE LASER MODELLING

In the copropagating case, the calculation of the photon distributions within an erbium ion doped fibre amplifier requires a solution of the set of two differential equations 7.27 with an initial boundary value of the pump and signal pump powers. However, when the pump wave is counter propagating with respect to the signal wave, the problem becomes an example of a two point boundary value problem. In such a case more advanced methods for the numerical integration of the ordinary differential equations must be applied, for example, the shooting method [16]. A situation whereby the pump wave is counter propagating with respect to the signal wave is also encountered in a fibre laser. In this section we consider, therefore, a Yb^{3+} silica glass fibre laser as an illustrating example (Figure 7.10).

Considering two levels, $^2F_{5/2}$ and $^2F_{7/2}$ of Yb^{3+} ions in the ground state, while pumping at 915 nm, results in a relatively simple representation of the energy level structure that can be described by a set of two equations [11]:

$$\begin{bmatrix} a_{11} & a_{12} \\ 1 & 1 \end{bmatrix} * \begin{bmatrix} N_1 \\ N_2 \end{bmatrix} = \begin{bmatrix} 0 \\ N \end{bmatrix} \tag{7.31}$$

In 7.31 the coefficients a_{xx} are as follows:

$$a_{11} = \frac{\Gamma_p \lambda_p \sigma_a(\lambda_p)}{Ahc}\left[P_p^+ + P_p^-\right] + \frac{\Gamma_s \lambda_p \sigma_a(\lambda_s)}{Ahc}\left[P_s^+ + P_s^-\right]$$

FIGURE 7.10   Schematic diagram showing Yb^{3+}-doped fibre configuration.

$$a_{12} = -\frac{\Gamma_p \lambda_p \sigma_e \left(\lambda_p\right)}{Ahc}\left[P_p^+ + P_p^-\right] - \frac{\Gamma_s \lambda_s \sigma_e \left(\lambda_s\right)}{Ahc}\left[P_s^+ + P_s^-\right] - \frac{1}{\tau}$$

where τ is the radiative lifetime of level $^2F_{5/2}$. $P_p^\pm$ and $P_s^\pm$ are the optical powers for the pump and, signal waves, respectively. The signs "+" and "−" refer to forward and backward travelling waves. A is the doping cross-section, h is Planck's constant, c is the speed of light in free space, $\lambda_{s/p}$ is signal/pump wavelength, $\Gamma_{s/p}$ is the signal/pump confinement factor, and $\sigma_{a/e}$ is the absorption/emission cross-section for the $^2F_{5/2}$–$^2F_{7/2}$ transition (Figure 7.11). It is noted that, again, the approximate formulae 7.26 were adopted to relate power with photon flux density.

Equation 7.31 is complemented by a set of ordinary differential equations that describe the evolution of the pump and signal power:

$$\frac{dP_p^\pm}{dz} = \mp\Gamma_p\left[\sigma_a\left(\lambda_p\right)N_2 - \sigma_e\left(\lambda_p\right)N_1\right]P_p^\pm \mp \alpha\, P_p^\pm \tag{7.32a}$$

$$\frac{dP_s^\pm}{dz} = \mp\Gamma_s\left[\sigma_a\left(\lambda_s\right)N_2 - \sigma_e\left(\lambda_s\right)N_1\right]P_s^\pm \mp \alpha\, P_s^\pm \tag{7.32b}$$

where α is the fibre attenuation coefficient. Because the pump and signal wavelengths do not differ significantly, no distinction between the pump and signal absorption coefficient is made. To find the solutions of 7.31 and 7.32, we need to impose the boundary conditions at the ends of the fibre. For the structure shown in Figure 7.10, these are for the pump and the signal:

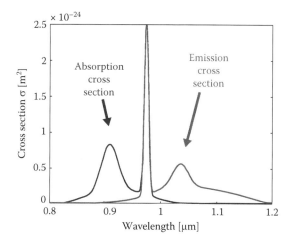

FIGURE 7.11 Yb^{3+} emission and absorption cross-sections in silica glass.

$$P_p^{+}\left(z = 0\right) = P_{pin}$$

$$P_p^{-}\left(z = L\right) = r_F P_p^{+}\left(z = L\right)$$

$$P^{+}(\lambda, z = 0) = r_{DM} P^{-}(\lambda, z = 0)$$

$$P^{-}(\lambda, z = L) = r_F P^{+}(\lambda, z = L) \tag{7.33}$$

where P_{pin} is the input pump power, r_F is the Fresnel fibre facet reflectivity, which for silica glass was assumed equal to 0.035, and r_{DM} is the dichroic mirror reflectivity at the signal wavelength of 1.037 µm.

The implementation of the shooting method combined with the fixed point method to the solutions of equations 7.31 and 7.32 results in Algorithm 7.3.

We implemented Algorithm 7.3 and used it for the simulation of the Yb³⁺ fibre laser structure from Figure 7.10. The simulation parameters are summarized in Table 7.1.

The low value of the pump wave confinement factor stems from the fact that a double clad fibre is used [17]. The emission and absorption cross-section spectra used in the simulations are presented in Figure 7.11 [18]. Figure 7.12 show the dependence of the output power for the signal and pump waves on the position z within the Yb³⁺-doped fibre for subsequent iterations of the shooting method. The results obtained in the second iteration of the shooting method are already nearly settled down, and there is only a minor correction introduced in the third and fourth iterations.

ALGORITHM 7.3   1D YB³⁺ DOPED FIBRE LASER MODEL USING THE SHOOTING METHOD WITH A FIXED POINT METHOD

1. Start
2. Set the values of all parameters
3. Set the longitudinal step Δz
4. Set the values of the initial values of $P_p^{\pm}$ and $P_s^{\pm}$ at z = 0
5. Use an ordinary differential equations integration scheme for 7.32 to calculate the values of $P_p^{\pm}$ and $P_s^{\pm}$ at left side of the fibre facet at z = L
6. From 7.36 calculate the values of $P_p^{\pm}$ and $P_s^{\pm}$ after reflection at z = L
7. Use an ordinary differential equations integration scheme for 7.32 to calculate the values of $P_p^{\pm}$ and $P_s^{\pm}$ at right side of the fibre facet at z = 0
8. From 7.33 calculate the new values of $P_p^{\pm}$ and $P_s^{\pm}$ at z = 0
9. If the difference between the new and initial values of $P_p^{\pm}$ and $P_s^{\pm}$ at z = 0 is larger than a predefined residual, then set the initial values of $P_p^{\pm}$ and $P_s^{\pm}$ at z = 0 to the newly calculated values (fixed point method) and go to 5
10. Stop

TABLE 7.1
Modelling Parameters for Yb³⁺ Fibre Laser

Quantity	Value	Unit
Yb^{3+}-ion concentration	6×10^{25}	m^{-3}
Rare earth doping cross-section A	3×10^{-10}	m^2
Fibre length	1	m
Fibre loss	0.01	1/m
Lifetime of level $^2F_{5/2}$ τ	0.84	ms
Reflectivity at $z = L$ for signal wave	0.035	
Reflectivity at $z = L$ for pump wave	0.035	
Reflectivity at $z = 0$ for signal wave	0.99	
Reflectivity at $z = 0$ for pump wave	0	
Confinement factor for signal wave	0.7	
Confinement factor for pump wave	0.01	
Pump wavelength	0.915	µm
Signal wavelength	1.037	µm

For the sake of completeness we provide as follows in the MATLAB script that was used to implement the algorithm 7.3:

```
clear % clears variables
clear global % clears global variables
format long e
% initial constants
I = sqrt(-1);% remember not to overwrite pi or i !
pi = 3.141592653589793e+000;

% setting wavelength grid storing pump and signal wavelengths
global LAM_GRID
LAM_PUMP = 0.915e-6;%pump wavelength [m]
LAM_SS = 1.037e-6;%signal wavelength [m]
LAM_GRID = [LAM_PUMP;LAM_SS;LAM_PUMP;LAM_SS];
% pump and signal wavelengths in vertical
% vector for forw(1st) & back(2nd) waves

Fib_Len = 1.;% fibre laser length [m]
Pp_forw = 20.;
%pump power forward [W] (input at left side, i.e. z = 0)
ROC_p = 0.035;%reflectivity for pump at output
ROC_s = 0.035;%reflectivity for signal at output
RHR_p = 0.0;%reflectivity for pump at dichroic mirror
RHR_s = 0.99;%reflectivity for signal at dichroic mirror

%initial photon distribution
Y0 = 0.1*Pp_forw*ones(size(LAM_GRID));
%initial photon distribution for x = 0
Y0(1) = (1.-RHR_p)*Pp_forw;
%setting initial value of forward pump
```

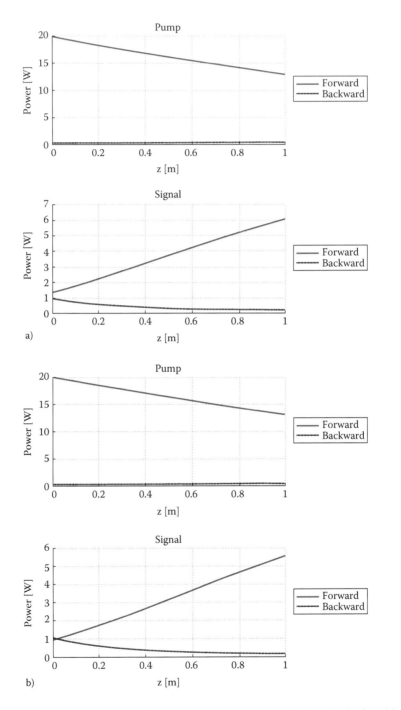

FIGURE 7.12 The dependence of the pump and signal power on the longitudinal position z within the fibre computed using Algorithm 7.3 after: a) one iteration, b) two iterations, c) three iterations, and d) four iterations of the shooting method. (*Continued*)

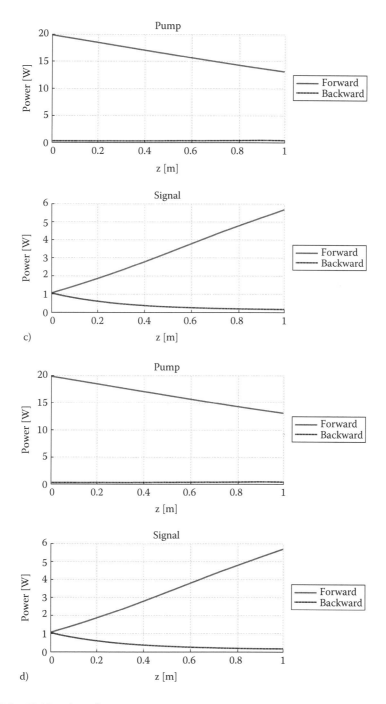

c)

d)

FIGURE 7.12 (*Continued*)

```
Y0(3) = 0.0;%settinig initial value of backward pump (at z = 0)

%calculation of the field distribution in the laser
cavity%%%%%%%%%%%%%%%%%%%% %%%%%%%%%%%%%%%%%%%%%%%%%%%%%%%%%%%%
%%%%%%%%%%%%%%%%%
'start'
options = odeset('RelTol',1e-3,'AbsTol',1.e-6*ones(size…
(LAM_GRID)));% accuracy params for ode
for j1 = 1:4

%forward integration
[t,y] = ode45(@one_step1,[0 Fib_Len],Y0,options);
%reflection at the right facet (z = Fib_len)
Y0 = y(size(y,1),:);
Y0(3) = ROC_p*Y0(1);
Y0(4) = ROC_s*Y0(2);
%backward integration
[t,y] = ode45(@one_step1,[Fib_Len 0],Y0,options);
%reflection at the left facet (z = 0)
Y0 = y(size(y,1),:);
Y0(1) = RHR_p*Y0(3)+(1.0-RHR_p)*Pp_forw;
Y0(2) = RHR_s*Y0(4);
end

function dydt = one_step1(t,y)
global LAM_GRID

pi = 3.141592653589793e+000;
h = 6.626e-34;%[Js]
v0 = 3e8;%[m/s]
Gamma_fibP = 0.01;%confinement factor for pump
Gamma_fibS = 0.7;%confinement factor for signal
A_dopRE = 3e-10;%rare earth doping cross section [m2]
tau_Yb = 840.e-6;% life time for Yb [s]
N_Yb = 6e25;%Yb dopant concentration [1/m3]
alfa_fib = 1.e-1;%fiber loss [1/m]

Gamma_fib = Gamma_fibS*ones(size(LAM_GRID));
Gamma_fib(1) = Gamma_fib(1)*Gamma_fibP/Gamma_fibS;
Gamma_fib(3) = Gamma_fib(3)*Gamma_fibP/Gamma_fibS;
%calculation of C coefficient = gamma*lam/(h*c*A_dopRE)
C_Yb = Gamma_fib.*LAM_GRID/(h*v0*A_dopRE);

%calculation of level populations: N2 and N1
N2_Yb = N_Yb*sum(C_Yb.*Yb_abs(LAM_GRID).*y)/…
(1/tau_Yb+sum(C_Yb.*(Yb_abs(LAM_GRID)+Yb_ems(LAM_GRID)).*y));
N1_Yb = N_Yb-N2_Yb;

%calculation of RHS' of coupled ODEs
dydt = Gamma_fib.*(Yb_abs(LAM_GRID)*N1_Yb-…
Yb_ems(LAM_GRID)*N2_Yb).*y;
```

```
dydt(1:2) = -dydt(1:2);
%introducing - sign for forward propagating waves
dydt(2) = dydt(2)-alfa_fib*y(2);
%adding fiber loss term to forw signal wave
dydt(4) = dydt(4)+alfa_fib*y(4);
%adding fiber loss term to back signal wave
```

The function dydt needs to be saved in the file one_step1.m to be compatible with the main program. Functions Yb_abs.m and Yb_ems.m are not provided but can be easily developed using Figure 7.11. We note that a standard MATLAB Runge-Kutta algorithm was used for the integration of the ordinary differential equations.

There is an extensive literature on the application of the one dimensional single wavelength steady state models to the modelling and design of various lanthanide ion doped fibre lasers. These include the erbium doped lasers [19], erbium-ytterbium co-doped lasers [20], thulium and holmium lasers [21], and more recently dysprosium, terbium and praseodymium lasers [22] that operate in the mid infrared wavelength region.

TIME DOMAIN MODELS

An important class of fibre laser and amplifier models are the time domain models. These models are usually based on the assumption that only fundamental mode is supported. As an illustrating example, let's consider an Yb^{3+}-doped amplifier. Again only two levels are involved: $^2F_{5/2}$ and $^2F_{7/2}$, whilst pumping at 915 nm, which results in a relatively simple two level representation of the energy level structure that can be described by a single ordinary differential equation:

$$\frac{dN_2}{dt} = a_{11}\left(N - N_2\right) + a_{12}N_2 \tag{7.34}$$

In 7.34 the coefficients a_{xx} are defined in the same way as in 7.28. Equation 7.34 is complemented by a set of ordinary differential equations that describe the evolution of the pump and signal power:

$$\frac{\partial P_p^\pm}{\partial z} \pm \frac{1}{v}\frac{\partial P_p^\pm}{\partial t} = \mp\Gamma_p\left[\sigma_a\left(\lambda_p\right)N_2 - \sigma_e\left(\lambda_p\right)N_1\right]P_p^\pm \mp a\ P_p^\pm \tag{7.35a}$$

$$\frac{\partial P(\lambda)^\pm}{\partial z} \pm \frac{1}{v}\frac{\partial P(\lambda)^\pm}{\partial t} = \mp\Gamma_s\left[\sigma_a\left(\lambda\right)N_2 - \sigma_e\left(\lambda\right)N_1\right]P(\lambda)^\pm \mp a\ P(\lambda)^\pm$$
$$+ 2\sigma_e\left(\lambda\right)N_2\frac{hc^2}{\lambda^3}\Delta\lambda \tag{7.35b}$$

Similarly, as in the examples considered in the sections "Fibre Amplifier Modelling" and "Fibre Laser Modelling," Equations 7.34 and 7.35 have to be complemented with the boundary conditions that provide the information on the device configuration.

The first step in solving 7.34 and 7.35 consists in converting the partial differential equation 7.35 into a set of ordinary differential equations. This can be achieved in two ways. Either the discretization is introduced along the z axis [23,24] or along the time axis [25]. Once this step is completed, the resulting set of first order ordinary differential equations can be solved by standard numerical methods. There are many examples that can be found in today's literature [17,23,25].

When very short pulses are considered, a more careful handling of the dispersion effects needs to be included in the model. This can be done by combining the standard time domain fibre device model with the techniques that will be discussed in Chapter 9 [26].

EXTRACTION OF MODELLING PARAMETERS

The accurate estimation of the lanthanide photonic device numerical simulation parameters is essential for predictive modelling. The key parameters that are needed for predictive modelling of lanthanide ion–doped photonic devices are emission and absorption cross-sections and the photoluminescence life time. The values of these parameters are obtained usually indirectly from measurements. The measurements can be performed on either a section of a lanthanide-doped fibre that is used for the realization of the photonic device or on a bulk glass sample. The bulk glass sample can be obtained either from the fibre preform or made separately. In principle, the most useful and predictive results are obtained by measuring emission and absorption cross sections and the photon life time directly for the fibre section. This, however, is usually not straightforward and requires a development of a dedicated measurement setup. Measuring the parameters using a bulk glass sample on the other hand, involves performing standard measurements that can be carried out in most optical laboratories, using standard measuring equipment. We outline the measurement procedures for the bulk glass sample first.

The basis for the extraction of the emission and absorption spectra and of the photoluminescence lifetime of lanthanide ion doped bulk glass samples are two standard measurements, that is, the photoluminescence measurement and the optical absorption measurement. To measure the absorption cross-section spectrum, a doped glass sample with two parallel flat surfaces needs to be prepared. The measurement itself is carried out typically using a Fourier transform infrared spectrophotometer (FTIR). The values of absorption cross-section can be calculated directly from the absorption measured by FTIR, if the sample thickness is known.

A typical experimental setup for photoluminescence spectrum measurement consists of a pump laser, chopper, photoluminescence collection optics, a long pass filter that prevents the scattered pump radiation from entering the monochromator, a monochromator, a photodetector, a lock-in amplifier, and a computer that controls all devices and collects the data samples through a data acquisition (DAQ) board (Figure 7.13).

The sample should be prepared in such a way that the effect of reabsorption is minimized as much as possible [27]. The lanthanide ion doping concentration should not be too high to avoid the upconversion effects. However, it should be also noted that a low dopant concentration results in a weak photoluminescence signal. The pump laser should operate at the wavelength that corresponds to one of the

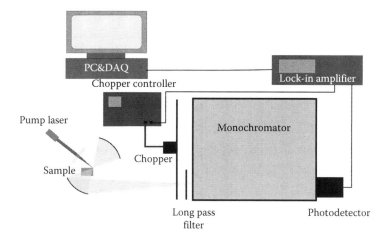

FIGURE 7.13 A schematic diagram of an experimental setup for photoluminescence measurement.

absorption bands of the given lanthanide ion. The optical signal is chopped after passing the chopper at the frequency that should be adjusted, taking into account the photoluminescence lifetime of the observed transition. Nearly the same setup can be used for the measurement of the photoluminescence lifetime. The output signal from the photo-detector should be in this case, however, connected to a scope via a trans-impedance amplifier instead of using a lock-in amplifier. The value of the photoluminescence lifetime is extracted from the exponentially decaying part of the waveform observed on the scope. Finally, we note that if FTIR is not available, the absorption cross section can be also measured using a modified setup of that used for the photoluminescence analysis. In this modified setup, the pump laser needs to be replaced with a white light source (Figure 7.14).

Once the absorption spectrum, photoluminescence spectrum, and photoluminescence lifetime are measured, the emission and absorption cross-sections and the photoluminescence lifetime corresponding to the radiative part of the transition can be obtained in a number of ways. As mentioned, the absorption cross-section can be directly obtained from the FTIR absorption measurement if the sample thickness is known, which is the case if the bulk glass sample is correctly prepared. The photoluminescence spectrum measurement on the other hand provides only the shape of the photoluminescence cross-section spectrum. It has to be therefore scaled to obtain the absolute values of the emission cross-section. For this purpose, either the Fuchbauer-Ladenburg theory or McCumber theory is applied [2]. Additionally, the data provided by the absorption measurement can be applied to investigate the efficiency of the radiative transitions in lanthanide ions using the Judd-Ofelt theory [2]. Furthermore, very often, only a limited range of measurements can be carried out in the laboratory using the available equipment. In such situations, Fuchtbauer-Ladenburg theory, McCumber theory, and Judd-Ofelt theory can be used to obtain all the required information. We list in the following the main useful formulae that stem from these three theories.

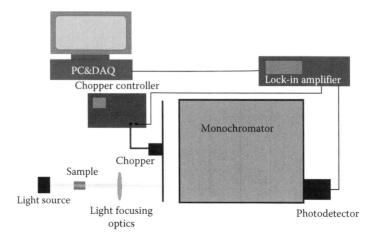

FIGURE 7.14 A schematic diagram of an experimental setup for absorption measurement.

We outline first the application of the Judd-Ofelt theory to a measured absorption cross-section spectrum. This technique allows the transition strength to be extracted from the absorption cross-section spectrum.

The equation that relates the absorption cross-section with the electrical dipole and magnetic dipole line strength has the following form:

$$
\int_{\text{band}} k(\lambda)\,d\lambda = \frac{8\pi^3 q^2}{3hc}\frac{N\lambda}{(2J+1)n_r^2}\left[\frac{n_r\left(n_r^2+2\right)^2}{9}S_{JJ'}^{\text{ed}} + n^3 S_{JJ'}^{\text{md}}\right] \tag{7.36a}
$$

in cgs units and in SI units:

$$
\int_{\text{band}} k(\lambda)\,d\lambda = \frac{8\pi^3 q^2}{3hc}\frac{N\lambda}{(2J+1)n_r^2}\left[\frac{n_r\left(n_r^2+2\right)^2}{9}S_{JJ'}^{\text{ed}} + n^3 S_{JJ'}^{\text{md}}\right]\frac{1}{4\pi\varepsilon_0} \tag{7.36b}
$$

whereby k is the absorption coefficient in cm^{-1}, q is elemental charge in Coulombs, h is the Planck's constant, c is the speed of light in free space, n_r is the refractive index of the bulk glass, J is the total angular momentum of the final state, and N is the lanthanide ion concentration, while $S_{JJ'}^{\text{ed}}$ and $S_{JJ'}^{\text{md}}$ denote respectively the electric and magnetic dipole strengths between the levels with total angular momentum J and J'. Following the Judd-Ofelt theory, the electric and magnetic dipole strengths are given by formulas:

$$
S_{JJ'}^{\text{ed}} = \sum_{t=2,4,6} \Omega_t \left|\left\langle 4f^n[S,L]J\left\|U^t\right\|4f^{n'}[S'L']J'\right\rangle\right|^2 \tag{7.37}
$$

$$S_{jj'}^{md} = \left(\frac{h}{4\pi mc}\right)\left\langle f^n \left[SL\right]J \|L + 2S\| f^n \left[S'L'\right]J'\right\rangle^2 \tag{7.38}$$

where Ω_t are the Judd-Ofelt parameters,

$$\left|\left\langle 4f^n \left[S,L\right]J \|U^t\| 4f^{n'} \left[S'L'\right]J'\right\rangle\right|$$

are reduced matrix elements between states J and J', m is the electron mass and

$$\left\langle f^n \left[SL\right]J \|L + 2S\| f^n \left[S'L'\right]J'\right\rangle$$

are reduced matrix elements of the operator $L + 2S$.

Substituting 7.37 and 7.38 into 7.36, one obtains a set of n equations for the Judd-Ofelt parameters Ω_t. Because the number of unknowns is three, this procedure leads to an overdetermined set of equations if more than three absorption bands are considered:

$$\begin{aligned}
F_n &= C_{JJ'}[\chi_{JJ'}^{ed} \cdot \sum_{t=2,4,6} (\Omega_t^{JJ'} \cdot U_{JJ'}^t) + \chi_{JJ'}^{md} \cdot S_{JJ'}^{md}]\frac{1}{4\pi\varepsilon_0} \\
&= C_{JJ'}[\chi_{JJ'}^{ed} \cdot (\Omega_2 \cdot U_{JJ'}^2 + \Omega_4 \cdot U_{JJ'}^4 + \Omega_6 \cdot U_{JJ'}^6) + \chi_{JJ'}^{md} \cdot S_{JJ'}^{md}]\frac{1}{4\pi\varepsilon_0} \\
&= (\Omega_2 \cdot M_{n1} + \Omega_4 \cdot M_{n2} + \Omega_6 \cdot M_{n3}) + S_n^{md} \\
&= M \cdot \Omega + S_n^{md}
\end{aligned} \tag{7.39}$$

where

$$F_n = \int_{band} k(\lambda)\, d\lambda$$

is calculated by integrating the area under the dependence of the light absorption coefficient on the wavelength for each band n corresponding to the transition from the initial J state to the final J' state. For the sake of clarity, we note that the index n is used to number all bands that are present in the FTIR absorption spectrum. The coefficient $C_{JJ'}$ is defined as:

$$C_{JJ'} = \frac{8\pi^3 q^2 N \lambda_{pk}}{3hc(2J+1)}\frac{1}{n_r^2}$$

Equation 7.39 can thus be compactly written as:

$$F = M\Omega - S^{\mathrm{md}}$$

where the matrix M is defined as:

$$M = \begin{bmatrix} M_{11} & M_{12} & M_{13} \\ M_{21} & M_{22} & M_{23} \\ \vdots & \vdots & \vdots \\ \vdots & \vdots & \vdots \\ M_{n1} & M_{n2} & M_{n3} \end{bmatrix}$$

$$M_{n1} = C_{JJ'} \chi_{JJ'}^{\mathrm{ed}} \frac{1}{4\pi\varepsilon_0} \cdot U_{JJ'}^2$$

$$M_{n2} = C_{JJ'} \chi_{JJ'}^{\mathrm{ed}} \frac{1}{4\pi\varepsilon_0} \cdot U_{JJ'}^4$$

$$M_{n3} = C_{JJ'} \chi_{JJ'}^{\mathrm{ed}} \frac{1}{4\pi\varepsilon_0} \cdot U_{JJ'}^6$$

the three-element vector Ω as

$$\Omega = \begin{bmatrix} \Omega_2 \\ \Omega_4 \\ \Omega_6 \end{bmatrix}$$

and n element vectors S^{ed}, S^{md}, and F as

$$S^{\mathrm{ed}} = \begin{bmatrix} S_1^{\mathrm{ed}} \\ S_2^{\mathrm{ed}} \\ \vdots \\ \vdots \\ S_n^{\mathrm{ed}} \end{bmatrix}$$

$$S^{\mathrm{md}} = \begin{bmatrix} S_1^{\mathrm{md}} \\ S_2^{\mathrm{md}} \\ \vdots \\ \vdots \\ S_n^{\mathrm{md}} \end{bmatrix}$$

$$F = \begin{bmatrix} F_1 \\ F_2 \\ \vdots \\ \vdots \\ F_n \end{bmatrix}$$

while

$$S_n^{md} = C_{JJ'} \chi_{JJ'}^{md} \frac{1}{4\pi\varepsilon_0} \cdot S_{JJ'}^{md}$$

and n is the number of transitions included in the fit, which is usually taken to equal the number of absorption bands that are present in the FTIR absorption spectrum.

From 7.39 the values of Ω_t in the least squares sense can be obtained from [28]:

$$\Omega = \left(M^T M\right)^{-1} M^T \left(F - S^{md}\right) \tag{7.40}$$

where the vector F contains the values of areas obtained from the integral on the left hand side of Equation 7.36.

Once the Judd-Ofelt parameters Ω_t are determined, the spontaneous radiative transition rate from level J to J' can be obtained from:

$$A_{JJ'} = \frac{64\pi^4 q^2}{3h(2J+1)\lambda^3} \left(\frac{n_r \left(n_r^2 + 2\right)^2}{9} S_{JJ'}^{ed} + n^3 S_{JJ'}^{md} \right) \tag{7.41a}$$

in cgs units and in SI units:

$$A_{JJ'} = \frac{64\pi^4 q^2}{3h(2J+1)\lambda^3} \left(\frac{n_r \left(n_r^2 + 2\right)^2}{9} S_{JJ'}^{ed} + n^3 S_{JJ'}^{md} \right) \frac{1}{4\pi\varepsilon_0} \tag{7.41b}$$

where J is the quantum number of the upper state. The total radiative life-time of the state J into all states J' is defined as the inverse of the sum of spontaneous radiative transition rates into all J' states:

$$\tau_J = \frac{1}{\sum_{J'} A_{JJ'}} \tag{7.42}$$

The branching ratio that gives the probability of an excited state to decay into the state with the quantum number J' is defined as:

$$\beta_{JJ'} = \tau_J A_{JJ'} \tag{7.43}$$

The electric dipole line strengths are conveniently obtained from

$$S_{JJ'}^{ed} = \Omega_2 \cdot U^{(2)} + \Omega_4 \cdot U^{(4)} + \Omega_6 \cdot U^{(6)}$$

The difference between the values of photon decay times observed experimentally and the ones that are predicted by 7.42 and 7.43 results from the fact that in a glass host the radiative transitions are accompanied by the nonradiative transitions. The nonradiative transitions result from the interaction of photons with the glass host phonons. The rate of the nonradiative transitions can be estimated from the formula [2]:

$$A_{nr} = B \exp\left(-a \Delta E\right) \left[1 - \exp\left(\frac{\hbar \omega}{k_B T}\right) \right]^{-N_{ph}} \tag{7.44}$$

where B and a are the host dependent constants, ΔE is the energy gap (in cm^{-1}) between the considered transition levels, $\hbar \omega$ is the phonon energy, k_B is the Boltzmann's constant, T is the temperature, and N_{ph} is the smallest integer number of phonons needed to bridge the energy gap.

The shape of the emission cross-section spectrum can be obtained from the absorption cross-section spectrum using the McCumber theory [2,29]:

$$s_e(\nu) = \sigma_a(\nu) \exp\left(-\frac{h\nu}{kT}\right) \tag{7.45}$$

Once the shape of the emission spectrum is known (either from the photoluminescence measurement or from 7.45), the absolute values can be obtained from [30]:

$$\sigma_e(\nu) = s_e(\nu) \exp\left(\frac{\varepsilon}{kT}\right) \tag{7.46}$$

where ε is a parameter that can be extracted using the procedure outlined in Miniscalo and Quimby [30]. McCumber theory allows also the radiative lifetime to be related to the emission cross-section [2]:

$$\frac{1}{\tau} = \frac{8\pi n_r^2}{c^2} \int \nu^2 \sigma_e(\nu) \, d\nu \tag{7.47}$$

Alternative formulae are provided by the Fuchtbauer-Ladenburg theory [2]:

$$\frac{1}{\tau} = \frac{8\pi n_r^2}{\lambda^2} \int \sigma_e(\nu)d\nu = \frac{g_1}{g_2}\frac{8\pi n_r^2}{\lambda^2}\int \sigma_a(\nu)d\nu \qquad (7.48)$$

where λ is the free space wavelength while g_1 and g_2 are the number of sublevels in the lower and upper levels, respectively.

It should be noted that both the Fuchtbauer-Ladenburg theory and the McCumber theory are based on a number of assumptions and therefore can introduce significant errors. The discussion of the limitations of both theories is available in references [2,27,31–34]. The McCumber analysis is also helpful in evaluating the gain spectral shape dependence on the temperature [35].

As an illustrating example, we will consider a chalcogenide glass sample doped with Tb^{3+}. The sample prepared for the FTIR absorption measurement is shown in Figure 7.15. The raw measurement result obtained from FTIR is shown in Figure 7.16 and the result obtained after the baseline extraction is shown in Figure 7.17. These results include the absorption between the ground state $^7F_6$ and the states ^7F_k where $k = 5, 4, 3, 2, 1,$ and 0. Using 7.40 we calculated the Judd-Ofelt parameters. The results of these calculations are: $\Omega_2 = 7.9760 \times 10^{-20}$ cm², $\Omega_4 = 6.0978 \times 10^{-20}$ cm², and $\Omega_6 = 2.3531 \times 10^{-20}$ cm². The reduced matrix elements used for this calculations were taken from Carnall et al. [36] and are collected together in Table 7.2. The values of integrals needed for evaluation of the left hand side of 7.36 are given in Table 7.3.

Once the values of Ω_2, Ω_4, and Ω_6 are known, the radiative lifetimes and branching ratios can be obtained from 7.42 and 7.43. The emission cross section for the relevant transitions can be obtained using the McCumber theory 7.45 whilst calculating the parameter ε in 7.46 using the procedure outlined in Miniscalo and Quimby [30] and

FIGURE 7.15 1000 ppm Tb^{3+}-doped chalcogenide glass sample prepared for FTIR absorption measurement. (Sample prepared by Z. Tang and Ł. Sójka in the Mid-IR Photonics Group, George Green Institute for Electromagnetics Research, the University of Nottingham, UK.)

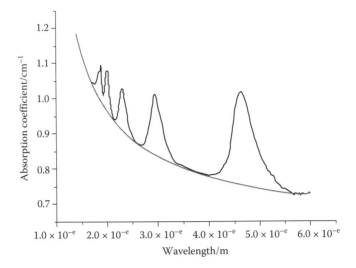

FIGURE 7.16 Measured FTIR absorption spectrum for the sample from Figure 7.15 with the red line showing the baseline. (Measurement obtained by Ł. Sójka and D. Furniss in the Mid-IR Photonics Group, George Green Institute for Electromagnetics Research, the University of Nottingham, UK.)

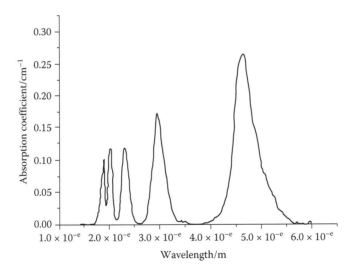

FIGURE 7.17 Absorption cross-section obtained after extracting the baseline from results presented in Figure 7.13. (Result obtained by Ł. Sójka, Mid-IR Photonics Group, George Green Institute for Electromagnetics Research, the University of Nottingham, UK.)

TABLE 7.2

Reduced Matrix Elements for Tb^{3+}

Transition	$[U2]^2$	$[U4]^2$	$[U6]^2$
$^7F_0 - ^7F_6$	0.0	0.0	0.1441
$^7F_1 - ^7F_6$	0.0	0.0	0.3761
$^7F_2 - ^7F_6$	0.0	0.0481	0.4695
$^7F_3 - ^7F_6$	0.0	0.2323	0.4129
$^7F_4 - ^7F_6$	0.0888	0.5159	0.2658
$^7F_5 - ^7F_6$	0.5377	0.6420	0.1178

Source: Carnall, W.T. et al. Crosswhithe, Energy level structure and transition probabilities in the spectra of the trivalent lanthanides in LaF$_3$. *Argonne National Laboratory report/CNRS Colloq.*, 1977. 255(65).

TABLE 7.3

Areas under the Absorption Curve from Figure 7.17 Expressed in nm/cm

n	Transition	Area [nm/cm]	λ_{pk} [nm]
1	$^7F_0 - ^7F_6$	5.23017	1857
2	$^7F_1 - ^7F_6$	4.37141	1891
3	$^7F_2 - ^7F_6$	11.6962	2025
4	$^7F_3 - ^7F_6$	20.0819	2309
5	$^7F_4 - ^7F_6$	48.995	2951
6	$^7F_5 - ^7F_6$	148.262	4633

Digonnet et al. [31]. Alternatively, the scaling of the spectral shape obtained from 7.46 can be carried out by applying either the formula 7.47 or 7.48.

In the case of fibre measurements, a typical setup is presented in Figure 7.18. It consists of a pump laser, broadband source, chopper, fibre coupler, monochromator, and pump and signal detection circuitry [33]. In this setup the dependence of the gain and loss on the pump power for a lanthanide ion–doped fibre section is directly measured. In the setup shown in Figure 7.15, the dependence of fluorescence from the side of the fibre on the pump power can also be measured [33]. The emission and absorption cross-sections are then derived from the measured gain and loss spectra. A more detailed description of the measurement techniques applied in the case of a fibre can be found in Desurvire [3].

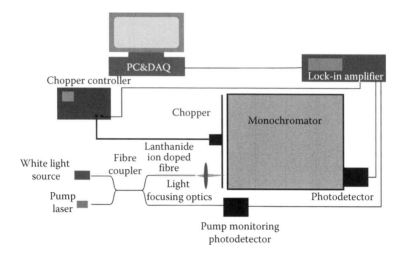

FIGURE 7.18 A schematic diagram of an experimental setup for optical characterization of lanthanide ion doped fibre.

LANTHANIDE ION INTERACTION EFFECTS

A small distance between lanthanide ions increases the probability of ion–ion interactions. Such a situation typically occurs at high dopant concentration. However, it can also be a signature of a strong inhomogeneous lanthanide ion distribution within the host material [37]. Interactions between the lanthanide ions have a major impact on the experimentally observed photoluminescence spectra and photoluminescence lifetimes, and hence also on the device operation that uses a lanthanide doped fibre. There are three main examples of interactions observed in lanthanide ion doped glasses: energy migration, cross relaxation, and cooperative upconversion.

Figure 7.19 shows schematically the energy migration process. Of the two lanthanide ions participating in the energy migration "ion 1" is initially in the upper energy state while "ion 2" is in the lower energy state. In the energy migration process, the energy is transferred from the ion in the higher energy state to the ion in the lower energy state. Therefore, finally "ion 1" is in the upper energy state while "ion 2" is in the lower energy state. The energy migration process may have an adverse influence on an optical amplifier operation because it enables the transfer of pump photon energy to ions that may be de-excited nonradiatively or to ions that do not interact with the incident signal light. Energy migration can also have a positive effect. For instance Er^{3+} doped fibres are co-doped with Yb^{3+} to facilitate the pumping process. The pump photons in such devices are absorbed primarily by Yb^{3+} ions, which have a large absorption cross-section, and transferred to Er^{3+} ions via the energy migration process. In the available literature, there are many more examples of using co-doping to improve optical device characteristics [38,39].

Figure 7.20 shows schematically the cooperative upconversion process. In the cooperative upconversion process initially "ion 1" is in the energy level 1 while "ion 2" in the energy level 2. After the interaction takes place, "ion 1" moves to the lowest considered

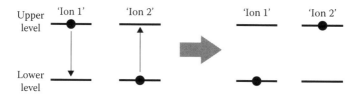

FIGURE 7.19 A schematic diagram of the energy migration process.

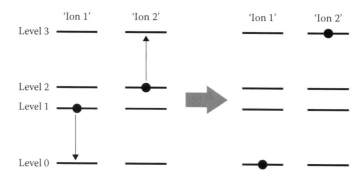

FIGURE 7.20 A schematic diagram of the upconversion process.

energy level while "ion 2" to the highest one. For the upconversion process to be efficient, the energy difference between levels 1 and 0 has to match closely the energy difference between the levels 3 and 2. The inverse process is referred to in the literature as cross-relaxation. In general, the upconversion processes have generally a negative impact on the optical device operation. For instance, after an upconversion event, the ion that ends up in the higher energy level can fall back nonradiatively to the lower energy level, thus wasting one pump photon. This is a typical situation in an Er^{3+} doped high phonon energy host [40]. Furthermore, upconversion processes lead to a reduction of the effective photon lifetime that leads, in turn, to a reduction of the device efficiency [2]. The upconversion processes can, however, be very useful in constructing short wavelength lanthanide ion doped fibre lasers [41] while the cross-relaxation process can be potentially used to enhance the quantum efficiency of Tm^{3+} fibre lasers beyond 100% [42].

Figure 7.21 shows two examples of lanthanide ion interactions in glass hosts. The upconversion shown in Figure 7.21a takes place in a silica glass host and involves four levels of Er^{3+} ion. The upconversion process is followed by the nonradiative transition of the upconverted ion to the $^4I_{13/2}$ level. If we consider that both ions have been brought to the $^4I_{13/2}$ level by two pump photons, the final state corresponds to a situation whereby the energy of one pump photon has been effectively lost. Such an interaction has therefore a negative impact on an erbium doped fibre amplifier operation. The cross-relaxation shown in Figure 7.21b takes place in a germinate glass and involves 4 levels of Tm^{3+}. After the cross-relaxation process takes place, two ions are in the $^3F_4$ level. If we consider that the first ion was brought to the highest level with

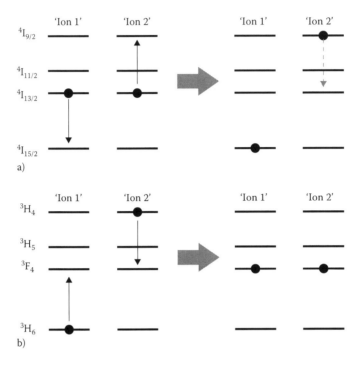

FIGURE 7.21 Schematic diagrams showing examples of a) an upconversion process and b) a cross-relaxation process.

a pump photon corresponding to 808 nm, we have a situation whereby one pump photon brings two ions into the $^3F_4$ level. Such a process has, therefore, been utilized for the realization of high-efficiency thulium lasers operating at 2 µm [42].

The theory of ion–ion interactions in solids has been developed initially by Dexter [43] and then refined by Kushida [44]. A rigorous study of energy transfer between lanthanide ions requires the fabrication of a set of samples with varying dopant concentration, and the determination of the dominant ion interaction mechanisms via comparison of experimentally observed and theoretically predicted photoluminescence lifetimes, for example, [45]. However, very often a simplified phenomenological approach is followed instead, for example, [46,47]. In the simplified approach, the time dependence of the photoluminescence decay is studied to derive a proportionality constant for a phenomenological term that is added to the rate equations to simulate the effect of the energy transfer between ions.

REFERENCES

1. Digonnet, M.J.F., *Rare-Earth-Doped Fiber Lasers and Amplifiers*. 2001, New York: Marcel Dekker. Inc.
2. Becker, P.C., N.A. Olsson, and J.R. Simpson, *Erbium Doped Fiber Amplifiers. Fundamentals and Technology*. 1999, London: Academic Press.
3. Desurvire, E., *Erbium-Doped Fibre Amplifiers, Principles and Applications*. 1993, New York: John Wiley & Sons.

4. Saleh, B.E.A. and M.C. Teich, *Fundamentals of Photonics*. 1991, New York: John Wiley & Sons Inc.

5. Ebeling, K.J., *Integrated Optoelectronics*. 1992, Berlin: Springer-Verlag.

6. Sobelman, I.I., *Atomic Spectra and Radiative Transitions*. 1996, London: Springer.

7. Judd, B.R., *Operator Techniques in Atomic Spectroscopy*. 1963, New York: McGraw-Hill.

8. Desurvire, E., *Erbium-Doped Fibre Amplifiers, Principles and Applications*. 2002, New York: John Wiley & Sons.

9. Janos, M. and S.C. Guy, Signal-induced refractive index changes in erbium-doped fiber amplifiers. *Journal of Lightwave Technology*, 1998. 16(4): p. 542–548.

10. Giles, C.R. and E. Desurvire, Modeling erbium-doped fiber amplifiers. *Journal of Lightwave Technology*, 1991. 9(2): p. 271–283.

11. Wang, Y., C.Q. Xu, and H. Po, Analysis of Raman and thermal effects in kilowatt fiber lasers. *Optics Communications*, 2004. 242(4–6): p. 487–502.

12. Pedersen, B., et al., The design of erbium-doped fiber amplifiers. *Journal of Lightwave Technology*, 1991. 9(9): p. 1105–1112.

13. Sorbello, G., S. Taccheo, and P. Laporta, Numerical modelling and experimental investigation of double-cladding erbium-ytterbium-doped fibre amplifiers. *Optical and Quantum Electronics*, 2001. 33(6): p. 599–619.

14. Gorjan, M., T. North, and M. Rochette, Model of the amplified spontaneous emission generation in thulium-doped silica fibers. *Journal of the Optical Society of America B-Optical Physics*, 2012. 29(10): p. 2886–2890.

15. Biallo, D., et al., Time domain analysis of optical amplification in Er3+ doped SiO2-TiO2 planar waveguide. *Optics Express*, 2005. 13(12): p. 4683–4692.

16. Press, W.H., S.A. Teukolsky, W.T. Vetterling, and B.P. Flannery, *Numerical Recipes in C++: The Art of Scientific Computing*. 2002, Cambridge: Cambridge University Press.

17. Wang, Y. and H. Po, Dynamic characteristics of double-clad fiber amplifiers for high-power pulse amplification. *Journal of Lightwave Technology*, 2003. 21(10): p. 2262–2270.

18. Pask, H.M., et al. Ytterbium-doped silica fiber lasers - versatile sources for the 1–1.2 micrometre region. *IEEE Journal of Selected Topics in Quantum Electronics*, 1995. 1(1): p. 2–13.

19. Jackson, S.D., M. Pollnau, and J. Li, Diode pumped erbium cascade fiber lasers. *IEEE Journal of Quantum Electronics*, 2011. 47(4): p. 471–478.

20. Yelen, K., L.M.B. Hickey, and M.N. Zervas, Experimentally verified modeling of erbium-ytterbium co-doped DFB fiber lasers. *Journal of Lightwave Technology*, 2005. 23(3): p. 1380–1392.

21. Evans, C.A., et al., Numerical rate equation modeling of a similar to 2.1-mu m - Tm3+/Ho3+ Co-doped tellurite fiber laser. *Journal of Lightwave Technology*, 2009. 27(19): p. 4280–4288.

22. Sojka, L., et al., Study of mid-infrared laser action in chalcogenide rare earth doped glass with Dy3+, Pr3+ and Tb3+. *Optical Materials Express*, 2012. 2(11): p. 1632–1640.

23. Canat, G., et al., Dynamics of high-power erbium-ytterbium fiber amplifiers. *Journal of the Optical Society of America B-Optical Physics*, 2005. 22(11): p. 2308–2318.

24. Eichhorn, M., Numerical modeling of Tm-doped double-clad fluoride fiber amplifiers. *IEEE Journal of Quantum Electronics*, 2005. 41(12): p. 1574–1581.

25. Zhang, Z., et al., Numerical analysis of stimulated inelastic scatterings in ytterbium-doped double-clad fiber amplifier with multi-ns-duration and multi-hundred-kW peak-power output. *Optics Communications*, 2009. 282(6): p. 1186–1190.

26. Huang, Z., et al., Combined numerical model of laser rate equation and Ginzburg-Landau equation for ytterbium-doped fiber amplifier. *Journal of the Optical Society of America B-Optical Physics*, 2012. 29(6): p. 1418–1423.

27. Martin, R.M. and R.S. Quimby, Experimental evidence of the validity of the McCumber theory relating emission and absorption for rare-earth glasses. *Journal of the Optical Society of America B*, 2006. 23(9): p. 1770–1775.

28. Evans, C.A., et al., Numerical rate equation modeling of a ~ 2.1 - mm - Tm^{3+}/Ho^{3+} co-doped tellurite fiber laser. *Journal of Lightwave Technology*, 2009. 27(19): p. 4280–4288.

29. McCumber, D.E., Einstein relations connecting broadband emission and absorption spectra. *Physical Review*, 1964. 136(4A): p. 954–957.

30. Miniscalo, W.J. and R.S. Quimby, General procedure for the analysis of Er^{3+} cross sections. *Optics Letters*, 1991. 16(4): p. 258–260.

31. Digonnet, M.J.F., E. Murphy-Chutorian, and D.G. Falquier, Fundamental limitations of the McCumber relation applied to Er-doped silica and other amorphous-host lasers. *IEEE Journal of Quantum Electronics*, 2002. 38(12): p. 1629–1637.

32. Quimby, R.S., Range of validity of McCumber theory in relating absorption and emission cross sections. *Journal of Applied Physics*, 2002. 92(1): p. 180–187.

33. Barnes, W.L., et al., Absorption and emission cross sections of Er^{3+} doped silica fibres. *IEEE Journal of Quantum Electronics*, 1991. 27(4): p. 1004–1010.

34. Foster, S. and A. Tikhomirov, In Defence of the McCumber relation for erbium-doped silica and other laser glasses. *IEEE Journal of Quantum Electronics*, 2009. 45(10): p. 1232–1239.

35. Bolshtyansky, M., P. Wysocki, and N. Conti, Model of temperature dependence for gain shape of erbium-doped fiber amplifier. *Journal of Lightwave Technology*, 2000. 18(11): p. 1533–1540.

36. Carnall, W.T., H. Crosswhite, and H.M. Crosswhithe, Energy level structure and transition probabilities in the spectra of the trivalent lanthanides in LaF_3. *Argonne National Laboratory report/CNRS Colloq.*, 1977. 255(65).

37. Kik, P.G. and A. Polman, Cooperative upconversion as the gain-limiting factor in Er doped miniature Al_2O_3 optical waveguide amplifiers. *Journal of Applied Physics*, 2003. 91(9): p. 5008–5012.

38. Choi, Y.G., et al., Comparative study of energy transfer from Er^{3+} to Ce^{3+} in tellurite and sulfide glasses under 980 nm excitation. *Journal of Applied Physics*, 2000. 88(7): p. 3832–3839.

39. Tsang, Y., et al., Tm^{3+}/Ho^{3+} codoped tellurite fiber laser. *Optics Letters*, 2008. 33(11): p. 1282–1284.

40. van den Hoven, G.N., et al., Upconversion in Er-implanted Al_2O_3 waveguides. *Journal of Applied Physics*, 1996. 79(3): p. 1258–1266.

41. Paschotta, R., et al., 230 mW of blue light from a thulium-doped upconversion fiber laser. *IEEE Journal of Selected Topics in Quantum Electronics*, 1997. 3(4): p. 1100–1102.

42. Wu, J., et al., Efficient thulium-doped 2 mm germanate fiber laser. *IEEE Photonics Technology Letters*, 1996. 18(2): p. 334–336.

43. Dexter, D.L., A theory of sensitized luminescence in solids. *Journal of Chemical Physics*, 1953. 21(5): p. 836–850.

44. Kushida, T., Energy transfer and cooperative optical transitions in rare-earth doped inorganic materials. I. Transition probability calculation. *Journal of the Physical Society of Japan*, 1973. 34(5): p. 1318–1326.

45. de Sousa, D.F. and L.A.O. Nunes, Microscopic and macroscopic parameters of energy transfer between Tm^{3+} ions in fluoroindogallate glasses. *Physical Review B*, 2002. 66(024207): p. 1–7.

46. van den Hoven, G.N., et al., Upconversion in Er-implanted Al_2O_3 waveguides. *Journal of Applied Physics*, 1996. 79(3): p. 1258–1266.

47. Park, S.H., et al., Energy transfer between Er3+ and Pr3+ in chalcogenide glasses for dual-wavelength fiber-optic amplifiers. *Journal of Applied Physics*, 2002. 91(11): p. 9072–9077.

8 Laser Diode Modelling

Laser diodes due to their compact structure, high reliability, and simple pumping mechanism have found applications in optical telecom, medicine, printing, and as pumps for solid state and fibre lasers. In this chapter we discuss the modelling and design of laser diodes (LDs). We start from a short introduction that facilitates the classification of various LD models and helps understand the benefits of applying a specific model to a particular problem. In the following sections, we discuss zero dimensional (0D), one-dimensional (1D), and multidimensional (LD) models. The discussion is supported by illustrating examples. It is assumed that the reader is familiar with the LD theory and with the physics of semiconductor optoelectronic devices [1–12]. This chapter also capitalises on the material covered in Chapters 3 through 7. Hence, for its best understanding, it is recommended that Chapters 3 through 7 are read beforehand. The material covered in this chapter is also aimed at facilitating the understanding of more advanced textbooks that are specifically dedicated to LD modelling [13–15].

INTRODUCTION

Numerous laser diode (LD) models can be found in literature and available as commercial design tools from several software developing companies. These models vary quite significantly in terms of the computational resources required, complexity, and accuracy. This fact is a consequence of a large variety of problems that are related to the LD modelling and design process. Hence, the selection of an appropriate model depends on many factors like, for instance, the type of LD, that is, vertical cavity surface emitting laser (VCSEL) diode or edge emitting laser diode (EELD). Furthermore, a LD can operate either as a stand-alone device or within an external cavity scheme. The type of the resonator is also an important factor, for example, Fabry-Perot, distributed feedback (DFB), or distributed Bragg reflectors (DBR) LD resonators. Considering the operating conditions, a LD can operate in a continuous wave (CW), in directly modulated (by current injection), pulsed, or self-pulsed mode (e.g. passive mode locking). Additionally, a design engineer may be focused only on a particular LD operating region of the L-I-V characteristics, for example, a near threshold behaviour, moderate output power operation, or a high output power operation. Finally, the choice of the appropriate model is influenced by the imposed set of target design parameters, for example, wall-plug efficiency, threshold current, and vertical and horizontal beam divergence.

All these circumstances in which an LD design engineer needs to operate render the appropriate model selection process a fairly complex task. To facilitate the model selection process, we introduce next a classification of LD models according to the model dimensionality.

As discussed in Chapter 1, photonic device models (including LD models) can be classed in a number of ways. In this chapter we order the presented material by differentiating between the models according to the number of spatial dimensions considered when numerically solving the equations that describe the underlying physical phenomena that accompany the process of light generation: that is, zero-dimensional (0D), one-dimensional (1D), two-dimensional (2D), and three-dimensional (3D) models. In a LD the optical gain is achieved through the photon interaction with electrical current carriers. The local photon and carrier concentrations depend also on the temperature distribution within the device. Therefore, a typical LD model consists of at least an electrical part and an optical part, which might also be complemented by a thermal solver that self-consistently calculates the temperature distribution within the device. This fact slightly complicates the classification of LD models according to the model dimensionality because an LD model can have a 2D optical part and a 3D electrical part, for instance. In this chapter to avoid confusion we always state the dimensionality of each part of the model unless all parts have the same dimensionality.

Usually, more accurate models include more spatial dimensions and are more demanding in terms of the computing resources required (Figure 8.1). In general, 0D models are less accurate than full 3D models. However, 0D models benefit from much better computational efficiency. LD models can also be classified according to the way that the time dimension is handled, that is, CW, time domain, and spectral models (Figure 8.1). Again, the CW models are usually more computationally efficient than the time domain and spectral models, but provide less information about a LD operating characteristics.

The 0D models therefore are typically used for an initial parameter space search and initial design. In the case of 0D models, the three spatial dimensions are reduced to one point using simplifying assumptions. The most often used assumption is that the spatial distribution of the LD operating mode can be calculated neglecting the refractive index perturbation coming from the gain distribution. The most common 0D LD models are based on the rate equations [2,11,12]. Advanced 0D models are based on the density matrix approach [3].

There are three types of 1D LD models: models that include either the longitudinal or one of the transverse dimensions. For the sake of clarity, Figure 8.2 defines

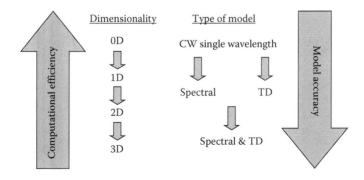

FIGURE 8.1   Single lasing mode reservoir model.

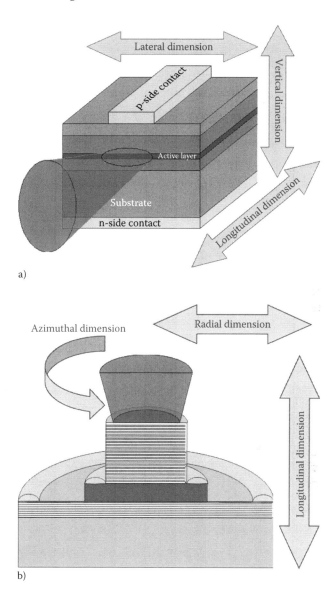

FIGURE 8.2   Schematic diagram of a) edge emitting LD and b) VCSEL.

the transverse and the longitudinal dimensions for an EELD and VCSEL. The longitudinal dimension is aligned with the direction of the beam propagation within the cavity. Longitudinal 1D models are particularly often used when studying LD mode locking whilst the vertical 1D models are used for an initial design of an EELD epitaxial structure.

In the case of the 2D models, the two most often used ones are the models that contain either both transverse dimensions or the horizontal and longitudinal dimensions.

The former models are mainly used for the design of the ridge waveguide based EELDs while the latter ones are used for the design of the high power tapered and broad area EELDs.

The 3D models are rather rarely used for the design of EELDs due to large dimensions of the LD structure and hence low computational efficiency of the models. Even quasi-3D models set a high level of demand in terms of the necessary computer memory and CPU time. For VCSELs, on the other hand, 3D models are much more efficient due to much smaller size of the cavity and hence are relatively often used in their design process.

In the following sections, we provide a more detailed description and several illustrating examples of 0D, 1D, and higher dimensional LD models.

0D LD MODELS

0D LD models are based on numerous simplifying assumptions that can hardly be justified in many cases of practical importance. Nonetheless, 0D LD models are very helpful in the design process as a computationally efficient tool for making an initial search of the parameter space before more advanced, accurate, and predictive models are employed for the detailed design. 0D laser models are also used in textbooks to explain the basic properties of the LDs [11,16,17]. As far as examples of particular applications are concerned, time domain 0D models were applied for the analysis of bistable lasers [18] and frequency domain 0D models were used for the initial design of the high speed lasers [19] and noise analysis [20]. The 0D models also have proven useful in analysing the behaviour of coupled-cavity lasers [21] and of injection [22] and phase locking lasers [23].

In this section we consider several examples of 0D rate equations based models and introduce a clear distinction between the CW, time domain, and spectral models. The 0D rate equations can be derived in a number of ways [16,24]. Here we follow the phenomenological approach, that is, using the reservoir model [16]. The reservoir model is based on an assumption that the process of light generated by a LD can be represented by two coupled reservoirs: a photon reservoir of the lasing mode and a current carrier reservoir (Figure 8.3).

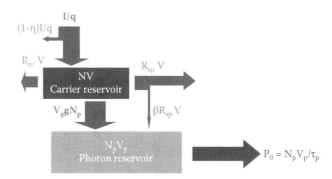

FIGURE 8.3 Single lasing mode reservoir model.

The carrier reservoir is fed by the LD bias current that injects the carriers into the quantum well (QW). Some carriers injected into the QW do not interact with the photons and hence are considered "lost." To account for this phenomenon, a current injection efficiency parameter η is included in the model. Carriers from the carrier reservoir are depleted through a number of channels. One of these channels feeds directly the photon reservoir through the process of the stimulated emission into the lasing mode. The carriers are also lost through various processes of nonradiative recombination and through the spontaneous emission. A part of the spontaneously emitted light, couples to the lasing optical mode while the rest feeds the other optical modes of the laser cavity and hence does not couple to the photon reservoir.

After the preceding discussion, we derive the rate equations for a QW EELD. We first consider the carrier rate equation. The carrier reservoir dimensions are given in Figure 8.4.

To obtain the carrier rate equation, we consider the temporal rate of change for the carrier density, the net carrier generation rate within the carrier reservoir, and the net carrier flux through the sides of the carrier reservoir (similarly, as in Chapters 5 and 6 when deriving in a phenomenological way the basic equations). Thus, we arrive at the following equation:

$$\iiint\limits_{V_c} \frac{dN}{dt} dV = -\iiint\limits_{V_c} \left(R_{sp} + R_{st} + R_{nr} \right) dV + \oiint\limits_{S_c} \frac{J}{q} dS \qquad (8.1)$$

In 8.1 N is the carrier concentration, V_c denotes the carrier reservoir volume, and S_c is its surface. R_{sp}, R_{st}, and R_{nr} are the rate of the spontaneous emission, stimulated emission, and nonradiative recombination, respectively. J and q denote respectively current density and the elemental charge so that the ratio of these two quantities provides the carrier flux density. We assume that N, R_{sp}, R_{st}, and R_{nr} are constant within

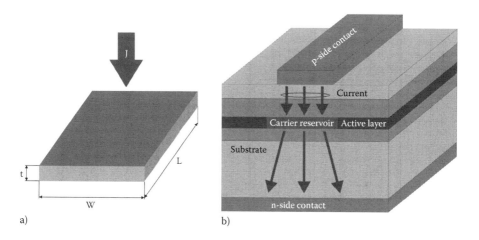

a)   b)

FIGURE 8.4   a) Carrier reservoir dimensions and b) a physical position of the carrier reservoir in a quantum well EELD.

the carrier reservoir. This is a fairly restrictive assumption and will be discussed in more detail in the following sections. Furthermore, we assume that J is constant on the surface of the carrier reservoir and can be obtained as the ratio of the LD bias current and the top surface of the carrier reservoir: $\eta I/(Lw)$ where L and w are, respectively the length and width of the carrier reservoir (Figure 8.4) and I is the LD bias current, providing carriers to the carrier reservoir, while η is the carrier injection efficiency. We note that the provided expression for the current density neglects the contribution of the carrier flux through the side walls of the carrier reservoir, which is a fair assumption for a typical QW EELD structure, considering the fact that the active region thickness (t in Figure 8.4) is much smaller than the transverse dimensions of the carrier reservoir (w and L in Figure 8.4). Thus, we obtain from 8.1 after calculating the integrals:

$$\frac{dN}{dt} = -\left(R_{sp} + R_{st} + R_{nr}\right) + \frac{\eta I}{qV_c} \tag{8.2}$$

Similarly, we can obtain the photon rate equation by considering the temporal photon density change rate and the net photon generation rate within the photon reservoir:

$$\iiint_{V_p} \frac{dN_p}{dt} dV = \iiint_{V_c} \left(\beta R_{sp} + R_{st}\right) dV - \iiint_{V_p} R_l dV \tag{8.3}$$

In 8.3, N_p stands for the photon density, V_p is the volume of the photon reservoir, β is the spontaneous emission coupling coefficient, and R_l is the rate of the photon loss from the photon reservoir, which is mainly due to leakage through the mirrors and the free carrier absorption. Again, assuming that the photon density and R_{sp}, R_{st}, and R_l are constant within the photon reservoir (or carrier reservoir as applicable) allows us to calculate easily the integrals in 8.3, which yields:

$$\frac{dN_p}{dt} = \Gamma\left(\beta R_{sp} + R_{st}\right) - R_l \tag{8.4}$$

where $\Gamma = V_c/V_p$ is the confinement factor.

Equations 8.2 and 8.4 have a fairly general form. To focus the discussion we reduce them to:

$$\frac{dN}{dt} = \frac{\eta I}{qV_c} - \frac{N}{\tau} - v_g g N_p$$

$$\frac{dN_p}{dt} = \left[\Gamma v_g g - \frac{1}{\tau_p}\right] N_p + \Gamma\beta\frac{N}{\tau} \tag{8.5}$$

where v_g is the photon group velocity in the cavity, g stands for gain, τ_p is the photon lifetime, and τ denotes the carrier lifetime. Equation 8.5 contains many additional approximations when compared with 8.2 and 8.4. However, this does not restrict the generality of the discussion of their numerical solutions that follows in the subsections on 0D modelling. An in-depth discussion of the approximations and parameters used in Equation 8.5 is available from standard reference books on LDs [11,12,16] and therefore we only provide a brief description in the following text.

In a QW EELD the length and thickness of the carrier reservoir can be assumed approximately equal to the length of the LD and the thickness of the QW, respectively. The width of the carrier reservoir is more difficult to estimate. In the first instance, it can be assumed equal to the p side contact width. For more accurate estimates, simulations of the current spreading in the LD cross-section have to be performed, using a drift-diffusion model described in Chapter 5.

The current carriers within a LD recombine through both radiative and nonradiative recombination processes. The nonradiative recombination rate is mainly the effect of the Auger recombination and Shockley-Read-Hall (SRH) recombination. The formulae that describes quantitatively these processes can be found in books on LD modelling, and in more detail in books on semiconductor physics and semiconductor device modelling e.g. [12]. In 8.5 we assumed that the nonradiative recombination is negligible.

The group velocity gives the velocity of photons in the lasing mode. Formally, group velocity is defined as the derivative of the propagation constant of the transverse mode with respect to the angular frequency. It is straightforward to calculate this parameter using a mode solver if the refractive index distribution in the LD transverse (transverse with respect to the direction of the lasing mode propagation) cross-section is known. A typical procedure would involve calculating the propagation constant at a number of angular frequencies around the operating wavelength of the LD using the numerical methods discussed in Chapter 3, interpolating the results, and calculating the derivative. To get an accurate result, the material dispersion of the refractive index in each LD layer should be included in the calculations.

Gain measured in inverse units of length is one of the most important parameters of a LD. It can be either calculated or measured using the Hakki-Paoli method. Usually a combined approach is preferred. Namely the spectral dependence of the gain curve is measured using the Hakki-Paoli method [25,26] and then it is used to calibrate the parameters for the gain calculations. The calculation of the gain is performed typically in two stages. Firstly, the dependence of the energy on the electron momentum is calculated using methods of quantum mechanics [27]. Then the golden rule is used to calculate the value of gain. The details of the gain calculations can be found for instance in references [3,16].

The confinement factor gives the measure of the overlap between the photon and carrier reservoirs. It can be defined as a ratio of the carrier reservoir volume over the photon reservoir volume. Such a definition, although giving some insight into the nature of this parameter, does not provide a direct procedure for its calculation. More detail on calculation of the confinement factor can be found in reference [12]. We discuss several practical methods for calculating the confinement factor in the following sections of this chapter.

Photon lifetime is a parameter that gives the measure of how quickly the laser cavity is depleted of photons once the pumping process stops (the bias current source is turned off). This parameter can be fairly simply estimated by considering the photon amplification process with the laser cavity (using the travelling wave approach) cf. Chapter 2 in [12].

The spontaneous emission coupling coefficient says which fraction of the spontaneous emission couples to the lasing mode. It is not straightforward to calculate this coefficient and there are a number of papers that use various approaches to solve this problem, cf. references in Section 4.4 of [12]. Alternatively, the spontaneous emission coupling coefficient can be treated as a fitting parameter to match the experimentally observed curvature of the light current characteristic near threshold. Additionally, in the spectral model, the spontaneous emission coupling coefficient can be used as a fitting parameter for matching the calculated output power spectral width with the measured LD spectrum. Such fitting should be done after fitting the shape of the gain spectrum to the measured gain spectrum because both the shape of the gain curve and the spontaneous emission coupling coefficient affect the shape of the output spectrum.

Another parameter used in a rate equations model is the rate of the spontaneous emission. Usually, a bimolecular recombination coefficient is used for this purpose [14]. Here we simply use a linear approximation $R_{sp} = N/\tau$ [11]. Finally, the rate of the stimulated emission was approximated by $R_{st} = v_g g N_p$ [11,12].

After this preliminary discussion of the rate equations, we will consider three types of 0D LD models: a CW model, a time domain model, and a spectral model.

0D CW MODEL

In the case of the CW model, we consider the solutions of the rate Equations 8.5 that are obtained under a DC bias current and are characterized by a constant value of the photon density. Under these assumptions Equation 8.5 reduces to:

$$0 = \frac{\eta I}{q V_c} - R - v_g g N_p$$

$$0 = \left[\Gamma v_g g - \frac{1}{\tau_p} \right] N_p + \Gamma \beta R_{sp} \tag{8.6}$$

Before proceeding with obtaining the solution of 8.6, we first recast these equations into the following form:

$$N = \frac{\dfrac{\eta I}{q V} + v_g a N_{tr} N_p}{\dfrac{1}{\tau} + v_g a N_p} \tag{8.7a}$$

$$N_{\mathrm{p}} = N_{\mathrm{p}} \exp \left\{ \left[\left(\Gamma v_g a N_{\mathrm{tr}} \left(\frac{N}{N_{\mathrm{tr}}} - 1 \right) + \Gamma \beta \frac{1}{\tau} \frac{N_{\mathrm{tr}}}{N_{\mathrm{p}}} \frac{N}{N_{tr}} - \frac{1}{\tau_{\mathrm{p}}} \right) \frac{2L}{v_g} \right] \right\} \qquad (8.7\mathrm{b})$$

where L is the cavity length (Figure 8.3). The importance of this form will become obvious when studying the more dimensional models in the next section. Equation 8.7 can be further recast into:

$$\frac{N}{N_{\mathrm{tr}}} = \frac{\dfrac{\tau \eta I}{q V N_{\mathrm{tr}}} + \tau v_g a N_{\mathrm{p}}}{1 + \tau v_g a N_{\mathrm{p}}} = \frac{E + A N_{\mathrm{p}}}{1 + A N_{\mathrm{p}}} \qquad (8.8\mathrm{a})$$

Substituting 8.8a into 8.7b yields:

$$N_{\mathrm{p}} = N_{\mathrm{p}} \exp \left\{ \left(B \left[\frac{E + A N_{\mathrm{p}}}{1 + A N_{\mathrm{p}}} - 1 \right] + \frac{C}{N_{\mathrm{p}}} \frac{E + A N_{\mathrm{p}}}{1 + A N_{\mathrm{p}}} - D \right) \right\} \qquad (8.8\mathrm{b})$$

where $A = \tau v_g a$, $B = 2L\Gamma a N_{\mathrm{tr}}$, $C = 2L\Gamma \beta N_{\mathrm{tr}}/(v_g \tau)$, $D = 2L/(v_g \tau_{\mathrm{p}})$, and $E = \tau I \eta/(N_{\mathrm{tr}} q V)$. Equation 8.8b is expressed using only five independent parameters, which is significantly less than in Equation 8.7b. Furthermore, N_{p} is the only unknown in 8.8b so this equation can be solved for N_{p}. Once N_{p} is known the carrier concentration can be obtained from 8.8a. Equation 8.8a can be solved by any numerical method that is routinely used to solve nonlinear algebraic equations, for example, the fixed point method (FPM) and Newton-Raphson method (NRM). It can be observed, however, that Equation 8.8b can also be solved analytically, which is obvious from recasting it into:

$$-DA\, N_{\mathrm{p}}^2 + \left[BE - B + CA - D \right] N_{\mathrm{p}} + CE = 0 \qquad (8.9)$$

The quadratic Equation 8.9 can be easily solved. Its zeroes are: $N_{\mathrm{p}} = \left(b \pm \sqrt{b^2 + 4DACE} \right)/(2DA)$ where $b = BE - B + CA - D$. Because the $DACE$ and CE products are always larger than zero and $|b| < \sqrt{b^2 + 4DACE}$, the value of N_{p} calculated with the plus sign in front of the square root is always positive while the one calculated with the minus sign is always negative. Therefore, the physically meaningful solution is obtained with the plus sign while the minus sign yields a spurious solution. We note also that an analytical solution of 8.8b could only be obtained because the fitting function for gain and recombination terms is linear. In the general case only numerical methods can be used to obtain a solution of 8.6. We consider in this section two representative numerical techniques for solving 8.8b.

First we consider the FPM. For this purpose we set up the following fixed point sequence for Equation 8.8b:

$$N_p^{n+1} = N_p^n \exp\left\{\left(B\left[\frac{E + AN_p^n}{1 + AN_p^n} - 1\right] + \frac{C}{N_p}\frac{E + AN_p^n}{1 + AN_p^n} - D\right)\right\} \qquad (8.10)$$

To gain a better understanding of the FPM, we first perform the analysis of the sequence 8.10. The necessary and sufficient condition for the convergence of a fixed point sequence is that the absolute value of the derivative of the right-hand side (RHS) of Equation 8.6 in the vicinity of the solution is less than 1 [28].

As an example we consider an $In_{0.2}Ga_{0.8}As/GaAs$ QW laser that has been designed for an operating wavelength near 980 nm (see the section "Time Domain Analysis" in reference [12]). We approximate the gain dependence on the carrier density using a linear model: $g = a(N - N_{tr})$ whereby the linear gain coefficient $a = 7.88 \times 10^{-16}$ cm^2 and the transparency carrier density $N_{tr} = 1.8 \times 10^{18}$/cm^3. The values of all modelling parameters used in this subsection are listed in Table 8.1.

Figure 8.5 shows the dependence of the RHS of Equation 8.10 on the photon density at selected values of the bias current for the LD with the parameter set given in Table 8.1.

The crossing point of the RHS of 8.6 with the line $N_p = N_p$ gives a solution of 8.10. It can be observed that the inclination of the line that is tangential to the graphical representation of 8.10 at the crossing point with the line $N_p = N_p$ increases with the bias current. In fact, the inclination of this line is larger than 45 degrees, even at very small values of the bias current. This indicates that the FPM converges to the solution only if the bias current is relatively low. This conclusion is confirmed by the results shown in Figure 8.6.

TABLE 8.1
Values of a Gallium Arsenide Laser Diode Parameters

v_g	$(3/4.7) \times 10^{10}$ cm/s	Group Velocity
d	8 nm	Quantum well (QW) width
w	2 µm	Waveguide width
L	0.25 mm	Cavity length
V	4×10^{-12} cm^3	Active cavity volume = $L \times w \times d$
Γ	0.032	Confinement factor
N_{tr}	1.8×10^{18} 1/cm^3	Transparency carrier density
a	7.88×10^{-16} cm^2	Linear gain coefficient
τ	2.71 ns	Carrier lifetime
τ_p	2.77 ps	Photon lifetime
λ	980 nm	Wavelength
β	0.869×10^{-4}	Spontaneous emission coefficient
η	1	Current injection efficiency
R	0.32	Facet reflectivity

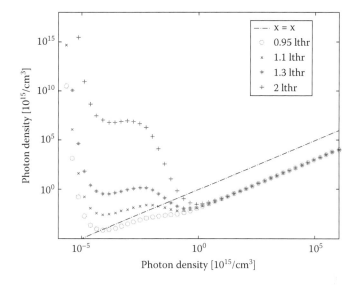

FIGURE 8.5 The plot of RHS of Equation 8.8b. (Sujecki, S., Stability of steady-state high-power semiconductor laser models. *Journal of Optical Society of America B*, 2007. 24(4): p. 1053–1060.)

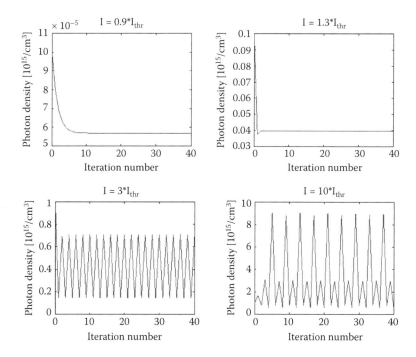

FIGURE 8.6 Dependence of the photon density on the iteration number for the fixed point method (FPM) using the sequence defined by the eqution 8.10. (Sujecki, S., Stability of steady-state high-power semiconductor laser models. *Journal of Optical Society of America B*, 2007. 24(4): p. 1053–1060.)

**ALGORITHM 8.1   THE FPM SOLUTION
OF THE RATE EQUATIONS 8.8**

1.   Start
2.   Set the modelling parameters
3.   Select the initial value of photon density N_p^0
4.   Obtain N_p^{n+1} from 8.11
5.   If $\left| \vec{N}^{i+1} - \vec{N}^i \right| > tolerance$ go to 4
6.   Stop

Figure 8.6 gives the dependence of the photon density N_p on the fixed point iteration number at four selected values of the bias current. Near the laser threshold, the
convergence of the FPM is fast. However, for larger values of the bias current, the FPM
does not converge, and instead oscillates around the solution. Unfortunately, the convergence is lost at quite small values of the bias current, so the FPM sequence defined
by Equation 8.10 is not of large practical importance. However, quite a simple modification of the FPM sequence 8.10 can secure the convergence for an arbitrary value of
the bias current. An introduction of a parameter p in 8.10 gives the following sequence:

$$N_p^{n+1} = N_p^n \exp\left\{ \left(B\left[\frac{E + AN_p^n}{1 + AN_p^n} - 1 \right] + \frac{C}{N_p} \frac{E + AN_p^n}{1 + AN_p^n} - D \right) p \right\} \qquad (8.11)$$

It can be easily verified that the introduction of the parameter p does not modify
the values of N_p that satisfy Equation 8.10. The implementation of the FPM based on
Equation 8.11 leads to the following simple algorithm.

The Algorithm 8.1 can be implemented using the following MATLAB script:

```
% semiconductor laser analysis using 0D rate equations
% 0D steady state analysis - fixed point method
clear
format long e
pi = 3.141592653589793e+000;

h = 6.626e-34;%[Js]
q = 1.609e-19;%[C]
V = 4e-12;%[cm3]
gamma = 0.032;%confinement factor
betasp = 1e-4;%spontaneous emission
vg = 3e10/3.9;%[cm/s]
v0 = 3e10;
lam = 1.3e-4;%[cm]
freq = v0/lam;%[1/s]
tau = 0.5e-9;%[s]
L = 0.025;% cavity length [cm]
```

```
Rf = 0.32;% mirror reflectivity front
Rb = 0.32;% mirror reflectivity back
taup = 2*L/(vg*log(1/(Rf*Rb)));
%[s] = 1/(alfam*vg);alfam = (ln(1/R))/L
neta = 0.7;%[current injection efficiency
Ntr = 2.1e+18;% [1/cm3]
alfam = 1/(vg*taup);
% cavity optical loss (no scattering loss)
I =.002;% total current [A]

alfam = 1/(vg*taup);%no scattering loss

a_gain = 8.;% 1/gain curvature coefficient [(nm*cm^1/2)]
a = 2.1e-15;% gain coefficient [cm2]
alam = a*ones(1,1);

iter_no = 40;% number of iterations
% iteration loop
a_e(1) = neta*I/(q*V);a_e(2) = -1/tau;a_e(3) = -vg*a;
a_e(4) = vg*a*Ntr;
b_e(1) = gamma*vg*a;b_e(2) = -gamma*vg*a*Ntr;
b_e(3) = gamma*betasp/tau;
b_e(4) = -1/taup;

x = 2.571210747598624e+015*ones(1,1);%initial photon density
suma = sum(vg*alam.*x);
y = (-a_e(1)-suma*Ntr)/(-suma+a_e(2));%initial carrier density

relax_par = 0.6;
for k = 1:iter_no
        x = exp((b_e(1)*y.*alam/a+b_e(2)*alam/a+b_e(3)*...
        y./x+b_e(4))*relax_par*L/vg).*x;
        photons(:,k) = x;
    suma = sum(vg*alam.*x);
    y = (-a_e(1)-suma*Ntr)/(-suma+a_e(2));
    carriers(k) = y;
end
```

For simplicity in the aforementioned code, we have limited the number of iterations rather than relying on the checking of a residual in step 4 of the Algorithm 8.1. However, a residual check can be easily introduced into the script if required.

Figure 8.7 shows the dependence of the photon density on the iteration number calculated using the FPM sequence defined by Equation 8.11. These results were obtained at 10 times the threshold current. Two values of the parameter p were used. For $p = 1$, which is equivalent to using the sequence 8.10, and $p = 0.5$. No convergence can be observed when $p = 1$. However, when $p = 0.5$ a convergent sequence is obtained. In fact, it can be seen from 8.11 that p reduces the coefficient by which photon density is multiplied and therefore also reduces the value of the derivative of the RHS of 8.11 near the crossing point with the line $N_p = N_p$. Hence, with 8.11 we obtained a FPM that can calculate solutions of the rate Equation 8.6. We note that

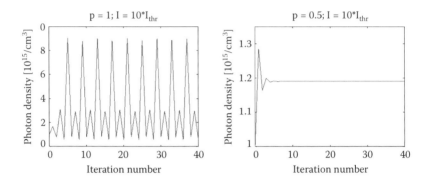

FIGURE 8.7   Dependence of the photon density on the iteration number for the FPM using the sequence defined by Equation 8.11.

the convergence rate depends critically on the arbitrary parameter p. Fortunately, it is not difficult to select an appropriate value of p by trial and error. This is because the range of p parameters is restricted to the interval $(0,1)$ only. Furthermore, for large values of p (near to 1), an oscillating sequence is obtained, whilst for small ones (near to 0) a very slowly converging sequence follows.

For comparison, we also solve Equation 8.8 using the Newton method (NM). Figure 8.8 shows the dependence of the photon density on the iteration number for the NM at two selected values of the bias current. Fast convergence is achieved both near the threshold and for large values of the bias current. This is to be expected because it is known that the quadratic convergence of NM is guaranteed if the initial guess is sufficiently near to the solution. Further away from the solution, the NM has a slower rate of convergence. This problem can be remedied by using a global Newton method as discussed in Chapter 6. However, it should be noted that when starting far away from the solution the NM algorithm can converge to a spurious solution. Equation 8.8b, in fact, is equal to zero at three points. Figure 8.9 shows a plot of the RHS of 8.8b including negative values of the photon density. The intersections with the line $N_p = N_p$ correspond to the zeroes of 8.8b. There is one solution for a photon density equal to zero; that is, a trivial solution of Equation 8.8b that was added implicitly while recasting Equation 8.6 into 8.8b, and the two solutions that have been considered while discussing Equation 8.9. Namely, there is one spurious solution for a negative value of the photon density; that is, a solution that is an artefact produced by the mathematical model. There is also a physically meaningful solution that is obtained for the positive value of the photon density. Consequently, the NM algorithm may either converge to a solution of Equation 8.8b that has no physical meaning or fail to converge if either a numerical overflow or underflow is encountered during the NR iteration process. The standard way of forcing the NM algorithm to converge towards the physically meaningful solution consists in introducing the slack variables [29].

The difficulties encountered while applying the NR method are illustrated in Figure 8.10. If the initial guess differs much from the solution, the NR method fails to converge. However, when the guess is brought nearer to the solution, the convergence rate is initially very slow and eventually a quadratic rate is attained. The FPM

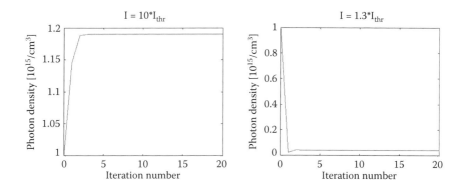

FIGURE 8.8 Dependence of the photon density on the iteration number for the Newton method.

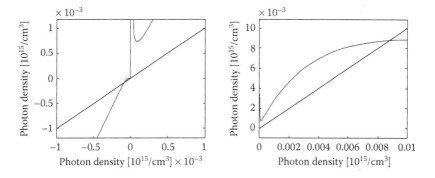

FIGURE 8.9 The plot of RHS of Equation 8.8b.

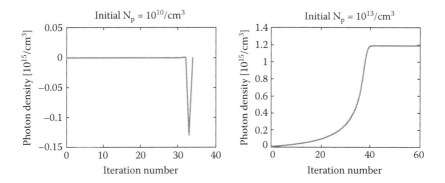

FIGURE 8.10 Dependence of the photon density on the iteration number for the Newton method at $I = 10 \times I_{th}$.

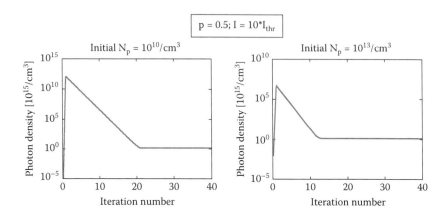

FIGURE 8.11   Dependence of the photon density on the iteration number for the FPM.

on the other hand is much less sensitive to the initial guess if the parameter p is selected optimally (Figure 8.11). It can be therefore concluded that the FPM is more robust than the NM. However, the rate of the FPM convergence strongly depends on the value of an arbitrary parameter p.

0D TIME DOMAIN MODEL

The CW model can only be used to study the characteristics of a LD at a constant bias current and if the output power does not vary with time. However, in many practical applications LDs are used in the pulse operating mode, using for instance either the direct modulation or mode locking [4]. To study the properties of a LD under such operating conditions, an application of the time domain modelling tools is necessary. In this section we consider a 0D time domain model. The time domain analysis consists in solving the two coupled ordinary differential Equation 8.5. As far as numerical analysis is concerned, this is a straightforward task, because the problem considered is an example of an initial value problem. For an illustration, we used an explicit Euler method to integrated the ordinary differential equations:

$$N(t_{n+1}) = N(t_n) + \Delta t \left(\frac{\eta I(t_n)}{qV} - \frac{N(t_n)}{\tau} - v_g a \left(N(t_n) - N_{tr} \right) N_p(t_n) \right)$$

$$N_p(t_{n+1}) = N_p(t_n) + \Delta t \left(\left[\Gamma v_g a \left(N(t_n) - N_{tr} \right) - \frac{1}{\tau_p} \right] N_p(t_n) + \Gamma \beta \frac{N(t_n)}{\tau} \right) \qquad (8.12)$$

where Δt is the time step. In 8.12, similarly as in 8.6, we used a linear gain approximation. The implementation of the Euler method to the solution of Equation 8.5 results in Algorithm 8.2.

ALGORITHM 8.2   THE EULER METHOD
SOLUTION OF THE RATE EQUATIONS 8.5

1.   Start
2.   Set the modelling parameters
3.   Calculate the initial value of photon density $N_p(t_0)$ and carrier density $N(t_0)$ at the given initial value of the bias current (using e.g. FPM)
4.   Obtain $N_p(t_{n+1})$ and $N(t_{n+1})$ from 8.12
5.   If $t_{n+1} < t_{final}$ go to 4
6.   Stop

Algorithm 8.2 can be implemented using the following MATLAB script:

```
% semiconductor laser analysis using 0D rate equations
% TD analysis
clear
format long e
pi = 3.141592653589793e+000;

h = 6.626e-34;%[Js]
q = 1.609e-19;%[C]
V = 4e-12;%[cm3]
gamma = 0.032;%confinement factor
betasp = 1e-4;%spontaneous emission
vg = 3e10/3.9;%[cm/s]
v0 = 3e10;
lam = 1.3e-4;%[cm]
freq = v0/lam;%[1/s]
tau =.5e-9;%[s]
L = 0.025;% cavity length [cm]
Rf = 0.32;% mirror reflectivity front
Rb = 0.32;% mirror reflectivity back
taup = 2*L/(vg*log(1/(Rf*Rb)))
%[s] = 1/(alfam*vg);alfam = (ln(1/R))/L
neta = 0.7;%[current injection efficiency
Ntr = 2.1e+18;% [1/cm3]
alfam = 1/(vg*taup);
% cavity optical loss (no scattering loss)
I =.006;% total current [A]

alfam = 1/(vg*taup);%no scattering loss

a_gain = 8.;% 1/gain curvature coefficient [(nm*cm^1/2)]
a = 2.1e-15;% gain coefficient [cm2]
alam = a*ones(1,1);
```

```
%%%%%%%%%%%%%%%%%%%%%%%%%%%%%%%%%%%%%%%%%%%%%%%%%%%%%%%%%%%%%%
% initial bias point calculation
iter_no = 40;% number of iterations
% iteration loop
a_e(1) = neta*I/(q*V);a_e(2) = -1/tau;a_e(3) = -vg*a;
a_e(4) = vg*a*Ntr;
b_e(1) = gamma*vg*a;b_e(2) = -gamma*vg*a*Ntr;
b_e(3) = gamma*betasp/tau;
b_e(4) = -1/taup;

x = 2.571210747598624e+015*ones(1,1);%initial photon density
suma = sum(vg*alam.*x);
y = (-a_e(1)-suma*Ntr)/(-suma+a_e(2));%initial carrier density

relax_par = 0.6;
for k = 1:iter_no
        x = exp((b_e(1)*y.*alam/a+b_e(2)*alam/a+b_e(3)…
        *y./x+b_e(4))*relax_par*L/vg).*x;
        photons(:,k) = x;
    suma = sum(vg*alam.*x);
    y = (-a_e(1)-suma*Ntr)/(-suma+a_e(2));
    carriers(k) = y;
    k
end

%%%%%%%%%%%%%%%%%%%%%%%%%%%%%%%%%%%%%%%%%%%%%%%%%%%%%%%%%%%%%%
% Time domain analysis
x = (photons(:,iter_no));% initial photon distribution
y = carriers(iter_no);% intitial carrier distribution
Num_of_TS = 10000;% Number of time steps
Current = 0.02*ones(1,Num_of_TS);%Driving current waveform
for j = 1:Num_of_TS/100
    Current(j) = I;
end

dt = 0.4e-12;%time step for Euler method
for mk = 1:Num_of_TS
    a_e(1) = neta*Current(mk)/(q*V);
    x = x+dt*((b_e(1)*y.*alam/a+b_e(2)*alam/
      a+b_e(4)).*x+b_e(3)*y);
    photons1(:,mk) = x;
    suma = sum(vg*alam.*x);
    y = y+dt*((a_e(1)+suma*Ntr)-y.*(suma-a_e(2)));
    carriers1(mk) = y;
end
```

The modelling parameters used in the simulations are summarised in Table 8.2 and refer to a GaInNAs/GaAs QW LD [30].

Using the Algorithm 8.2, we calculated the dependence of the photon and carrier density on time during turning on of the LD bias current. The calculated current

TABLE 8.2
Values of a GaInNAs/GaAs QW Laser Diode Parameters

v_g	$(3/3.9)*10^{10}$ cm/s	Group Velocity
d	8 nm	QW width
w	2 mm	Waveguide width
L	0.25 mm	Cavity length
V	$4*10^{-12}$ cm^3	Active cavity volume $= L*w*d$
Γ	0.032	Confinement factor
N_{tr}	$2.1*10^{18}$ 1/cm^3	Transparency carrier density
a	$2.1*10^{-15}$ cm^2	Linear gain coefficient
τ	0.5 ns	Carrier lifetime (assumed)
τ_p	2.85 ps	Photon lifetime
λ	1300 nm	Wavelength
β	$1*10^{-4}$	Spontaneous emission coefficient
η	0.7	Current injection efficiency
R	0.32	Facet reflectivity
$\Delta\lambda$	0.87 nm	Resonant wavelength spacing

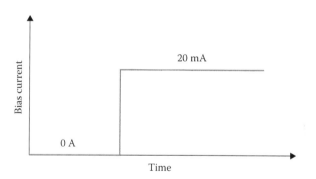

FIGURE 8.12   The bias current waveform for the results shown in Figure 8.13.

waveform is shown in Figure 8.12 while the photon and carrier density waveforms are presented in Figure 8.13. The time step was 0.2 ps. The results obtained show the expected turn on delay and ringing effects cf. Chapter 5 in [12]. A very short time step required by the Euler method can be easily relaxed by using more sophisticated integration algorithms for the ordinary differential equations [31].

0D SPECTRAL MODEL

In a typical LD spectrum usually, several longitudinal modes lase simultaneously. For studying the dependence of the LD output power spectrum on the bias current, neither of the models presented in the previous sections is appropriate. For this

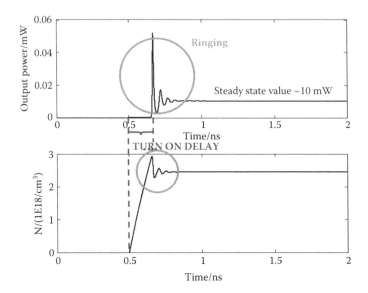

FIGURE 8.13   The photon and carrier density waveforms.

purpose the rate Equation 8.5 has to be extended to account for multiple longitudinal modes [11,32]:

$$\frac{dN}{dt} = \frac{\eta I}{qV} - \frac{N}{\tau} - v_g \sum_n \left[a*(N - N_{tr}) - dg^2 * (\lambda_p - \lambda_n)^2 \right] * N_p(\lambda_n)$$

$$\frac{dN_p(\lambda_n)}{dt} = \left[\Gamma v_g * \left(a*(N - N_{tr}) - dg^2 * (\lambda_p - \lambda_n)^2 \right) - \frac{1}{\tau_p} \right] * N_p(\lambda_n) + \Gamma\beta\frac{N}{\tau} \quad (8.13)$$

In 8.13 the linear gain approximation has been extended to include the gain dependence on the wavelength: $g_n = a(N - N_{tr}) - (dg)^2*(\lambda_p - \lambda_n)^2$ whereby dg defines the gain curvature near the gain peak wavelength λ_p and λ_n, which is the operating wavelength of the nth longitudinal mode. Such a formula for gain implies a quadratic dependence of gain on the wavelength near the peak gain, which is well justified if the LD output spectrum is narrow [11]. The index "n" in 8.13 is used to denote the longitudinal modes. For instance, if seven longitudinal modes are considered, then the index n runs from –3 to 3 (Figure 8.14).

In the steady state, the derivatives on the LHS of 8.13 are equal to zero and the following set of equations is obtained:

$$0 = \frac{\eta I}{qV} - \frac{N}{\tau} - v_g \sum_n \left[a*(N - N_{tr}) - dg^2 * (\lambda_p - \lambda_n)^2 \right] * N_p(\lambda_n)$$

$$0 = \left[\Gamma v_g * \left(a*(N - N_{tr}) - dg^2 * (\lambda_p - \lambda_n)^2 \right) - \frac{1}{\tau_p} \right] * N_p(\lambda_n) + \Gamma\beta\frac{N}{\tau} \quad (8.14)$$

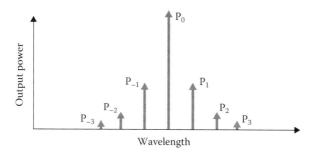

FIGURE 8.14 A schematic diagram of a LD output power spectrum.

Similarly, as in the Section "0D CW Model," Equation 8.14 can be recast in the following form:

$$0 = N - \frac{\frac{\eta I}{qV} + v_g \sum_n \left[aN_{tr} - dg^2 * \left(\lambda_p - \lambda_n \right)^2 \right] * N_p \left(\lambda_n \right)}{\frac{1}{\tau} + v_g \sum_n a * N_p \left(\lambda_n \right)}$$

$$0 = N_p \left(\lambda_n \right) - N_p \left(\lambda_n \right) * \exp \left\{ \left[\left[\Gamma v_g * \left(a * \left(N - N_{tr} \right) - dg^2 * \left(\lambda_p - \lambda_n \right)^2 \right) - \frac{1}{\tau_p} \right] \right. \right.$$

$$\left. + \Gamma \beta \frac{1}{\tau} \frac{1}{N_p \left(\lambda_n \right)} \right) \frac{2L}{v_g} \right\} \tag{8.15}$$

Equation 8.15 can be solved using a similar FPM sequence as in the Section "0D CW Model":

$$N^{i+1} = \frac{\frac{\eta I}{qV} + v_g \sum_n \left[aN_{tr} - dg^2 * \left(\lambda_p - \lambda_n \right)^2 \right] * N_p^i \left(\lambda_n \right)}{\frac{1}{\tau} + v_g \sum_n a * N_p^i \left(\lambda_n \right)}$$

$$N_p^{i+1} \left(\lambda_n \right) = N_p^i \left(\lambda_n \right) * \exp \left\{ \left[\left[\Gamma v_g * \left(a * \left(N^i - N_{tr} \right) - dg^2 * \left(\lambda_p - \lambda_n \right)^2 \right) - \frac{1}{\tau_p} \right] \right. \right.$$

$$\left. + \Gamma \beta \frac{1}{\tau} \frac{1}{N_p^i \left(\lambda_n \right)} \right) \frac{2Lp}{v_g} \right\} \tag{8.16}$$

Equation 8.16 can be conveniently represented in the following form:

$$N^{i+1} = G_0\left(N^i, N_p^i\left(\lambda_1\right),..., N_p^i\left(\lambda_n\right)\right)$$

$$N_p^{i+1}\left(\lambda_n\right) = G_n\left(N^i, N_p^i\left(\lambda_1\right),..., N_p^i\left(\lambda_n\right)\right)$$

which can yet be presented in a more compact fashion:

$$\vec{N}^{i+1} = \vec{G}\left(\vec{N}^i\right) \tag{8.17}$$

whereby the vectors $\vec{G}$ and $\vec{N}^i$ are defined as follows:

$$\vec{G}\left(\vec{N}^i\right) = \begin{bmatrix} G_0\left(\vec{N}^i\right) \\ G_1\left(\vec{N}^i\right) \\ \vdots \\ G_n\left(\vec{N}^i\right) \end{bmatrix}$$

$$\vec{N}^i = \begin{bmatrix} N_0^i \\ N_1^i \\ \vdots \\ N_n^i \end{bmatrix} = \begin{bmatrix} N^i \\ N_p^i\left(\lambda_1\right) \\ \vdots \\ N_p^i\left(\lambda_n\right) \end{bmatrix}$$

Equation 8.15 can be also solved using the NRM as an alternative technique to 8.16. Using the notation applied in 8.17 NRM yields the following sequence for calculating iteratively a solution of 8.15:

$$\vec{N}^{i+1} = \vec{N}^i + \Delta\vec{N}^i \tag{8.18}$$

$\Delta\vec{N}^i$ is obtained from:

$$J^i\Delta\vec{N}^i = -\vec{F}^i\left(\vec{N}^i\right)$$

and

$$\vec{F}^i\left(\vec{N}^i\right) = -\vec{N}^i + \vec{G}\left(\vec{N}^i\right) = 0$$

while J is the Jacobian matrix:

$$J = \begin{bmatrix} \dfrac{\partial F_0}{\partial N_0} & \dfrac{\partial F_0}{\partial N_1} & \cdots & \dfrac{\partial F_0}{\partial N_n} \\[2mm] \dfrac{\partial F_1}{\partial N_0} & \dfrac{\partial F_1}{\partial N_1} & \cdots & \dfrac{\partial F_1}{\partial N_n} \\[2mm] \vdots & \vdots & \ddots & \vdots \\[2mm] \dfrac{\partial F_n}{\partial N_0} & \dfrac{\partial F_n}{\partial N_1} & \cdots & \dfrac{\partial F_n}{\partial N_n} \end{bmatrix}$$

The Jacobian matrix can be conveniently calculated using finite differences [33]:

$$J \approx \begin{bmatrix} \dfrac{F_0\left(\vec{N}+\Delta N_0\right)-F_0\left(\vec{N}\right)}{\Delta N_0} & \dfrac{F_0\left(\vec{N}+\Delta N_1\right)-F_0\left(\vec{N}\right)}{\Delta N_1} & \cdots & \dfrac{F_0\left(\vec{N}+\Delta N_n\right)-F_0\left(\vec{N}\right)}{\Delta N_n} \\[3mm] \dfrac{F_1\left(\vec{N}+\Delta N_0\right)-F_1\left(\vec{N}\right)}{\Delta N_0} & \dfrac{F_1\left(\vec{N}+\Delta N_1\right)-F_1\left(\vec{N}\right)}{\Delta N_1} & \cdots & \dfrac{F_1\left(\vec{N}+\Delta N_n\right)-F_1\left(\vec{N}\right)}{\Delta N_n} \\[3mm] \vdots & \vdots & \ddots & \vdots \\[3mm] \dfrac{F_n\left(\vec{N}+\Delta N_0\right)-F_n\left(\vec{N}\right)}{\Delta N_0} & \dfrac{F_n\left(\vec{N}+\Delta N_1\right)-F_n\left(\vec{N}\right)}{\Delta N_1} & \cdots & \dfrac{F_n\left(\vec{N}+\Delta N_n\right)-F_n\left(\vec{N}\right)}{\Delta N_n} \end{bmatrix}$$

$$(8.19)$$

The addition of a scalar ΔN_n to the vector $\vec{N}$ is understood as adding a scalar ΔN_n to the nth element of the vector $\vec{N}$. The Jacobian matrix 8.19 has been evaluated following the guidance provided in reference [33], that is, by varying each element of $\vec{N}$ by 0.01%. The form of the Jacobian matrix 8.19 allows an arbitrary function (or a lookup table) to be incorporated into the model to calculate the gain and recombination rate dependence on the carrier density.

Figure 8.15 compares the light current characteristics obtained using the FPM considering only one wavelength, that is, Equation 8.8 with the result obtained using the spectral model that is based on the fixed point sequence 8.16. We used the same LD parameters as in the subsection "0D Time Domain Model." The gain parameter $dg = 1/(8 \text{ nm cm}^{1/2})$. We used the following formula for calculating the wavelength spacing [16]: $\Delta\lambda = \lambda^2/(2*n_g*L)$. The threshold current and slope efficiency, calculated by both methods, is generally very similar. However, near the threshold there is a significant difference. In the case of the spectral model, the output power before threshold is noticeably larger than zero. Figure 8.16 shows the dependence of the photon density on the FPM iteration number and the output power spectrum calculated after performing 10,000 FPM iterations. These results show that the convergence of the FPM is very slow.

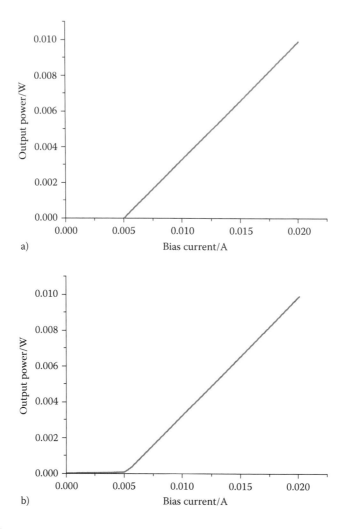

FIGURE 8.15 Light current characteristics calculated the a) CW single wavelength model and b) CW spectral model.

The convergence rate can be significantly improved by the application of the NRM 8.18. However, the NRM is very sensitive to the initial guess. Therefore, we use an algorithm that combines the FPM with the NM. The FPM is used to get an initial guess for the NRM. The NRM then refines the solution within several iterations. This leads to Algorithm 8.3.

Figure 8.17 shows the dependence of the photon density on the iteration step and the calculated output spectrum. The initial guess was calculated using 10,000 iterations of the FPM, which corresponds to the results shown in Figure 8.16. The results

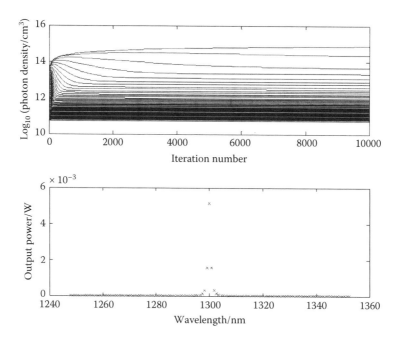

FIGURE 8.16 The dependence of the photon density on the iteration number and the wavelength dependence of the output power obtained after 10,000 fixed point iterations. 121 wavelengths were included in this simulation and the bias current is 20 mA.

ALGORITHM 8.3   THE NRM SOLUTION OF THE RATE EQUATIONS 8.14

1. Start
2. Set modelling parameters
3. Calculate the initial value of $\vec{N}^i = \vec{N}^0$ using a fixed number of FPM 8.16 iterations
4. Initiate NM using the initial guess obtained from FPM
5. Obtain N_p^{n+1} from 8.18
6. If $\left| \vec{N}^{i+1} - \vec{N}^i \right| > tolerance$ go to 5
7. Stop

calculated by the NRM settled down within 10 iterations. It can be observed that the output spectrum obtained after 10,000 iterations of the FPM (Figure 8.16) is still far from the converged one (Figure 8.17) that was obtained by refining the FPM results using the NRM.

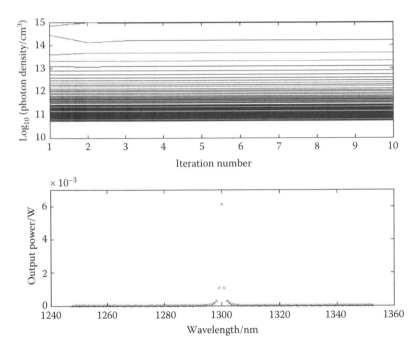

FIGURE 8.17 The dependence of the photon density on the iteration number and the wavelength dependence of the output power calculated using the Newton-Raphson method. In this simulation, 121 wavelengths were included and the bias current is 20 mA.

The last example considered in this subsection is a 0D time domain spectral model. The application of the Euler method to 8.13 yields:

$$N|_{t0+\Delta t} = N|_{t0} + \Delta t \left(\frac{\eta I|_{t0}}{qV} - \frac{N|_{t0}}{\tau} \right.$$

$$\left. - v_g \sum_n \left[a * \left(N|_{t0} - N_{tr} \right) - dg^2 * \left(\lambda_p - \lambda_n \right)^2 \right] * N_p|_{t0} \left(\lambda_n \right) \right)$$

$$N_p|_{t0+\Delta t} \left(\lambda_n \right) = N_p|_{t0} \left(\lambda_n \right) + \Delta t \left[\Gamma v_g * \left(a * \left(N|_{t0} - N_{tr} \right) - dg^2 * \left(\lambda_p - \lambda_n \right)^2 \right) - \frac{1}{\tau_p} \right]$$

$$* N_p|_{t0} \left(\lambda_n \right) + \Gamma \beta \frac{N|_{t0}}{\tau} \tag{8.20}$$

The calculation of the solution of 8.13 using 8.20 subject to known initial conditions results in an algorithm that effectively mimics Algorithm 8.2, with the only difference being in the number of equations that need to be considered. The initial

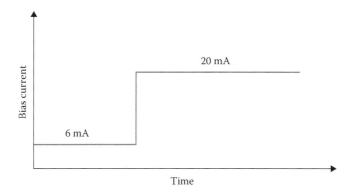

FIGURE 8.18   The bias current waveform for the results shown in Figures 8.18 and 8.19.

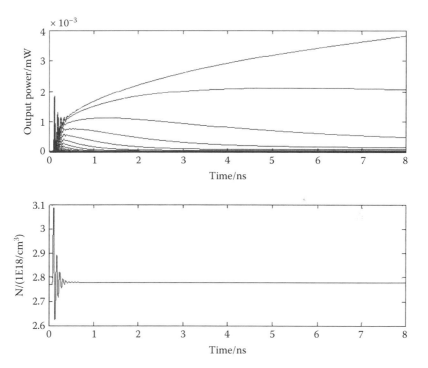

FIGURE 8.19   The longitudinal mode resolved photon density and carrier density waveforms.

values of the photon and carrier densities that are consistent with the given bias current are obtained using Algorithm 8.3.

Figures 8.19 and 8.20 show the dependence of the photon and carrier densities on time subject to the bias current waveform given in Figure 8.18. These results were obtained with the modelling parameters listed in Table 8.2. The calculation step was 0.4 ps while the values of the wavelength spacing and the gain parameter

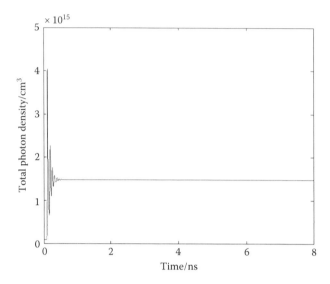

FIGURE 8.20   The total photon density waveform.

dg were the same as for the steady state spectral model. The total photon density and the carrier density settle down within several nanoseconds as predicted by the single wavelength model. However, the output spectrum evolves much slower, cf. Figures 8.19 and 8.20.

In summary, the spectral model allows analysing the LD output spectrum and its time evolution. This however, is obtained at a significant increase of the computational effort when compared with the single wavelength models. In the next section, we examine the validity of the assumption that the photon and carrier distribution within the laser cavity is independent of the spatial position. For this purpose we derive a 1D LD model.

1D LASER DIODE MODELS

There are three types of the 1D LD models depending on which dimension is explicitly included in the calculations (Figure 8.2). The vertical dimension models are usually applied to design the epitaxial structure of the LD. The lateral dimension models were used in studying the properties of the broad area lasers [34], lasers arrays [35], and VCSELs [36]. The steady-state longitudinal dimension models proved useful when analysing DBR lasers [37] and for studying the spatial hole burning [38]. 1D time domain models found a particular application in the study of the pulsed operation of LDs: the Q-switching, gain switching, passive and the active mode locking [4]. The longitudinal 1D models are also used extensively in the analysis of the laser amplifiers [10,39–44]. Typically, the travelling wave approach is followed in these applications when calculating the photon distribution within the laser cavity [45–51]. More recently, however, the delay differential equation approach is applied instead, because it allows for much more efficient calculations [52–54].

In this section we select a 1D CW longitudinal model as an illustrating example. To derive the equation for the photon flux evolution within the LD cavity we assume that the gain is provided by a thin QW layer. Furthermore, the gain region introduces only a small perturbation $\Delta n(x,y)$ to the refractive index distribution $n(x,y)$ of the optical waveguide cf. Chapter 6 in [12] while the optical beam trapped in the LD cavity propagates in the fundamental transverse mode of the unperturbed waveguide. We limit the considerations to the scalar case only. We apply the travelling wave approximation and decompose the inner cavity field distribution into the sum of the forward and backward propagating waves.

Let's consider first the equation for the evolution of the forward propagating wave. Taking advantage of all the assumptions we can represent, the field distribution of the forward propagating wave in the product form: $\Psi(x,y,z) = A_f(z)*F(x,y)*e^{j(\omega t-\beta z)}$ where $F(x,y)$ is the fundamental mode distribution of the unperturbed waveguide, $A_f(z)$ is a slowly changing function of z, and β is the fundamental mode propagation constant. To define $\Psi(x,y,z)$ uniquely, we impose the condition:

$$\iint F^*(x,y)F(x,y)\,dxdy = I_N$$

on $F(x,y)$ where "*" stands for the complex conjugate. We substitute $\Psi(x,y,z)$ into the scalar wave Equation (2.1), and after following the same steps as in the section "Fibre Amplifier Modelling," we obtain (cf. Equation 7.19):

$$\frac{dP_f(z)}{dz} = \hat{\Gamma}(z)P_f(z) \tag{8.21a}$$

where

$$\hat{\Gamma}(z) = \frac{k_0}{\beta}\iint F^*(x,y)\ n\ (x,y,z)g_m(x,y,z)F(x,y)\,dxdy\Big/I_N \tag{8.21b}$$

In the same way, the equation describing the evolution of the backward propagating wave can be derived. The equations describing the forward and backward wave evolution can be complemented by either the ambipolar equation 6.60 or the drift-diffusion Equation 6.11, to form a set of equations that can be solved self-consistently subject to the known boundary conditions. The solution of such set of equations would necessitate calculating the carrier concentrations at a given longitudinal position, in the entire transverse plane, similarly to Algorithm 7.1. Therefore, to obtain a 1D LD model, further simplifying assumptions are unavoidable. A 1D model can be obtained if at a given longitudinal position the material gain within the carrier reservoir is assumed constant, that is: $g_m(x,y,z) = g_m(z)$ and zero outside of it. This allows simplifying 8.21b to:

$$\hat{\Gamma}(z) = \frac{k_0 n_{qw}g_m(z)}{\beta}\iint\limits_{S_{ct}} F^*(x,y)\ F(x,y)\,dxdy\Big/I_N \tag{8.22}$$

where n_{qw} is the value of the refractive index in the QW layer and the surface of S_{ct} is the surface of the transverse cross section of the carrier reservoir at a given longitudinal position (Figure 8.4). Equation 8.22 can be further recast into:

$$\hat{\Gamma}(z) = \Gamma g_m(z) \tag{8.23}$$

where the z independent confinement factor Γ is expressed as:

$$\Gamma = \frac{k_0 n_{qw}}{\beta} \iint\limits_{S_{ct}} F^*(x,y)\ F(x,y)\,dxdy \Big/ I_N \tag{8.24}$$

Equation 8.24 reduces to

$$\Gamma = \iint\limits_{S_{ct}} F^*(x,y)\ F(x,y)\,dxdy \Big/ I_N \tag{8.25}$$

if the mode effective index value does not differ significantly from the QW refractive index.

Introducing 8.23 into 8.21a yields:

$$\frac{dP_f(z)}{dz} = \Gamma g_m(z) P_f(z) \tag{8.26a}$$

Similarly, for the backward propagating wave we obtain the equation:

$$\frac{dP_b(z)}{dz} = \Gamma g_m(z) P_b(z) \tag{8.26b}$$

The assumption of the constant value of gain within the carrier reservoir cross section is consistent with assuming that the carrier density is constant within the carrier reservoir in a transverse plane at a given longitudinal position. Hence, we complement Equation 8.26 by a 1D version of the ambipolar Equation 6.60 and obtain the following set of equations:

$$\begin{cases} \dfrac{dP_f(z)}{dz} = \Gamma g_m(z) P_f(z) \\[2mm] \dfrac{dP_b(z)}{dz} = \Gamma g_m(z) P_b(z) \\[2mm] \dfrac{d^2 N(z)}{dz^2} + \dfrac{N(z)}{\tau} + v_g g_m(z) N_p(z) + \dfrac{J(z)}{qd} = 0 \end{cases} \tag{8.27}$$

In 8.27 we made the same approximations to the recombination rates as in the 0D model case. To solve 8.27 we need to know the relationship between the backward and forward wave propagating powers and the photon density N_p. This can be obtained from Equations 7.6 and 7.26 (Figure 8.4):

$$N_{p,f}(z) = \Gamma P_f(z) / \left(v_g twh\nu\right)$$
$$N_{p,b}(z) = \Gamma P_b(z) / \left(v_g twh\nu\right) \tag{8.28}$$

where v_g is the modal group velocity while the carrier reservoir transverse cross section surface is equal to tw (Figure 8.4). Adding the photon density of the forward and backward propagating waves yields the total photon density N_p:

$$N_p(z) = N_{p,f}(z) + N_{p,b}(z) \tag{8.29}$$

Using 8.28 one can recast the equations for the forward and backward propagating waves in 8.27 into:

$$\left|\begin{array}{l} \dfrac{dN_{p,f}(z)}{dz} = \Gamma g_m(z) N_{p,f}(z) \\[3mm] \dfrac{dN_{p,b}(z)}{dz} = \Gamma g_m(z) N_{p,b}(z) \end{array}\right. \tag{8.30}$$

When studying the characteristics of EELDs, it is customary to neglect the longitudinal carrier diffusion. Thus Equation 8.30 is complemented by:

$$\frac{N(z)}{\tau} + v_g g_m(z) N_p(z) + \frac{J(z)}{qd} = 0 \tag{8.31}$$

Equation 8.30 needs to be solved subject to the following boundary conditions:

$$N_{p,b}(z = 0) = R_R N_{p,f}(z = 0) \tag{8.32a}$$

$$N_{p,f}(z = L) = R_F N_{p,b}(z = L) \tag{8.32b}$$

where R_F and R_R are the front facet and rear facet reflectivity, respectively.

Thus, in principle, Equations 8.30 and 8.31 can be solved using the algorithms presented in Chapter 7. However, in the context of LDs, another algorithm has found more wide spread application. To derive this algorithm, we introduce first a longitudinal discretisation (Figure 8.21).

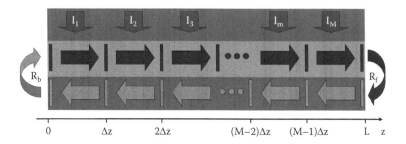

FIGURE 8.21 Schematic illustration of longitudinal discretisation process.

Equation 8.31 can be locally reduced to the ordinary differential equations with constant coefficients by introducing the longitudinal discretisation (Figure 8.21). Thus, locally within each longitudinal section, the material gain is constant:

$$\begin{cases} \dfrac{dN_{p,f}(z)}{dz} = \Gamma g_m(z_0) N_{p,f}(z) \\[2ex] \dfrac{dN_{p,b}(z)}{dz} = \Gamma g_m(z_0) N_{p,b}(z) \end{cases} \tag{8.33}$$

Hence, Equation 8.33 can locally be solved analytically:

$$N_{p,f}(z_0 + \Delta z) = \exp\big(g_m(z_0)\Delta z\big)\ N_{p,f}(z_0) \tag{8.34a}$$

$$N_{p,b}(z_0 - \Delta z) = \exp\big(-g_m(z_0)\Delta z\big)\ N_{p,b}(z_0) \tag{8.34b}$$

Consistently, with the introduced longitudinal discretisation for a ridge wave-guide EELD, Equation 8.31 can be further simplified:

$$\frac{N(z)}{\tau} + v_g g_m(z) N_p(z) + \frac{I}{q\Delta z w d\ M} = 0 \tag{8.35}$$

where M is the total number of longitudinal subsection introduced during the discretisation (Figure 8.21). Equation 8.35 allows the carrier concentration to be calculated consistently subject to the known values of the photon density.

We now solve Equation 8.34 combined with 8.35 using the coupled solution method (CSM) [55]. The application of CSM to the solution of 8.34 and 8.35 results in the Algorithm 8.4.

ALGORITHM 8.4   THE CSM SOLUTION
OF EQUATIONS 8.34 AND 8.35

1. Start
2. Set modelling parameters
3. Provide the initial value for N and $N_{p,f}$ at $z = 0$,
4. Set $N_{p,b} = 0$ for all z
5. Applying 8.34a calculate $N_{p,f}$ at $z = z + \Delta z$ using most recent values of $N_{p,b}$
6. Applying 8.35 calculate N at current z position
7. If $z < L$ goto 5
8. Apply 8.32b
9. Applying 8.34b calculate $N_{p,b}$ at $z = z - \Delta z$ using most recent values of $N_{p,f}$
10. Applying 8.35 calculate N at current z position
11. If $z > L$ goto 9
12. Apply 8.32a
13. If the difference between the new and initial value of $N_{p,f}$ at $z = 0$ is larger than a predefined residual, then set the value of $N_{p,f}$ at $z = 0$ to the newly calculated value and go to 5
14. Stop

TABLE 8.3
Values of a GaAs QW Laser Diode Parameters

Front facet reflectivity	$R_f = 0.1\%$
Rear facet reflectivity	$R_b = 95\%$
Wavelength	$l = 980$ nm
Transparency carrier density	$N_{tr} = 2 \times 10^{18}$ [1/cm^3]
Gain cross section	$a = 4 \times 10^{-15}$ [cm^2]
Group velocity	$v_g = 10^{10}$ [cm/s]
Confinement factor	$\Gamma = 0.006$
Quantum well width	$W_{QW} = 10 \times 10^{-7}$ [cm]
Laser cavity length	$L = 0.2$ [cm]
Laser cavity volume	$V = 6 \times 10^{-10}$ [cm^3]
Carrier life time	$\tau = 0.7$ [ns]
Initial photon density	10^{15} [1/cm^3]

The initial value of the photon density in the forward propagating wave and the carrier density at $z = 0$ can be consistently calculated using the 0D model discussed in section "0D CW Model."

Table 8.4 shows the summary of 1D CSM simulations for 732 nm high power LD. The CSM method converges for bias currents from threshold up to 10 times the threshold current (Table 8.4). The simulation parameters are summarised in Table 8.3 [56].

TABLE 8.4

Summary of CSM Simulations Obtained with 40 Longitudinal slices; x–Method Failed to Converge, √ - Method Converged

Bias current [I_{th}] =	0.95	1.015	1.1	1.2–10
CSM	√	√	√	√

There is one important property of the CSM which should be taken into account when applying this algorithm. Namely, the stability of the CSM and the rate of convergence depend significantly on the number of the longitudinal slices [56]. In fact, the CSM algorithm can fail to converge to the solution below a certain minimal number of the longitudinal slices M. This minimal number of the longitudinal slices is dependent on the parameters of a particular LD. Figure 8.21 shows the dependence of the photon density associated with the forward propagating wave, at the rear facet, on the iteration (round trip) number for the LD structure studied (Table 8.3). The bias current is 10 times larger than the threshold current (i.e., equals approximately 3.4 A). When the number of longitudinal slices is small, the CSM fails to converge and enters oscillatory cycles similar to the ones observed in the 0D FPM case with $p = 1$. However, by increasing the number of longitudinal slices, the convergence of the CSM algorithm can be imposed (Figure 8.22).

Finally, in Figure 8.23 we show the dependence of the photon density in the forward and backward propagating waves on the longitudinal position. Clearly, there is a two order of magnitude of difference between the photon densities carried by both waves. Furthermore, the photon density strongly depends on the longitudinal position, and its distribution is not symmetrical with respect to the centre of the cavity $z = L/2$.

The 1D spatial LD analysis gives more information about the LD operation than the 0D analysis. However, among other things, it fails to predict the output beam parameters for more complicated laser cavity structures. Hence, nowadays multidimensional models are used for the design of lasers diodes. We discuss these models in the next section.

MULTIDIMENSIONAL LD MODELS

Multidimensional LD models usually contain an optical, electrical, and thermal part, and also include an advanced gain model. Furthemore, the electrical part of the model may include fairly complex interactions between the confined and unconfined carriers [57,58]. The spatial dimensionality of each part of the model can also be different. In terms of computational efficiency, the multidimensional models are less efficient than the 0D and 1D models. Therefore, multidimensional models are mostly used in the final stage of the design process, to refine the key specific design parameters of a LD.

Multidimensional models are particularly applicable to VCSELs due to their relatively small spatial dimensions, when compared with EELDs. If possible, the multidimensional VCSEL models rely on the rotational symmetry of the laser structure and solve the electrical and heat diffusion equations along the radial direction [59–61].

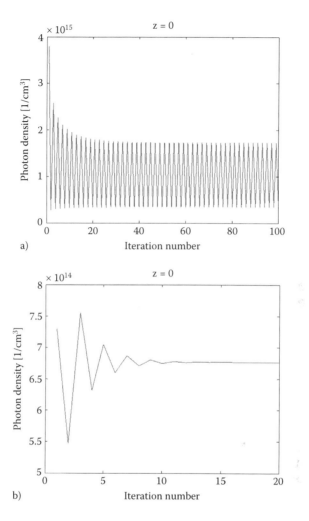

FIGURE 8.22 The dependence of the photon density at $z = 0$ on the CSM iteration number at the bias current $I = 10\,I_{thr}$ obtained using a) four longitudinal slices and b) 10 longitudinal slices. (Sujecki, S., Stability of steady-state high-power semiconductor laser models. *Journal of Optical Society of America B*, 2007. 24(4): p. 1053–1060.)

More advanced models also solve the electrical and heat diffusion equations along the longitudinal dimension [62,63]. When the rotational symmetry is not present, 3D VCSEL models have to be applied. Such models rely either on carrying out the modal decomposition for the optical field and solving for the carrier and temperature distributions, using a suitable expansion into Bessel and trigonometric functions products [64,65], or by applying directly the FDTD method [66].

A fairly often used multidimensional model that is applicable to ridge waveguide EELDs is obtained by complementing the drift diffusion equations 6.11 by the photon rate Equation 8.4 [29]. The optical field distribution is calculated using the mode

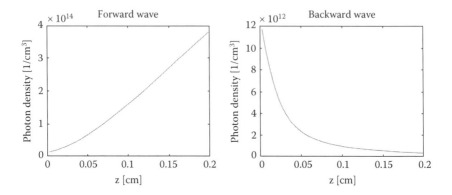

FIGURE 8.23 The dependence of the photon density in the backward and forward waves on the longitudinal position at the bias current $I = 2I_{thr}$ obtained using CSM with 100 longitudinal slices.

solvers discussed in Chapter 3. It is therefore fairly straightforward to include vectorial mode solvers in such LD models [67]. The interactions between the confined and unconfined carriers in QW lasers can be added using the quantum carrier capture-escape model [57,58].

When studying the optical beam evolution of the high power broad area or tapered lasers operating either as stand-alone devices or in an external cavity scheme, usually models consisting of at least a 2D electrical and optical part, calculating directly the carrier and photon distributions along the lateral and longitudinal dimensions are used [68–70]. The optical part of the model relies on representing the internal cavity photon distribution as a superposition of two counter-propagating optical waves. Such models were applied for optimizing the beam quality of a tapered cavity LD at low powers [71], the influence of the carrier and temperature distributions [72,73], spontaneous emission induced filamentation processes [74], the spatio-temporal dynamics of carriers and photons [75], and the DFB provided by intracavity gratings [76]. The steady state 2D high power LD models are relatively computationally efficient. However, to include the analysis of the temporal dynamics, time domain models need to be applied [77–79]. Unfortunately, due to low computational efficiency, it is not practical to include drift-diffusion equations in the time domain models of high power lasers. This can be done, however, by applying the quasi-3D approximation in steady state models [55,80].

Next, we give an example of a quasi 3D high power tapered laser model. The discussion of its structure gives the understanding of the complexity level that is involved in developing multidimensional LD models. This model combines a 3D electrical part that relies on the solution of the drift-diffusion equations 6.11 while neglecting the longitudinal carrier diffusion and the 2D optical part. The quasi 3D model is also complemented by a 3D model of heat diffusion.

A tapered laser consists of a straight waveguide section and a tapered amplifier section, as shown in the schematic diagram in Figure 8.24. The straight waveguide section (sometimes accompanied by a beam spoiler) plays a key role and serves as a modal filter, ensuring that the tapered amplifier is only excited by the fundamental

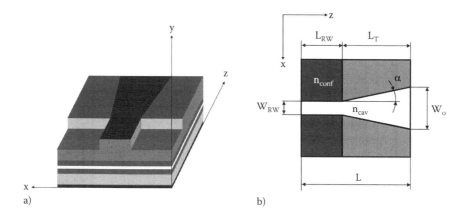

FIGURE 8.24   a) Tapered laser structure and b) its 2D effective index representation.

transverse mode of the straight waveguide. The tapered amplifier section allows the optical beam to expand gradually to lower the optical power density. This minimizes the intensity of detrimental effects (e.g., spatial hole burning), allows an increase in the total output power, and increases the catastrophic optical mirror damage (COMD) threshold.

The tapered lasers have the cavity length of several millimeters (i.e. more than 1,000 wavelengths). Thus, the use of direct 3D Maxwell equation solvers results in prohibitively long calculations times. On the other hand, the 0D and 1D rate equation models are not sufficient for the design of high power tapered edge emitting lasers because these models are unable to reproduce the phenomena that lead to the degradation of beam quality, that is, nonlinear interactions between electrical carriers, phonons, and photons, which result in self-focusing, spatial hole burning, and filamentation. The only way to meet these conflicting constraints is the application of the beam propagation method (BPM), cf. Chapter 4, which allows the beam propagation within the laser cavity to be studied directly by representing the intra-cavity optical field distribution as a superposition of two counterpropagating waves. Furthermore, the application of the BPM allows for a detailed analysis of steady-state nonlinear photon-carrier-temperature interactions and the identification of the origin of filamentation and spatial hole burning effects.

To derive a quasi-3D model, we will first derive a 2D optical model. We take advantage of the experimentally confirmed fact that the edge emitting semiconductor lasers produce light with a fixed linear polarisation; that is, the polarisation coupling can be usually neglected in the first-order approximation and instead of solving the vectorial wave equation the semivectorial equation can be solved. Furthermore, if the advantage of a planar shallow etch rib loaded waveguide structure is taken the effective index method can be applied to speed up the calculations (Figure 8.24). This approximation neglects vertical diffraction, but vertical diffraction is not expected to play a significant role in a well-designed and fabricated high power EELDs. Thus, for the x-polarized mode, the wave propagation within the laser cavity is described by either Equation 4.6c or 4.6d. Similarly, for the y-polarized mode we can use either 4.6a or 4.6b (cf. Figure 8.24).

Decomposing the intracavity field into the superposition of the forward and backward propagating wave yields (cf. 4.9):

$$\frac{\partial F_f}{\partial z} = -j\sqrt{L + \omega^2 \mu_0 \varepsilon} \; F_f \qquad (8.36a)$$

$$\frac{\partial F_b}{\partial z} = j\sqrt{L + \omega^2 \mu_0 \varepsilon} \; F_b \qquad (8.36b)$$

where F_f and F_b stand for the relevant component (cf. section "Introduction") of the electromagnetic field distribution of the forward and backward propagating wave. For the solution of 8.36, in principle, any BPM algorithm discussed in the Chapter 4 can be used. However, the matrix expansion FD-BPM algorithms were found so far most effective [55].

The transverse distribution of the effective index depends on the carrier and temperature distribution within the cavity. When combined with the FD-BPM algorithm, the calculation of the effective index distribution results in a fairly significant numerical overhead since the value of effective index needs to be calculated at every node of the transverse FD mesh. However, the effective index perturbation introduced by the carrier and temperature distribution within the cavity is small when compared with the absolute value of the effective index of the unperturbed waveguide. It is therefore convenient to calculate first the effective index distribution without the perturbations resulting from the carrier and temperature distribution and then include the effect of varying carrier density and local temperature using the vertical confinement factor Γ_v. Γ_v can be calculated using the perturbation theory (cf. appendix 14 in [12]). As an illustrating example, we derive Γ_v for the TE mode in the case of the electric field formulation. Without the perturbation the transverse component of the electric field and the propagation constant satisfies the equation (cf. Equation 3.39a):

$$\frac{\partial^2 Y}{\partial y^2} + \omega^2 \mu_0 \varepsilon_0 n^2 Y = \beta^2 Y \qquad (8.37)$$

where n is the refractive index distribution of the "unperturbed waveguide."

We now calculate the effect of the small refractive index perturbation Δn on the field distribution Y and the propagation constant β. The refractive index perturbation introduces a small perturbation of the field distribution ΔY and of the propagation constant $\Delta \beta$. Thus the perturbed field satisfies the equation:

$$\frac{\partial^2 (Y + \Delta Y)}{\partial y^2} + \omega^2 \mu_0 \varepsilon_0 (n + \Delta n)^2 (Y + \Delta Y) = (\beta + \Delta \beta)^2 (Y + \Delta Y) \qquad (8.38)$$

Dropping second order perturbation terms and using 8.37 results in

$$\frac{\partial^2 \Delta Y}{\partial y^2} + \omega^2 \mu_0 \varepsilon_0 n^2 \Delta Y - \beta^2 \Delta Y + 2\omega^2 \mu_0 \varepsilon_0 n \Delta n Y = 2\beta \Delta \beta Y \qquad (8.39)$$

Premultiplying by Y^* where "*" stands for complex conjugate and integrating over y gives:

$$\int Y^* \left(\frac{\partial^2 \Delta Y}{\partial y^2} + \omega^2 \mu_0 \varepsilon_0 n^2 \Delta Y - \beta^2 \Delta Y \right) dy + 2\omega^2 \mu_0 \varepsilon_0 \int n \Delta n Y^* Y dy = 2\beta \Delta \beta \int Y^* Y dy$$

$$(8.40)$$

If ΔY and Y tend to zero when y tends to infinity, the following equality holds:

$$\int Y^* \frac{\partial^2 \Delta Y}{\partial y^2} dy = \int \Delta Y \frac{\partial^2 Y^*}{\partial y^2} dy$$

If n is purely real, Y^* fulfils 8.37 and thus we obtain

$$\int Y^* \left(\frac{\partial^2 \Delta Y}{\partial y^2} + \omega^2 \mu_0 \varepsilon_0 n^2 \Delta Y - \beta^2 \Delta Y \right) dy = \int \Delta Y \left(\frac{\partial^2 Y^*}{\partial y^2} + \omega^2 \mu_0 \varepsilon_0 n^2 Y^* - \beta^2 Y^* \right) dy = 0$$

Therefore 8.40 yields:

$$\omega^2 \mu_0 \varepsilon_0 \int n \Delta n Y^* Y dy = \beta \Delta \beta \int Y^* Y dy \qquad (8.41)$$

If the refractive index perturbation comes from a QW region only, then from 8.41 we obtain the formula for $\Delta \beta$:

$$\Delta \beta = \frac{\omega^2 \mu_0 \varepsilon_0 n_{qw} \displaystyle\int_{qw} \Delta n Y^* Y dy}{\beta \displaystyle\int Y^* Y dy} \qquad (8.42)$$

where n_{qw} is the value of the refractive index in the QW. Assuming that Δn does not depend on y within the QW yields the final formula:

$$\Delta \beta = \frac{\omega^2 \mu_0 \varepsilon_0 n_{qw} \Gamma_y \Delta n}{\beta} \qquad (8.43)$$

while the vertical confinement factor is defined as:

$$\Gamma_v = \frac{\int_{qw} Y^* Y dy}{\int Y^* Y dy} \tag{8.44}$$

The $\Delta\beta$ calculated by 8.43 can be then used to calculate the effective index distribution for 8.36:

$$n_{eff} = \frac{\beta + \Delta\beta}{\omega\sqrt{\mu_0 \varepsilon_0}} \tag{8.45}$$

Thus β and Γ_v need to be calculated only once at the beginning of the simulation and then 8.45 can be used to calculate efficiently the effective index distribution.

In order to calculate the effective index distribution from 8.45, one needs to know the refractive index perturbation Δn in 8.43, which can only be determined once the temperature and carrier distributions are known. The carrier concentrations in the laser cavity are determined by solving self-consistently the current continuity equations for electrons and holes together with the Poisson Equation (cf. Equations 6.6 and 6.11):

$$div\ \vec{J}_n - q\left(R_{nr} + R_{sp} + F_n^{qw}\right) = 0 \tag{8.46a}$$

$$div\ \vec{J}_p + q\left(R_{nr} + R_{sp} + F_p^{qw}\right) = 0 \tag{8.46b}$$

$$div\left(\varepsilon_s grad\ \varphi\right) + q\left(p - n + N_D - N_A\right) = 0 \tag{8.46c}$$

where R_{nr} and R_{sp} are, respectively, the nonradiative and spontaneous recombination rates in bulk regions. j_n and j_p are electron and hole current densities, respectively; p and n stand for electron and hole concentrations; and N_D and N_A denote donor and acteptor concentrations, while φ stands for electric potential distribution. The terms F_n^{qw} and F_p^{qw} appearing in the continuity equations, account for the carrier capture–escape rates between the confined and unconfined states and they link Equation 8.46 with the continuity equations for each QW:

$$\int_{qw} F_n^{qw} dv - R_{nr}^{qw} + R_{sp}^{qw} + R_{st} = 0$$

$$\int_{qw} F_p^{qw} dv - R_{nr}^{qw} + R_{sp}^{qw} + R_{st} = 0,$$

where R_{nr}^{qw}, R_{sp}^{qw}, and R_{st} denote the nonradiative, spontaneous, and stimulated recombination rates in QW regions.

The temperature distribution in the cavity is calculated by solving the heat conduction equation (cf. Equation 5.5):

$$\nabla \left(-\kappa \nabla T \right) = Q \qquad (8.47)$$

where κ is the thermal conductivity and W is the heat generation rate. Thereafter, the carriers are assumed to be in thermal equilibrium with the lattice.

The current densities in Equation 8.46 include standard drift-diffusion terms together with the term arising from thermal gradients. Fermi-Dirac statistics is considered for 2D carriers in QW, and Boltzmann statistics for carriers in bulk regions. A parabolic band approximation is used for the gain after fitting the band parameters to complete band-mixing calculations for the valence band at the beginning of the simulation. The local material gain is calculated with a Lorentzian linewidth broadening function at the lasing wavelength as a function of the local temperature, and electron and hole concentrations. Spontaneous emission, Shockley-Read-Hall, and Auger recombination rates, are calculated using standard expressions [12], taking into account temperature dependencies. Free carrier absorption and band gap renormalisation are included by means of empirical formulations [12]. The thermal model considers Joule, nonradiative recombination, and absorption by free carriers as local heat sources in Equation 5.3, together with the so-called excess power needed to fulfil energy conservation [81]. More detail on electrical and thermal models are given in [81,82].

Near the gain maximum the real part of the refractive index perturbation of the QW is proportional to the square root of the carrier density [83]. This was confirmed experimentally [81], and therefore this dependence was included in the model. The temperature dependence of the real part of the refractive index is also included using phenomenological expressions fitted to experimental data [81].

The standard algorithm used for solving electrical, thermal, and optical parts of the model is the CSM, which was discussed in the section "1D Laser Diode Models" [84]. The application of the CSM algorithm to a high power tapered LD is illustrated in Figure 8.25. In the algorithm the analysis is initiated at the rear facet. The forward propagating wave optical field is propagated using the BPM towards the front facet. Equations 8.46 and 8.47 are solved for each longitudinal position in 2D (including both transverse dimensions, Figure 8.2a) using the latest update of the photon distribution to obtain new carrier, potential, and temperature distributions in the cavity. These new carrier and temperature profiles are used to update the lateral gain and refractive index distributions, which are then used in Equation 8.36a to obtain a new photon distribution at the next longitudinal position. After calculating the distribution of the wave reflected at the front facet, the propagation is continued for the backward propagating wave using 8.36b. This process is repeated until the photon distribution of the wave reflected field at the back facet stops changing after each subsequent round-trip. Thus, the phase of the incoming wave has not been considered in the convergence criterion. This simplification

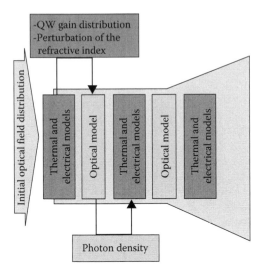

FIGURE 8.25 Flow chart of coupled solution method. (Sujecki, S., et al., Nonlinear properties of tapered laser cavities. *IEEE Journal of Selected Topics in Quantum Electronics*, 2003. 9(3): pp. 823–834.)

relies again on an assumption that within the intermodal wavelength distance the refractive index and optical gain do not change significantly. Thus, the problem arises of how to select the wavelength at which the calculations are carried out. In this work the emission wavelength was taken to be that of the gain maximum at each current.

As an example, a high power tapered LD structure, operating at 732 nm, was chosen. The epitaxy for a typical 732 nm GaAs-based laser is given in Table 8.5. The cavity structures studied have a 4° gain-guided taper and a 3 μm wide ridge waveguide width. The lengths of the ridge waveguide and tapered gain sections are 0.75 mm and 1.25 mm, respectively. The rear facet reflectivity is 95%. The outer slab effective index changes with etch depth. The ridge waveguide section is deeply etched and has refractive contrast of 0.02. The tapered section is gain guided. The presented results have been published in Sujecki et al. [55]. More details on the presented model can be found in Borruel et al. [87]. The model presented in this section was also successfully applied to analyse tapered lasers with patterned contacts [88], design clarinet lasers [89], study external cavity lasers, and to interpret results obtained by n-contact imaging [90].

The material parameters used in the simulations were taken from standard references [50,51], when available. Typical parameters for GaAs were used also in the QW when GaAsP data were not found. Table 8.6 collects the most important parameters used as default in the simulations, which were, unchanged throughout the simulations.

Facet reflectivity calculations were performed using a free space radiation method (for unguided waves in the coating) coupled with the finite difference beam propagation method (for the guided waves in the semiconductor) [85]. These calculations indicate that the use of simple reflection coefficients is satisfactory for the structures studied here and only need to be calculated accurately once at the beginning of the

TABLE 8.5

Epitaxial Structure of the Laser Cavity

Layer	Material	Thickness [nm]	Doping [cm⁻¹]	Refractive Index
p-contact metal				
p-contact layer	GaAs		p+	3.741
p-cladding	$Al_{0.70}Ga_{0.30}As$		p	3.229
p-waveguide	$Al_{0.65}Ga_{0.35}As$	500	p	3.263
QW	$GaAs_{0.67}P_{0.33}$	9	n	3.585
n-waveguide	$Al_{0.65}Ga_{0.35}As$	500	n	3.263
n-cladding	$Al_{0.70}Ga_{0.30}As$		n	3.229
n-buffer+ substrate	GaAs		n	3.741
n-contact metal				

TABLE 8.6

Simulation Parameters

Material Properties

Band Offset $\Delta E_c/\Delta E_g$ (QW)	0.32	
Intraband relaxation time	0.1	ps
Electron capture time		ps
Hole capture time	1.2	ps
Auger coefficient	1.5E-30 (at RT)	cm^6s^{-1}
Band-gap renorm. coefficient	−0.00225e-05	eV
Free carrier absorption cross-section	3E-18 (n); 7E-18 (p)	cm^2

BPM Parameters

Calculation area	2000×400	μm^2
Lateral step	0.1–0.4	μm
Longitudinal step	1–2	μm

simulation. We have also found that the use of the Padé (1,1) wide angle scheme [86] (cf. Chapter 4) suffices to properly treat the beam propagation in the tapers studied. The analysis of beam spoilers has also been included by using the simplified approach employed by Marjoulis [73], in which the fields impinging on the beam spoilers are completely absorbed.

The comparison between experiments and simulations included a two-step fitting procedure. In the first step, the measured threshold currents and slope efficiencies of a set of broad area lasers fabricated using the same epitaxial structure were compared with one-dimensional simulations performed with the laser simulator described in reference [82]. The fitting parameters were the trap density in the confinement region and the internal scattering losses. In the second step, the widths of the calculated and experimental far field patterns for the tapered laser under study were compared. The proportionality constant relating the carrier induced index perturbation and the carrier density was used as a fit parameter, and a good matching was found using

-2.7×10^{-11} [cm$^{3/2}$] for the epitaxial structure analysed in this chapter. Simulations were also performed for different laser geometries (taper angle, cavity length) with the same epitaxial structure.

In Figures 8.26 and 8.27, we show two examples of high power tapered laser simulations that were obtained using the quasi-3D tapered laser model. These examples demonstrate the ability of the model to predict the optical beam parameters at various values of the bias current.

Figure 8.26 shows a comparison between experimentally and numerically obtained light-current characteristics, M^2, power in the central lobe of the virtual source and astigmatism. Good agreement is obtained between experiment and simulation for the light–current characteristics yielding both the same threshold current and slope efficiency. Similarly, good agreement is observed between the measured and calculated values of power in the central lobe of the virtual source and the astigmatism.

Figure 8.27 shows the numerical and experimental near field patterns at selected output powers. The near-field patterns agree quite well. The experimental and numerical results both show the appearance of side lobes and filamentation at larger output powers along with a gradual narrowing of the near-field pattern.

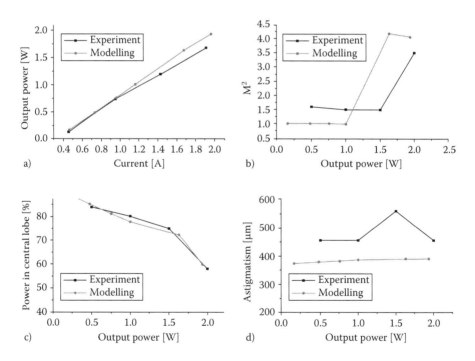

FIGURE 8.26 Light-current characteristics, current-voltage dependence and the dependence of beam divergence, M^2, power in central lobe of the virtual source, astigmatism, and beam waist on the output power. $L = 2$ mm, $L_{RW} = 0.75$ mm, $W_{RW} = 3$ μm, $W_o = 90.3$ μm, $R_{RW} = 95\%$, $R_o = 1\%$, and $\Delta n_{eff} = 0.008$. (Sujecki, S., et al., Nonlinear properties of tapered laser cavities. *IEEE Journal of Selected Topics in Quantum Electronics*, 2003. 9(3): p. 823–834.)

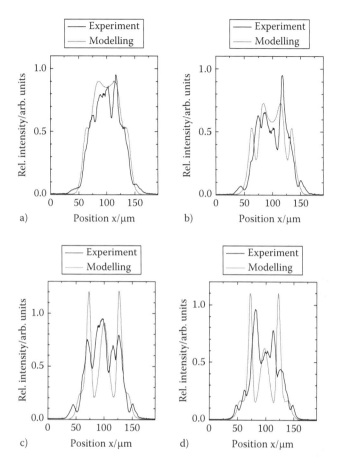

FIGURE 8.27 Near field power distributions, (the same geometrical parameters as in Figure 8.26), $R_{RW} = 95\%$, $R_o = 1\%$, a) $P = 0.5$ W, b) $P = 1$ W, c) $P = 1.5$ W, d) $P = 2$ W. (Sujecki, S., et al., Nonlinear properties of tapered laser cavities. *IEEE Journal of Selected Topics in Quantum Electronics*, 2003. 9(3): p. 823–834.)

REFERENCES

1. Numai, T., *Laser Diodes and their Applications to Communications and Information Processing*. 2010, New Jersey: John Wiley & Sons, Inc., IEEE Press.
2. Numai, T., *Fundamentals of Semiconductor Lasers*. 2004, New York: Springer-Verlag.
3. Chow, W.W. and S. Koch, *Semiconductor-Laser Fundamentals: Physics of the Gain Materials*. 1999, Berlin: Springer.
4. Vasil'ev, P., *Ultrafast Diode Lasers: Fundamentals and Applications*. 1995, Boston: Artech House.
5. Kawaguchi, H., *Bistabilities and Nonlinearities in Laser Diodes*. 1994, Boston: Artech House.
6. Agrawal, G.P. and N.K. Dutta, *Semiconductor Lasers*. 1993, New York: Van Nostrand Reinhold.
7. Zory, P.S., *Quantum Well Lasers*. 1993, Boston: Academic Press.
8. Thompson, G.H.B., *Physics of Semiconductor Laser Devices*. 1980, Chichester: Wiley.

9. Carroll, J.E., *Distributed Feedback Semiconductor Lasers*. 1998, London: SPIE Press.

10. Ghafouri-Shiraz, H., *The Principles of Semiconductor Laser Diodes and Amplifiers*. 2004, London: Imperial College Press.

11. Ebeling, K.J., *Integrated Optoelectronics*. 1992, Berlin: Springer-Verlag.

12. Coldren, L.A. and S.W. Corzine, Microwave and Optical Engineering. *Diode Lasers and Photonic Integrated Circuits*. ed. K. Chang. 1995, New York: Wiley.

13. Piprek, J., *Semiconductor Optoelectronic Devices Intorduction to Physics and Simulation*. 2003, San Diego: Academic Press.

14. Piprek, J., ed., *Optoelectronic Devices, Advanced Simulation and Analysis*. 2010, New York: Springer.

15. Piprek, J., ed., *Nitride Semiconductor Devices: Principles and Simulation*. 2007, Weinheim: Wiley VCH Verlag GmbH& Co. KGaA.

16. Coldren, L.A. and S.W. Corzine, *Diode Lasers and Photonic Integrated Circuits*. 1995, New York: Wiley.

17. Petermann, K., *Laser Diode Modulation and Noise*. 1991, London: Kluwer Academic Publishers.

18. Kawaguchi, H., Bistable laser diodes and their applications: State of the art. *IEEE Journal of Selected Topics in Quantum Electronics*, 1997. 3(5): p. 1254–1270.

19. Olshansky, R., et al., Frequency-response of 1.3-mi InGaAsP high-speed semiconductor-lasers. *IEEE Journal of Quantum Electronics*, 1987. 23(9): p. 1410–1418.

20. Harder, C., et al., Noise equivalent-circuit of a semiconductor-laser diode. *IEEE Journal of Quantum Electronics*, 1982. 18(3): p. 333–337.

21. Marcuse, D. and T.P. Lee, Rate-equation model of a coupled-cavity laser. *IEEE Journal of Quantum Electronics*, 1984. 20(2): p. 166–176.

22. Kurtz, R.M., et al., Mutual injection locking: A new architecture for high-power solid-state laser arrays. *IEEE Journal of Selected Topics in Quantum Electronics*, 2005. 11(3): p. 578–586.

23. Khurgin, J.B., I. Vurgaftman, and J.R. Meyer, Analysis of phase locking in diffraction-coupled arrays of semiconductor lasers with gain/index coupling. *IEEE Journal of Quantum Electronics*, 2005. 41(8): p. 1065–1074.

24. Marcuse, D., Classical derivation of the laser rate-equation. *IEEE Journal of Quantum Electronics*, 1983. 19(8): p. 1228–1231.

25. Hakki, B.W. and T.L. Paoli, Gain spectra in GaAs double-heterostructure injection lasers. *Journal of Applied Physics*, 1975. 46(3): p. 1299–1306.

26. Cassidy, D.T., Technique for measurement of the gain spectra of semiconductor diode-lasers. *Journal of Applied Physics*, 1984. 56(11): p. 3096–3099.

27. Bir, G.L. and G.E. Pikus, *Symmetry and Strain-Induced Effects in Semiconductors*. 1974, New York: Wiley.

28. Collatz, L., *The Numerical Treatment of Differential Equations*. 1960, Berlin: Springer-Verlag.

29. Alam, M.A., et al., Simulation of semiconductor quantum well lasers. *IEEE Transactions on Electron Devices*, 2000. 47(10): p. 1917–1925.

30. Lim, J.J., et al., Simulation of double quantum well GaInNAs laser diodes. *IET Optoelectronics*, 2007. 1(6): p. 259–265.

31. Press, W.H., Teukolsky, S.A., Vetterling, W.T., Flannery, B.P., *Numerical Recipes in C++: The Art of Scientific Computing*. 2002, Cambridge: Cambridge University Press.

32. Lee, T.P., et al., Short-cavity InGaAsP injection-lasers—dependence of mode spectra and single-longitudinal-mode power on cavity length. *IEEE Journal of Quantum Electronics*, 1982. 18(7): p. 1101–1113.

33. LeVeque, R.J. and Z. Li, The immersed interface method for elliptic equations with discontinuous coefficients and singular sources. *SIAM Journal on Numerical Analysis* 1994. 31(4): p. 1019–1044.

34. Lang, R.J., J. Salzman, and A. Yariv, Modal-analysis of semiconductor-lasers with nonplanar mirrors. *IEEE Journal of Quantum Electronics*, 1986. 22(3): p. 463–470.

35. Twu, Y., et al., Mode characteristics of phase-locked semiconductor-laser arrays at and above threshold. *IEEE Journal of Quantum Electronics*, 1987. 23(6): p. 788–803.

36. Tee, C.W., et al., Transient response of ARROW VCSELs. *IEEE Journal of Quantum Electronics*, 2005. 41(2): p. 140–147.

37. Izhaky, N. and A.A. Hardy, Analysis of DBR lasers and MOPA with saturable absorption gratings. *IEEE Journal of Quantum Electronics*, 1997. 33(7): p. 1149–1155.

38. Ryvkin, B.S. and E.A. Avrutin, Spatial hole burning in high-power edge-emitting lasers: A simple analytical model and the effect on laser performance. *Journal of Applied Physics*, 2011. 109(4).

39. Marcuse, D., Computer-model of an injection-laser amplifier. *IEEE Journal of Quantum Electronics*, 1983. 19(1): p. 63–73.

40. Agrawal, G.P. and N.A. Olsson, Self-phase modulation and spectral broadening of optical pulses in semiconductor-laser amplifiers. *IEEE Journal of Quantum Electronics*, 1989. 25(11): p. 2297–2306.

41. Agrawal, G.P., Population pulsations and nondegenerate 4-wave mixing in semiconductor-lasers and amplifiers. *Journal of the Optical Society of America B-Optical Physics*, 1988. 5(1): p. 147–159.

42. Tang, J.M. and K.A. Shore, Carrier diffusion and depletion effects on multiwave mixing in semiconductor lasers. *IEEE Journal of Selected Topics in Quantum Electronics*, 1997. 3(5): p. 1280–1286.

43. Koltchanov, I., et al., Gain dispersion and saturation effects in four-wave mixing in semiconductor laser amplifiers. *IEEE Journal of Quantum Electronics*, 1996. 32(4): p. 712–720.

44. Connelly, M.J., *Semiconductor Optical Amplifiers*. 2004, London: Kluver Academic Publishers.

45. Schell, M., et al., Fundamental limits of sub-ps pulse generation by active-mode locking of semiconductor-lasers—The spectral gain width and the facet reflectivities. *IEEE Journal of Quantum Electronics*, 1991. 27(6): p. 1661–1668.

46. Yang, W. and A. Gopinath, Study of passive-mode locking of semiconductor-lasers using time-domain modeling. *Applied Physics Letters*, 1993. 63(20): p. 2717–2719.

47. Jones, D.J., et al., Dynamics of monolithic passively mode-locked semiconductor-lasers. *IEEE Journal of Quantum Electronics*, 1995. 31(6): p. 1051–1058.

48. Wong, Y.L. and J.E. Carroll, A traveling-wave rate-equation analysis for semiconductor-lasers. *Solid-State Electronics*, 1987. 30(1): p. 13–19.

49. Chi, J.W.D., L. Chao, and M.K. Rao, Time-domain large-signal investigation on nonlinear interactions between an optical pulse and semiconductor waveguides. *IEEE Journal of Quantum Electronics*, 2001. 37(10): p. 1329–1336.

50. Vankwikelberge, P., G. Morthier, and R. Baets, CLADISS - a longitudinal multimode model for the analysis of the static, dynamic, and stochastic-behavior of diode-lasers with distributed feedback. *IEEE Journal of Quantum Electronics*, 1990. 26(10): p. 1728–1741.

51. Zhang, L.M., et al., Dynamic analysis of radiation and side-mode suppression in a 2nd-order dfb laser using time-domain large-signal traveling-wave model. *IEEE Journal of Quantum Electronics*, 1994. 30(6): p. 1389–1395.

52. Vladimirov, A.G., D. Turaev, and G. Kozyreff, Delay differential equations for mode-locked semiconductor lasers. *Optics Letters*, 2004. 29(11): p. 1221–1223.

53. Rossetti, M., P. Bardella, and I. Montrosset, Modeling Passive Mode-Locking in Quantum Dot Lasers: A Comparison Between a Finite-Difference Traveling-Wave Model and a Delayed Differential Equation Approach. *IEEE Journal of Quantum Electronics*, 2011. 47(5): p. 569–576.

54. Javaloyes, J. and S. Balle, Multimode dynamics in bidirectional laser cavities by folding space into time delay. *Optics Express*, 2012. 20(8): p. 8496–8502.

55. Sujecki, S., et al., Nonlinear properties of tapered laser cavities. *IEEE Journal of Selected Topics in Quantum Electronics*, 2003. 9(3): p. 823–834.

56. Sujecki, S., Stability of steady-state high-power semiconductor laser models. *Journal of Optical Society of America B*, 2007. 24(4): p. 1053–1060.

57. Grupen, M. and K. Hess, Simulation of carrier transport and nonlinearities in quantum-well laser diodes. *IEEE Journal of Quantum Electronics*, 1998. 34(1): p. 120–140.

58. Keating, T., et al., Temperature dependence of electrical and optical modulation responses of quantum-well lasers. *IEEE Journal of Quantum Electronics*, 1999. 35(10): p. 1526–1534.

59. Hadley, G.R., et al., Comprehensive numerical modeling of vertical-cavity surface-emitting lasers. *IEEE Journal of Quantum Electronics*, 1996. 32(4): p. 607–616.

60. Gustavsson, J.S., et al., A comprehensive model for the modal dynamics of vertical-cavity surface-emitting lasers. *IEEE Journal of Quantum Electronics*, 2002. 38(2): p. 203–212.

61. Wenzel, H. and H.J. Wunsche, The effective frequency method in the analysis of vertical-cavity surface-emitting lasers. *IEEE Journal of Quantum Electronics*, 1997. 33(7): p. 1156–1162.

62. Streiff, M., et al., A comprehensive VCSEL device simulator. *IEEE Journal of Selected Topics in Quantum Electronics*, 2003. 9(3): p. 879–891.

63. Conradi, O., S. Helfert, and R. Pregla, Comprehensive modeling of vertical-cavity laser-diodes by the method of lines. *IEEE Journal of Quantum Electronics*, 2001. 37(7): p. 928–935.

64. Debernardi, P., HOT-VELM: A comprehensive and efficient code for fully vectorial and 3-D hot-cavity VCSEL simulation. *IEEE Journal of Quantum Electronics*, 2009. 45(8): p. 969–982.

65. Jungo, M.X., D. Erni, and W. Bachtold, VISTAS: A comprehensive system-oriented spatiotemporal VCSEL model. *IEEE Journal of Selected Topics in Quantum Electronics*, 2003. 9(3): p. 939–948.

66. Lee, T.W., et al., Modal characteristics of ARROW-type vertical-cavity surface-emitting lasers. *IEEE Photonics Technology Letters*, 2001. 13(8): p. 770–772.

67. Grupen, M., et al., Coupling the electronic and optical problems in semiconductor quantum well laser simulations. *SPIE Proceedings*, 1994. 2146: p. 133–147.

68. Champagne, Y., S. Mailhot, and N. McCarthy, Numerical procedure for the lateral-mode analysis of broad-area semiconductor-lasers with an external-cavity. *IEEE Journal of Quantum Electronics*, 1995. 31(5): p. 795–810.

69. Wolff, S., et al., Fourier-optical transverse mode selection in external-cavity broad-area lasers: Experimental and numerical results. *IEEE Journal of Quantum Electronics*, 2003. 39(3): p. 448–458.

70. Lang, R.J., A.G. Larsson, and J.G. Cody, Lateral-modes of broad area semiconductor-lasers—Theory and experiment. *IEEE Journal of Quantum Electronics*, 1991. 27(3): p. 312–320.

71. Sujecki, S., et al., Optical properties of tapered laser cavities. *IEE Proceeding Journal*, 2003. 150(3): p. 246–252.

72. Williams, K.A., et al., Design of High—Brightness Tapered Laser Arrays. *IEEE Journal of Selected Topics in Quantum Electronics*, 1999. 5(3): p. 822–831.

73. Mariojouls, S., et al., Modelling of the Performance of High-Brightness Tapered Lasers. *SPIE Proceedings*, 2000. 3944: p. 395–406.

74. Ramanujan, S. and H.G. Winful, Spontaneous emission induced filamentation in flared amplifiers. *IEEE Journal of Quantum Electronics*, 1996. 32(5): p. 784–789.

75. Gering, E., O. Hess, and R. Wallenstein, Modelling of the Performance of High-Power diode amplifier systems with on opto-thermal microscopic spatio-temporal theory. *IEEE Journal of Quantum Electronics*, 1999. 35(3): p. 320–331.

76. Yu, S.F., A quasi-three-dimensional large-signal dynamic model of distributed feedback lasers. *IEEE Journal of Quantum Electronics*, 1996. 32(3): p. 424–432.

77. Ning, C.Z., et al., A first-principles fully space-time resolved model of a semiconductor laser. *Quantum and Semiclassical Optics*, 1997. 9(5): p. 681–691.

78. Williams, K.A., et al., Design of high-brightness tapered laser arrays. *IEEE Journal of Selected Topics in Quantum Electronics*, 1999. 5(3): p. 822–831.

79. Hess, O., S.W. Koch, and J.V. Moloney, Filamentation and beam-propagation in broad-area semiconductor-lasers. *IEEE Journal of Quantum Electronics*, 1995. 31(1): p. 35–43.

80. Borruel, L., et al., Quasi-3-D simulation of high-brightness tapered lasers. *IEEE Journal of Quantum Electronics*, 2004. 40(5): p. 463–472.

81. Borruel, L., et al., A Selfconsistent Electrical, Thermal and Optical Model of High Brightness Tapered Lasers. *SPIE Proceedings,* 2002. 4646: p. 355–366.

82. Arias, J., et al., One-Dimensional Simulation of High Power Laser Diode Structures. In *NUSOD*. 2002. Zurich. p. 8–9.

83. Wenzel, H., G. Erbert, and P.M. Enders, Improved theory of the refractive-index change in quantum-well lasers. *IEEE Journal of Selected Topics in Quantum Electronics*, 1999. 5(3): p. 637–642.

84. Wykes, J.G., et al., Convergence behaviour of coupled electromagnetic/thermal/electronic high power laser models. In *IEE Computation in Electromagnetics Conference*. 2002. Bournemouth, UK. p. 91–92.

85. Benson, T.M., P. Sewell, and A. Vukovic. Interfacing the Finite Difference Beam Propagation Method with Fast Semi-Analytic Techniques. In *OSA Integrated Photonics Research Symposium*. Quebec, Canada. 2000. p. 48–50.

86. Anada, T., et al., Very-wide-angle beam propagation methods for integrated optical circuits. *IEICE Transactions on Electronics*, 1999. E82-C(7): p. 1154–1158.

87. Borruel, L., et al., Quasi 3D simulation of high brightness tapered lasers. *IEEE Journal of Quantum Electronics*, 2004. 40(5): p. 463–472.

88. Borruel, L., et al., Modelling of patterned contacts in tapered lasers. *IEEE Journal of Quantum Electronics*, 2004. 40(10): p. 1384–1388.

89. Borruel, L., et al., Clarinet laser: Semiconductor laser design for high-brightness applications. *Applied Physics Letters*, 2005. 87(10): p. 101104.

90. Lim, J.J., et al., Design and simulation of next-generation high-power, high-brightness laser diodes. *IEEE Journal of Selected Topics in Quantum Electronics*, 2009. 15(3): p. 993–1008.

9 Pulse Propagation in Optical Fibres

The understanding of optical pulse evolution during the propagation in a light transmitting medium is essential for the design of the optical telecommunication links [1,2], supercontinuum sources [3], and self-pulsing mode locked lasers [4]. In this chapter we discuss the numerical techniques used in the modelling of optical pulse propagation in optical fibres. In particular we give a detailed description of the split-step Fourier method (SSFM). This relatively simple algorithm is routinely used while studying pulse propagation in optical fibres. We start with a short introduction in section "Introduction" after which we derive the nonlinear Schrödinger equation (NSE) in section "Propagation of Optical Pulses in Fibres." In the third section, we discuss the application of the SSFM to the solution of the NSE and outline how this numerical technique can be extended so that it is applicable to simulations of the supercontinuum generation.

INTRODUCTION

When analysing optical nonlinear phenomena one should clearly differentiate between the mathematical model used for the quantitative description of the observed beam evolution and the physical model that is used to explain the physical origin of the particular nonlinear phenomenon. The basis for the former gives the following equation that describes the dependence of the electrical polarisation vector on the electric field vector [5]:

$$\vec{P} = \varepsilon_0 \left(\chi^{(1)} \cdot \vec{E} + \chi^{(2)} : \vec{E}\vec{E} + \chi^{(3)} \vdots \vec{E}\vec{E}\vec{E} + \cdots \right) \tag{9.1}$$

Equation 9.1 gives a very general form of the dependence of the electrical polarization vector on the electrical field vector for systems that are time shift invariant. $\chi^{(n)}$ is the electrical susceptibility tensor that depends on time but also can be dependent on the electrical field vector and spatial variables. The first term in 9.1 gives the standard linear dependence of the polarisation on the electric field. The second term gives the Pockels effect and the third one gives the Kerr effect. It should be noted that 9.1 gives only the mathematical description of the physical phenomena and does not explain the physical origin. The latter aspect of the nonlinear optical phenomena is usually approached using either quantum electrodynamics or a phenomenological description that relies on the insight gained through a combination of experimental and theoretical analysis with the help of intuition aided by the general knowledge that stems from quantum electromagnetic theory. For instance, from the

mathematical description point of view, the Raman gain effect is accounted for by
the imaginary part of the third order nonlinear coefficient whereas the physical origin for it can be explained as a result of an interaction between the pump wave and
signal wave with the mediation of optical phonons. An advanced discussion of nonlinear optical phenomena can be found in standard textbooks [5–8]. In this chapter
we focus on the study of optical pulse propagation in optical waveguides using the
mathematical description of the underlying nonlinear phenomena. We discuss the
modelling methods for studying optical pulse propagation in fibres. In particular,
we develop a numerical modelling tool for studying the pulse propagation in optical
fibres using the SSFM.

PROPAGATION OF OPTICAL PULSES IN FIBRES

During the propagation along a typical telecom silica fibre, an optical pulse shape
may undergo an evolution that results from the interaction of the electromagnetic
field of the pulse with the medium, that is, the silica glass fibre. This evolution may
be a desired effect, when the design of a supercontinuum source is intended, or unde-
sired, for example, when a fibre is used as a transmission medium for long distance
optical telecom systems.

In the following we will derive the equation that governs the evolution of an
optical pulse within an optical waveguide. The derivation can be found in many
textbooks [1,2,9]. However, we carry out this derivation here in a manner that is
consistent with the preceding material. An equation for the pulse envelope evolution
can be derived starting from the Maxwell's equations:

$$\nabla \times \vec{E} = -\frac{\partial \vec{B}}{\partial t} \tag{9.2a}$$

$$\nabla \times \vec{H} = \frac{\partial \vec{D}}{\partial t} \tag{9.2b}$$

with the following conditions that impose the uniqueness of the solution:

$$\nabla \cdot \vec{D} = 0 \tag{9.3a}$$

$$\nabla \cdot \vec{B} = 0 \tag{9.3b}$$

and the material equations:

$$\vec{D} = \varepsilon_0 \vec{E} + \vec{P}, \tag{9.4a}$$

$$\vec{B} = \mu_0 \vec{H} \tag{9.4b}$$

The polarisation vector has two parts:

$$\vec{P} = \vec{P}_L + \vec{P}_{NL},$$ (9.5)

The linear part is given as:

$$\vec{P}_L(\vec{r},t) = \varepsilon_0 \int_{-\infty}^{\infty} \chi^{(1)}(\vec{r},\tau)\vec{E}(\vec{r},t-\tau)\,d\tau$$ (9.6a)

while the nonlinear part:

$$\vec{P}_{NL}(\vec{r},t) = \varepsilon_0 \int_{-\infty}^{\infty} \chi^{(3)}(\vec{r},\tau_1,\tau_2,\tau_3) : \vec{E}(\vec{r},t-\tau_1)\vec{E}(\vec{r},t-\tau_2)\vec{E}(\vec{r},t-\tau_3)\,d\tau_1 d\tau_2 d\tau_3$$ (9.6b)

where $\chi^{(1)}$ is a rank 2 tensor while $\chi^{(3)}$ is a rank 4 tensor. The nonlinear part consists only of the third order nonlinearity, cf. 9.1.

Taking the curl of 9.2a and substituting 9.2b on the right hand side yields:

$$\nabla \times \nabla \times \vec{E} = -\frac{1}{c_0^2}\frac{\partial^2 \vec{E}}{\partial t^2} - \mu_0 \frac{\partial^2 \vec{P}}{\partial t^2}$$ (9.7)

where $c_0 = 1/\sqrt{\mu_0\varepsilon_0}$. Because $\nabla \times \nabla \times \vec{E} = \nabla(\nabla \cdot \vec{E}) - \nabla^2\vec{E}$, Equation 9.7 can be recast into the form:

$$-\nabla^2\vec{E} + \nabla(\nabla\vec{E}) = -\frac{1}{c_0^2}\frac{\partial^2 \vec{E}}{\partial t^2} - \mu_0 \frac{\partial^2 \vec{P}}{\partial t^2}$$ (9.8)

We assume that the pulse propagates in an optical waveguide that is aligned with the z-direction (cf. Figure 3.1) and consequently $\chi^{(1)}$ is independent of z, while we assume that the contribution of $\chi^{(3)}$ to the transverse distribution of $\chi^{(1)}$ is negligible. Hence, $\chi^{(3)}$ is effectively considered to be a function of z only. This assumption is fairly well justified when studying optical pulse propagation in optical transmission fibres (Section 4.7 in reference [1]), but results in neglecting optical beam focusing caused by the transverse distribution of $\chi^{(3)}$, and hence for instance the phenomenon of spatial solitons [10]. We also assume that the electric field distribution has only one transverse component that is aligned along the x direction and neglect the polarization coupling. For $\chi^{(3)}$ we assume also that the nonlinear response of the medium is instantaneous [1,2]. We can therefore recast 9.6 into:

$$P_{Lx}(\vec{r},t) = \varepsilon_0 \int_{-\infty}^{\infty} \chi_{xx}^{(1)}(x,y,t-\tau)E_x(\vec{r},\tau)\,d\tau$$ (9.9a)

$$P_{NLx}\left(\vec{r},t\right) = \varepsilon_0 \chi_{xxxx}^{(3)}\left(z\right) E_x\left(\vec{r},t\right) E_x\left(\vec{r},t\right) E_x\left(\vec{r},t\right) \tag{9.9b}$$

In (9.9) $\chi_{xx}^{(1)}$ and $\chi_{xxxx}^{(3)}$ are the elements of the respective polarisation tensors that relate the corresponding components of the electric field and polarization vectors, that is, these numbers are scalars, cf. Section 4.3 in reference [1]. Further, from 9.4 we obtain the following equation:

$$\frac{\partial E_x\left(\vec{r},\tau\right)}{\partial x} = \int_{-\infty}^{\infty} \frac{1}{1+\chi_{xx}^{(1)}\left(x,y,t-\tau\right)} \frac{\partial \chi_{xx}^{(1)}\left(x,y,t-\tau\right) E_x\left(\vec{r},\tau\right)}{\partial x} d\tau \tag{9.10}$$

Using 9.9 and 9.10 in 9.8 yields the equation for the x component of the electric field vector:

$$\frac{\partial^2 E_x}{\partial y^2} + \frac{\partial^2 E_x}{\partial x^2} + \frac{\partial}{\partial x}\int_{-\infty}^{\infty} \frac{1}{1+\chi_{xx}^{(1)}\left(x,y,t-\tau\right)} \frac{\partial \chi_{xx}^{(1)}\left(x,y,t-\tau\right) E_x\left(\vec{r},\tau\right)}{\partial x} d\tau + \frac{\partial^2 E_x}{\partial z^2}$$

$$= \mu_0 \varepsilon_0 \frac{\partial^2 E_x}{\partial t^2} + \mu_0 \frac{\partial^2 P_x}{\partial t^2} \tag{9.11}$$

We do not use in 9.11 the assumption that $\nabla \cdot \vec{E} \approx 0$ [1,2,9], which allows for a consistent inclusion of a polarized approximation that is applicable to a much wider range of optical waveguides than the scalar one. Now we express the solution in the following way:

$$E_x\left(\vec{r},t\right) = \frac{1}{2}\left(\bar{E}_x\left(\vec{r},t\right)\exp\left(-j\omega_0 t\right) + \bar{E}_x^*\left(\vec{r},t\right)\exp\left(j\omega_0 t\right)\right) \tag{9.12}$$

The convention implied by 9.12 was selected to be consistent with all editions of reference [2]. This implies, however, that it is not consistent with other chapters of this book. Because the derivations that follow are presented in a fairly detailed fashion, this hopefully should not confuse the reader. The reader should also be aware that the derivation presented here is only a simplified version of that available in reference [2].

Substituting 9.12 into 9.11 while including 9.9 results in the following equation for $\bar{E}_x\left(\vec{r},t\right)$:

$$\exp\left(-j\omega_0 t\right)\left(\frac{\partial^2 \bar{E}_x}{\partial y^2} + \frac{\partial^2 \bar{E}_x}{\partial x^2}\right.$$

$$+ \frac{\partial}{\partial x}\int_{-\infty}^{\infty} \frac{1}{1+\chi_{xx}^{(1)}\left(x,y,t-\tau\right)} \frac{\partial \chi_{xx}^{(1)}\left(x,y,t-\tau\right) \bar{E}_x\left(\vec{r},\tau\right)}{\partial x} d\tau + \frac{\partial^2 \bar{E}_x}{\partial z^2}\right)$$

$$= \mu_0 \varepsilon_0 \left(\frac{\partial^2}{\partial t^2} \left[\bar{E}_x \exp\left(-j\omega_0 t\right) \right] + \frac{\partial^2}{\partial t^2} \int_{-\infty}^{\infty} \chi_{xx}^{(1)} \left(t - \tau\right) \bar{E}_x \exp\left(-j\omega_0 t\right) d\tau \right.$$

$$\left. + \chi_{xxxx}^{(3)} \frac{3}{4} \frac{\partial^2}{\partial t^2} \left[\left| \bar{E}_x \right|^2 \bar{E}_x \exp\left(-j\omega_0 t\right) \right] \right) \tag{9.13}$$

In the last term on the right hand side of 9.13, we included only the contributions which come with the $\exp(-j\omega_0 t)$ phase factor. For the Fourier transforms:

$$\tilde{\chi}_{xx}^{(1)} \left(x, y, \omega\right) = \int_{-\infty}^{\infty} \chi_{xx}^{(1)} \left(x, y, t\right) \exp\left(j\omega t\right) dt \tag{9.14a}$$

$$\tilde{E}_x \left(\vec{r}, \omega\right) = \int_{-\infty}^{\infty} \bar{E}_x \left(\vec{r}, t\right) \exp\left(j\omega t\right) dt \tag{9.14b}$$

equation 9.13 implies the following equation:

$$\frac{\partial^2 \tilde{E}_x \left(\vec{r}, \omega - \omega_0\right)}{\partial y^2} + \frac{\partial}{\partial x} \left(\frac{1}{\varepsilon_r} \frac{\partial\left(\varepsilon_r \tilde{E}_x \left(\vec{r}, \omega - \omega_0\right)\right)}{\partial x} \right) + \frac{\partial^2 \tilde{E}_x \left(\vec{r}, \omega - \omega_0\right)}{\partial z^2}$$

$$= -\mu_0 \varepsilon_0 \varepsilon_r \omega^2 \tilde{E}_x \left(\vec{r}, \omega - \omega_0\right) - \mu_0 \varepsilon_0 \chi_{xxxx}^{(3)} \frac{3}{4} \omega^2 \int_{-\infty}^{\infty} \left| \bar{E}_x \right|^2 \bar{E}_x \exp\left(-j\omega_0 t\right) \exp\left(j\omega t\right) dt \tag{9.15}$$

where $\varepsilon_r = 1 + \chi_{xx}^{(1)} \left(x, y, \omega\right)$. Now we extract from $\tilde{E}_x \left(\vec{r}, \omega - \omega_0\right)$ a quickly varying phase factor with respect to z:

$$\tilde{E}_x \left(\vec{r}, \omega - \omega_0\right) = \tilde{E}_x' \left(x, y, \omega\right) \exp(j\beta_0 z) \tilde{B} \left(z, \omega - \omega_0\right) \tag{9.16}$$

In 9.16 we assumed that transverse field distribution can be represented as the transverse field distribution of the fundamental mode of the optical waveguide while β_0 is the corresponding propagation constant at the angular frequency ω_0. To ensure the uniqueness of the representation 9.16, we impose the condition:

$$\iint_S \left| \tilde{E}_x \left(x, y, \omega_0\right) \right|^2 dx dy = 1$$

where S denotes the transverse plane. Substituting 9.16 into 9.15 and using 3.14a and applying the slowly varying (w.r.t z) envelope approximation for $\tilde{B}(z, \omega - \omega_0)$ yields:

$$\tilde{E}'_x \exp\left(j\beta_0 z\right)\left[\left(\beta^2 - \beta_0^2\right)\tilde{B} + 2j\beta_0 \frac{\partial \tilde{B}}{\partial z}\right]$$

$$= -\mu_0 \varepsilon_0 \chi_{xxxx}^{(3)} \frac{3}{4}\omega^2 \int_{-\infty}^{\infty} \left|\bar{E}_x\right|^2 \bar{E}_x \exp\left(-j\omega_0 t\right)\exp\left(j\omega t\right)dt \qquad (9.17)$$

where β is the dependent on ω fundamental mode propagation constant obtained by solving:

$$\frac{\partial^2 \tilde{E}_x\left(\vec{r}, \omega - \omega_0\right)}{\partial y^2} + \frac{\partial}{\partial x}\left(\frac{1}{\varepsilon_r}\frac{\partial\left(\varepsilon_r \tilde{E}_x\left(\vec{r}, \omega - \omega_0\right)\right)}{\partial x}\right) + \mu_0 \varepsilon_0 \varepsilon_r \omega^2 \tilde{E}_x\left(\vec{r}, \omega - \omega_0\right)$$

$$= \beta^2 \tilde{E}_x\left(\vec{r}, \omega - \omega_0\right)$$

which is equivalent to solving 3.14a for the fundamental mode while neglecting the nonlinear perturbation of the polarisation. Now we use approximation:

$$\beta^2 - \beta_0^2 = \left(\beta + \beta_0\right)\left(\beta - \beta_0\right) \approx 2\beta_0\left(\beta - \beta_0\right) \qquad (9.18)$$

and use a Taylor series expansion for β:

$$\beta - \beta_0 = \frac{\partial\beta}{\partial\omega}\bigg|_{\omega=\omega_0}\left(\omega - \omega_0\right) + \frac{1}{2}\frac{\partial^2\beta}{\partial\omega^2}\bigg|_{\omega=\omega_0}\left(\omega - \omega_0\right)^2 + \cdots \qquad (9.19)$$

We substitute 9.18 into 9.17 and use the expansion 9.19 while on the right hand side we approximate ω with ω_0:

$$\tilde{E}'_x \exp\left(j\beta_0 z\right)\left[2\beta_0\left(\beta_1\left(\omega - \omega_0\right) + \frac{1}{2}\beta_2\left(\omega - \omega_0\right)^2 + \cdots\right)\tilde{B} + 2j\beta_0 \frac{\partial\tilde{B}}{\partial z}\right]$$

$$= -\mu_0 \varepsilon_0 \chi_{xxxx}^{(3)} \frac{3}{4}\omega_0^2 \int_{-\infty}^{\infty} \left|\bar{E}_x\right|^2 \bar{E}_x \exp\left(-j\omega_0 t\right)\exp\left(j\omega t\right)dt \qquad (9.20)$$

where

$$\beta_1 = \frac{\partial\beta}{\partial\omega}\bigg|_{\omega=\omega_0} \quad ; \quad \beta_2 = \frac{\partial^2\beta}{\partial\omega^2}\bigg|_{\omega=\omega_0} .$$

Now from 9.20 in the time domain we obtain:

$$\tilde{E}'_x \exp\left(j\beta_0 z\right)\left(2\beta_0\left(j\beta_1 \frac{\partial}{\partial t} - \frac{1}{2}\beta_2 \frac{\partial^2}{\partial t^2} + \cdots\right)B + 2j\beta_0 \frac{\partial B}{\partial z}\right) = -\mu_0\varepsilon_0\chi^{(3)}_{xxxx}\frac{3}{4}\omega_0^2\left|\bar{E}_x\right|^2\bar{E}_x$$

(9.21)

where

$$B(z,t) = \int_{-\infty}^{\infty} \tilde{B}(z,\omega)\exp\left(-j\omega t\right)df$$

(9.22)

Consistently with 9.22, and also assuming that $\tilde{E}'_x(x,y,\omega) \approx \tilde{E}'_x(x,y,\omega_0)$, in time domain, 9.16 implies:

$$\bar{E}_x(\vec{r},t) = \tilde{E}'_x(x,y,\omega_0)\exp(j\beta_0 z)B(z,t)$$

(9.23)

Substituting 9.23 into 9.21 yields:

$$\tilde{E}'_x \exp\left(j\beta_0 z\right)\left(2\beta_0\left(j\beta_1 \frac{\partial}{\partial t} - \frac{1}{2}\beta_2 \frac{\partial^2}{\partial t^2} + \cdots\right)B + 2j\beta_0 \frac{\partial B}{\partial z}\right)$$

$$= -\mu_0\varepsilon_0\chi^{(3)}_{xxxx}\frac{3}{4}\omega_0^2\left|\tilde{E}'_x\right|^2\tilde{E}'_x|B|^2 B\exp\left(j\beta_0 z\right)$$

(9.24)

We multiply both sides of 9.24 by $\tilde{E}'^*_x$ and integrate in the transverse plane S both sides of 9.24, which yields:

$$2\beta_0\left(j\beta_1 \frac{\partial}{\partial t} - \frac{1}{2}\beta_2 \frac{\partial^2}{\partial t^2} + \cdots\right)B + 2j\beta_0 \frac{\partial B}{\partial z} = -\mu_0\varepsilon_0\chi^{(3)}_{xxxx}\frac{3}{4}\omega_0^2\Gamma|B|^2 B \quad (9.25)$$

where Γ is given as:

$$\Gamma = \frac{\displaystyle\iint_S \left|\tilde{E}_x(x,y,\omega_0)\right|^4 dxdy}{\displaystyle\iint_S \left|\tilde{E}_x(x,y,\omega_0)\right|^2 dxdy}$$

(9.26)

Now 9.25 can be conveniently expressed as:

$$\left(\beta_1 \frac{\partial}{\partial t} + \frac{j}{2}\beta_2 \frac{\partial^2}{\partial t^2} + \cdots\right)B + \frac{\partial B}{\partial z} = j\gamma'|B|^2 B \tag{9.27}$$

where

$$\gamma' = \frac{3}{8}\frac{\mu_0 \varepsilon_0 \omega_0^2}{\beta_0}\chi_{xxxx}^{(3)}(z) \approx \Gamma k_0 n_2', \; n_2' = \frac{3}{8n}\chi_{xxxx}^{(3)}(z), \text{ and } n(\text{core}) = \sqrt{1 + \chi_{xx}^{(1)}(\text{core})}$$

while $\chi_{xx}^{(1)}(\text{core})$ is the value of $\chi_{xx}^{(1)}$ in the optical waveguide core. In 9.27 B is expressed in units of [V] while n_2' is given in the same units as $\chi_{xxxx}^{(2)}$, namely, [m²/V²]. Consequently, the units of γ' are [1/(mV²)]. One can also observe that n_2' is the Kerr effect coefficient defined via the equation:

$$n(|E|) = n + n_2'|E|^2 \tag{9.28}$$

It is, however, more convenient to define n_2 in units of [m²/W]:

$$n(I) = n + n_2 I \tag{9.29}$$

where I is the optical intensity of the beam in units of [W/m²]. Using the concept of wave impedance $Z = \sqrt{\mu_0/(\varepsilon_0 \varepsilon_w)}$ and the relationship $I = |E|^2/Z$ (cf. Chapter 19 in reference [11], Chapter 2 in reference [2], and formula 3.57 in section "Optical Fibres"), one can recast 9.28 into:

$$n(|E|) = n + n_2 ZI \tag{9.30}$$

Now we can recast 9.27 into:

$$\left(\beta_1 \frac{\partial}{\partial t} + \frac{j}{2}\beta_2 \frac{\partial^2}{\partial t^2} + \cdots\right)A + \frac{\partial A}{\partial z} = j\gamma|A|^2 A \tag{9.31}$$

where $A = \dfrac{B}{\sqrt{Z}}$ and $\gamma \approx \Gamma k_0 n_2 = \gamma' Z$ while $n_2 = Z\dfrac{3}{8n}\chi_{xxxx}^{(3)}(z) = n_2' Z$. Now $|A|^2$ has the units of [W], which is more convenient when carrying out the calculations, while Γ and γ have the units of [1/m²] and [m⁻¹W⁻¹)], respectively. For the sake of completeness, we note that in 9.23 $\tilde{E}_x'(x, y, \omega_0)$ has the units of [1/m].

Finally, there is some discrepancy left with respect to the selection of the refractive index value n. If the fundamental mode propagates in a weakly guiding step index fibre, then according to the preceding discussion, and the discussion given in Chapter 3 (Effective Index Method section), the effective index of the fundamental mode at the frequency ω_0 is the most appropriate value.

Introducing the change of variables: $z = z$ and $T = t - z/v_g = t - \beta_1 z$ and retaining in the Taylor expansion terms up to the third order one obtains the following equation:

$$\frac{\partial A}{\partial z} = \left(-\frac{j}{2}\beta_2 \frac{\partial^2}{\partial T^2} + \frac{1}{6}\beta_3 \frac{\partial^3}{\partial T^3} \right) A + \left(j\gamma |A|^2 \right) A \tag{9.32}$$

Equation 9.32 is used extensively to study the optical pulse propagation in telecommunication fibres. In the next section, we give the details of the Split Step Fourier Method (SSFM) that is routinely used for its solution. A fairly recent survey of other numerical techniques that can be used for the solution of 9.32 is given in reference [2]. When the Taylor expansion includes only the terms up to the order 2, Equation 9.32 is referred to as the NSE.

SPLIT-STEP FOURIER METHOD

Equation 9.32 can be recast into the following form:

$$\frac{\partial A}{\partial z} = (D + N)A \tag{9.33}$$

whereby the operators D and N are given by:

$$D = -\frac{j}{2}\beta_2 \frac{\partial^2}{\partial T^2} + \frac{1}{6}\beta_3 \frac{\partial^3}{\partial T^3}; \quad N = j\gamma |A|^2 \tag{9.34}$$

Now, if we carry out the expansion of $A(z)$ for a given value of $z = z_0$, we obtain:

$$A(z_0 + \Delta z) = A(z_0) + \frac{dA(z)}{dz}\bigg|_{z0} \Delta z + \frac{d^2 A(z)}{dz^2}\bigg|_{z0} \Delta z^2 + \cdots \tag{9.35}$$

However, from 9.33 we can observe that the action of the operator d/dz on A is equivalent to the action of the operator $(D+N)$ on A. Hence 9.35 can be recast in the following form:

$$A(z_0 + \Delta z) = A(z_0) + (D + N)A(z_0)\Delta z + (D + N)^2 A(z_0)\Delta z \Delta z^2 + \cdots \tag{9.36}$$

which has the same form as the expansion of an exponential function. Hence, can be formally written in a concise form:

$$A(z_0 + \Delta z) = e^{(D+N)\Delta z} A(z_0) \tag{9.37}$$

and 9.36 is the definition of the exponential operator.

Equation 9.37 allows one to calculate the time dependence of A at the position $z_0 + \Delta z$ if the distribution $A(z_0)$ is known. Therefore, if applied repetitively, this equation can be used to calculate the time dependence of the envelope function A at a point along the fibre if the initial time dependence of A at a given point z_0 is given. The only problem that is left is how to calculate the result of the exponential operator acting on the envelope function. Several methods have been developed for this purpose. Obviously, the most straightforward one would be to use the definition of the exponential operator 9.36. However, this still requires some special measures for evaluating the differential operators within D. In order to circumvent these problems the Split Step Fourier Method (SSFM) has been developed. The first step in the implementation of SSFM is the splitting of the exponential operator:

$$A(z_0 + \Delta z) = e^{N\Delta z} e^{D\Delta z} A(z_0)$$

Such splitting does not incur any error if the operators $N\Delta z$ and $D\Delta z$ commute. This unfortunately is not the case. However, if Δz is small, the resulting error is negligible and the operator splitting can still be applied. Furthermore, it has been shown that the error is further reduced if the operator splitting is performed in a symmetrised form [2]:

$$A(z_0 + \Delta z) = e^{D\frac{\Delta z}{2}} e^{N\Delta z} e^{D\frac{\Delta z}{2}} A(z_0) \tag{9.38}$$

The final observation that leads to an efficient algorithm is that the D operator becomes diagonal in the Fourier space, or in other words the application of this operator to a function is equivalent to a multiplication by a constant (or another function). This can be shown by carrying out the Fourier transform (w.r.t. to time) for both sides of the following equation:

$$\frac{\partial A(z,T)}{\partial z} = DA = \left(-\frac{j}{2} \beta_2 \frac{\partial^2}{\partial T^2} + \frac{1}{6} \beta_3 \frac{\partial^3}{\partial T^3} \right) A(z,T) \tag{9.39}$$

which yields:

$$\frac{\partial A(z,\omega)}{\partial z} = j\left(\frac{1}{2} \beta_2 \omega^2 - \frac{1}{6} \beta_3 \omega^3 \right) A(z,\omega) \tag{9.40}$$

ALGORITHM 9.1   SSFM FOR THE PROPAGATION OF OPTICAL PULSES IN FIBRES

1. Start
2. Set the values of γ, β_1, β_2, and Δz
3. Select N for 2^N equidistant sampling points
4. Initialize: $z = 0$ and $A\ (x, z = 0) = A_0(x)$
5. Calculate the FFT of $A\ (x, z)$: $B\ (x, z) = \Im\big(A\ (x, z)\big)$
6. Calculate $B\left(x, z + \dfrac{1}{2}\Delta z\right) = B\ (x, z) e^{\frac{1}{2}D\Delta z}$
7. Calculate the inverse FFT of
$$B\left(x, z + \frac{1}{2}\Delta z\right): A\left(x, z + \frac{1}{2}\Delta z\right) = \Im^{-1}\left(B\left(x, z + \frac{1}{2}\Delta z\right)\right)$$
8. Calculate $C\left(x, z + \dfrac{1}{2}\Delta z\right) = A\left(x, z + \dfrac{1}{2}\Delta z\right) e^{N\Delta z}$
9. Calculate the FFT of
$$C\left(x, z + \frac{1}{2}\Delta z\right): D\left(x, z + \frac{1}{2}\Delta z\right) = \Im\left(C\left(x, z + \frac{1}{2}\Delta z\right)\right)$$
10. Calculate $D\ (x, z + \Delta z) = D\left(x, z + \dfrac{1}{2}\Delta z\right) e^{\frac{1}{2}D\Delta z}$
11. Calculate the inverse FFT of
$$D\ (x, z + \Delta z): A\ (x, z + \Delta z) = \Im^{-1}\big(D\ (x, z + \Delta z)\big)$$
12. If end of fibre not reached $A\ (x, z) = A\ (x, z + \Delta z)$ go to 5
13. Stop

and hence:

$$A\left(z_0 + \Delta z, \omega\right) = e^{j\left(\frac{1}{2}\beta_2\omega^2 - \frac{1}{6}\beta_3\omega^3\right)\Delta z} A\left(z_0, \omega\right) \tag{9.41}$$

which is straightforward to carry out.

Summarizing the discussion, Algorithm 9.1 can be used for calculating the evolution of the envelope function of an optical beam while propagating along an optical fibre. The algorithm 9.1 can be easily implemented using MATLAB script.

```
% program calculates propagation of an optical pulse in a
% fibre using split step FFT method
clear % clears variables
clear global % clears global variables
format long e
```

```
% initial constants
i = sqrt(-1);% remember not to overwrite pi or i !
pi = 3.141592653589793e+000;

% fibre parameters
B2 = -1.00; % |beta2| [ps^2/km]
B3 = 0.0;% |beta3| [ps^3/km]
G = 2;% gamma [1/(W*km)]
P0 = 0.5;% pulse peak power [W]

% iput data
dz =.1; % z step [km]
NoSamp = 2^10;% number of samples

wind_width = 50;% numerical window width[ps]
dt = wind_width/NoSamp;

t_init = dt*(0:NoSamp-1);% t values
t = t_init-wind_width/2;

beam_width = 1.;% beam half width[ps]
E_init = sqrt(P0)./cosh(t/beam_width);% initial field

veka = (0:NoSamp/2);
vekb = ((-NoSamp/2+1):1:-1);
vekab = [veka vekb];% setting frequencies from 0:pi and -pi:~0
freq = vekab*2*pi/wind_width;% calculates the values of omega

E = E_init;
E = fft(E);
% main propagation loop
N_steps = 500;
for j = 1:N_steps
ffun = E.*(exp(i*B2*freq.^2*dz/4-i*B3*freq.^3*dz/12));
% dispersion half step
E1 = ifft(ffun);
E1 = E1.*exp(i*G*abs(E1).^2*dz); % nonlinear step
E = fft(E1).*(exp(i*B2*freq.^2*dz/4-i*B3*freq.^3*dz/12));
% dispersion half step
power(j) = sum(abs(ffun).^2);
end
```

Similarly to the FFT method presented in Chapter 2, the SSFM relies on repeated calculation of an FFT and inverse FFT. In the above code, instead of shifting the FFT spectrum each step, we calculated the dependence of frequency on the sample number in such a way that the vector containing the frequency values can be directly multiplied with the vector containing the FFT samples. Figure 9.1 elucidates the way in which the vector containing the frequency values is constructed. A MATLAB code that implements the SSFM to the solution of NSE can be found in the fifth edition of the reference [2].

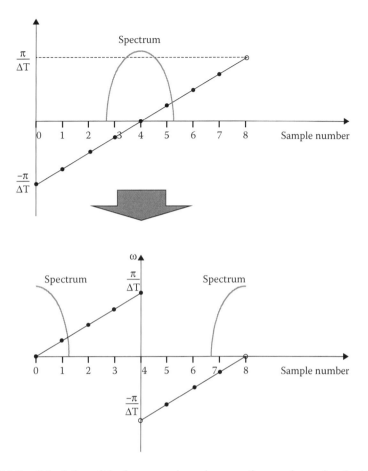

FIGURE 9.1 Calculation of the frequency dependence on the sample number. In this illustrating example, it was assumed that the total number of samples equals 2^3.

Figure 9.2 shows the power distribution for the input pulse and the output pulse after propagating 50 km down a single mode fibre. The fibre parameters β_2, β_3, and γ are 0 ps²/km, 0.1 ps³/km, and 2/(W km), respectively. These are typical values of the parameters for a single mode silica fibre at the wavelength of approximately 1.3 μm, where the material dispersion, which β_2 is related to, is equal to zero. This makes the derivative of the dispersion to dominate and hence the observed asymmetry in the output pulse. The propagation step used in the simulation was 0.1 km while 2^{10} samples were used along the time dimension. The initial pulse amplitude distribution was Gaussian:

$$A(z=0,t) = \sqrt{P_0}\, e^{-\left(\frac{t}{2\,T_0}\right)^2}$$

whereby the pulse-width $T_0 = 1$ ps and $P_0 = 1$ mW.

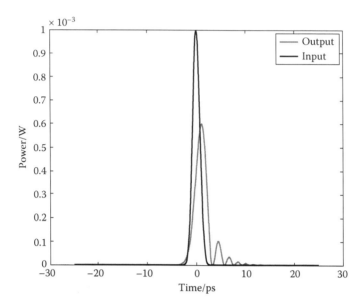

FIGURE 9.2 Gaussian pulse propagation in a fibre over the distance of 50 km. The pulse width is 1 ps while the peak power equals 1 mW. The fibre parameters β_2, β_3, and γ are 0 ps²/km, 0.1 ps³/km, and 2/(W km), respectively. The propagation step used in the simulation was 0.1 km while 2^{10} samples were used along the time dimension.

Figure 9.3 shows the fundamental soliton propagation in a fibre over the distance of 50 km. In this example the fibre parameters β_2, β_3, and γ are −1 ps²/km, 0 ps³/km, and 2/(W km), respectively. These parameter values are typical for a single mode silica fibre at a wavelength of approximately 1.55 μm. The propagation step used in the simulation was 0.1 km while 2^{10} samples were used along the time dimension. The initial pulse amplitude distribution was corresponding to the fundamental soliton:

$$A\left(z = 0, t\right) = \frac{\sqrt{P_0}}{\cosh\left(\dfrac{t}{T_0}\right)}$$

whereby the pulse-width $T_0 = 1$ ps and $P_0 = 0.5$ W. As expected in the absence of fibre loss, the pulse shape remains unchanged.

Similarly, to the case of the FFT algorithm presented in the Chapter 2, appropriate care must be taken to ensure that a sufficient sampling rate and window size are selected. However, unlike the algorithm presented in Chapter 2, the results obtained using Algorithm 9.1 depend on the longitudinal step size. This is a consequence of splitting the exponential operator in Equation 9.38. The error associated with this approximation can be minimized by using a sufficiently small value of

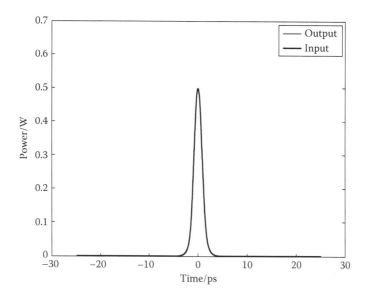

FIGURE 9.3 Secant pulse propagation in a fibre over the distance of 50 km. The pulse width is 1 ps while the peak power equals 0.5 W. The fibre parameters β_2, β_3, and γ are -1 ps²/km, 0 ps³/km, and 2/(W km), respectively. The propagation step used in the simulation was 0.1 km while 2^{10} samples were used along the time dimension.

the longitudinal step size. To illustrate this point, we recalculated the results from Figure 9.3 using a longer longitudinal step, that is, $\Delta z = 1$ km. The results of this simulation are shown in Figure 9.4. The output pulse changed shape, which is a purely numerical artefact introduced by the exponential operator splitting error. Interestingly, the results shown in Figure 9.2 can be reproduced using one longitudinal step equal to 50 km. This is because the pulse power is very small and hence $e^{N\Delta z} \approx 1$ and effectively commutes with $e^{D\frac{\Delta z}{2}}$ for any value of longitudinal step. Similarly, when $e^{D\frac{\Delta z}{2}} \approx 1$ holds, the SSFM can complete calculations within a single longitudinal step.

The efficiency of the SSFM can be improved by applying a variable longitudinal step [12]. The SSFM can be also applied to study four wave mixing in optical fibres [13] and higher order nonlinear effects [2]. More recently a Runge-Kutta algorithm was used to avoid the operator splitting. This allows for using a longer propagation step Δz [14]. When studying the supercontinuum generation in optical fibres, it is necessary to include the Raman effect [3]. The Raman effect can be described by a general third order nonlinearity [8]:

$$\vec{P}_{NL}(\vec{r},t) = \int_{-\infty}^{\infty} \chi^{(3)}(\tau_1,\tau_2,\tau_3)\vec{E}(\vec{r},t-\tau_1)\vec{E}(\vec{r},t-\tau_2)\vec{E}(\vec{r},t-\tau_3)\,d\tau_1 d\tau_2 d\tau_3 \quad (9.42)$$

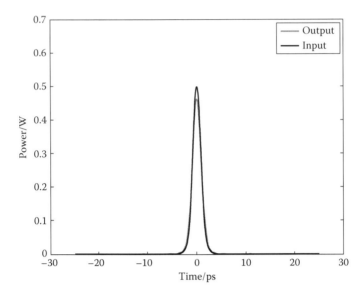

FIGURE 9.4 Secant pulse propagation in a fibre over the distance of 50 km. The pulse width is 1 ps while the peak power equals 0.5 W. The fibre parameters β_2, β_3, and γ are −1 ps²/km, 0 ps³/km, and 2/(W km), respectively. The propagation step used in the simulation was 1 km while 2^{10} samples were used along the time dimension.

In reference [15] it was shown that for the Raman effect the third order suscepti-bility can be represented in the following way:

$$\chi^{(3)}\left(t_1,t_2,t_3\right)=\chi_R\left(t_1\right)\delta\left(t_1-t_2\right)\delta\left(t_3\right)$$

(9.43)

where $\delta(t)$ is the Dirac delta distribution. Substituting 9.43 into 9.42 yields:

$$\vec{P}_{NL}\left(\vec{r},t\right)=\vec{E}\left(\vec{r},t\right)\int_{-\infty}^{\infty}\chi_R\left(\tau\right)\vec{E}\left(\vec{r},t-\tau\right)\vec{E}\left(\vec{r},t-\tau\right)d\tau$$

(9.44)

For the x component of the electric field, $\chi_R(t)$ in the silica glass can be approxi-mated in the following way:

$$\chi_{Rx}\left(t\right)=\varepsilon_0\chi_{xxxx}^{(3)}\left\{\left(1-f_R\right)\delta\left(t\right)+f_R h_R\left(t\right)\right\}$$

(9.45)

whereas $h_R(t)$ can be approximated by the following function [16]:

$$h_R\left(t\right)=1(t)\left(\tau_1^{-2}+\tau_2^{-2}\right)\tau_1\exp\left(-t/\tau_2\right)\sin\left(t/\tau_1\right)$$

(9.46)

where $1(t)$ is a unit step function. More detail on the modelling of the Raman gain spectrum in silica glass optical fibres can be found in references [17,18].

Substituting 9.46 into 9.8 leads to the generalized nonlinear Schroedinger equation (GNSE) [2,3,19,20]. The GNSE, subject to a fairly moderate extension, can be also numerically solved by the SSFM. Several versions of such an algorithm have been proposed [16,21]. To improve the efficiency, a variable longitudinal step and a Runge-Kutta algorithm was more recently proposed to avoid operator splitting when solving GNSE [22,23]. The details of the application of the SSFM to the solution of the GNSE with an example MATLAB script is provided in reference [24].

REFERENCES

1. Schneider, T., *Nonlinear Optics in Telecommunications*. 2004, Heidelberg: Springer.
2. Agrawal, G., *Nonlinear Fibre Optics*. 2013, London: Academic Press.
3. Dudley, M.J. and J.R. Taylor, *Supercontinuum Generation in Optical Fibres*. 2010, Cambridge: Cambridge University Press.
4. Vasil'ev, P., *Ultrafast Diode Lasers*. 1995, London: Artech House.
5. Bloembergen, N., *Nonlinear Optics*. 1996, London: World Scientific Publishing.
6. Shen, Y.R., *The Principles of Nonlinear Optics*. 2003, New Jersey: Wiley-Interscience.
7. Boyd, R.W., *Nonlinear Optics*. 2008, San Diego: Elsevier.
8. Sauter, E.G., *Nonlinear Optics*. 1996, New York: Wiley.
9. Kivshar, Y.S. and G. Agrawal, *Optical Solitons*. 2003, London: Academic Press.
10. Trillo, S. and W. Torruellas, *Spatial Solitons*. 2001, Berlin: Springer.
11. Saleh, B.E.A. and M.C. Teich, *Fundamentals of Photonics*. 1991, New York: John Wiley & Sons Inc..
12. Sinkin, O.V., et al., Optimization of the split-step Fourier method in modeling optical-fiber communications systems. *Journal of Lightwave Technology* 2003. 21(1): p. 61–68.
13. Bosco, G., et al., Suppression of spurious tones induced by the split-step method in fiber systems simulation. *IEEE Photonics Technology Letters*, 2000. 12(5): p. 489–491.
14. Zhang, Z., L. Chen, and X. Bao, A fourth-order Runge-Kutta in the interaction picture method for numerically solving the coupled nonlinear Schrodinger equation. *Optics Express*, 2010. 18(8): p. 8261–8276.
15. Zheltikov, A.M., The Raman effect in femto- and attosecond physics. *Physics-Uspekhi*, 2011. 54(1): p. 29–51.
16. Blow, K.J. and D. Wood, Theoretical description of transient stimulated Raman-scattering in optical fibers. *IEEE Journal of Quantum Electronics*, 1989. 25(12): p. 2665–2673.
17. Stolen, R.H., et al., Raman response function of silica-core fibers. *Journal of the Optical Society of America B-Optical Physics*, 1989. 6(6): p. 1159–1166.
18. Hollenbeck, D. and C.D. Cantrell, Multiple-vibrational-mode model for fiber-optic Raman gain spectrum and response function. *Journal of the Optical Society of America B-Optical Physics*, 2002. 19(12): p. 2886–2892.
19. Laegsgaard, J., Mode profile dispersion in the generalized nonlinear Schrodinger equation. *Optics Express*, 2007. 15(24): p. 16110–16123.
20. Laegsgaard, J., Modeling of nonlinear propagation in fiber tapers. *Journal of the Optical Society of America B-Optical Physics*, 2012. 29(11): p. 3183–3191.
21. Cristiani, I., et al., Dispersive wave generation by solitons in microstructured optical fibers. *Optics Express*, 2004. 12(1): p. 124–135.

22. Hult, J., A fourth-order Runge-Kutta in the interaction picture method for, simulating supercontinuum generation in optical fibers. *Journal of Lightwave Technology*, 2007. 25(12): p. 3770–3775.
23. Heidt, A.M., Efficient adaptive step size method for the simulation of supercontinuum generation in optical fibers. *Journal of Lightwave Technology*, 2009. 27(18): p. 3984–3991.
24. Travers, J.C., M.H. Frosz, and M.J. Dudley, Nonlinear fibre optics overview, In *Supercontinuum Generation in Optical Fibres*, M.J. Dudley and J.R. Taylor, Editors. 2010, Cambridge University Press: Cambridge.

Index

Printed and bound by CPI Group (UK) Ltd, Croydon, CR0 4YY

22/10/2024

01777621-0017